MATHEMATICAL PHYSICS

MATHEMATICAL PHYSICS

Hilary. D. Brewster

Oxford Book Company

Jaipur ▪ New Delhi

Published By

Oxford Book Company

Regd. Off. "Bony Residency", Gate No. 2
Opp. Tilak Public School, Vishweswariya Nagar,
Gopalpura Road, Jaipur - 302018, India
Ph: 0141 - 2761280, Telefax: 2761381
email - oxfordbookjpr@gmail.com
website - oxfordbookcompany.com

Delhi Office: 102, 1st Floor, Satyam House,
4327/3, Ansari Road, Darya Ganj,
New Delhi - 110002, India
Ph: 011-45652440

Published in India, 2025

ISBN : 9789380179025

Price : ₹ 2250/-

Typeset by Shivangi Computers, Jaipur

Printed and bound in India by Thomson Press Ltd.,
New Delhi

Preface

This book is intended to provide an account of those parts of pure mathematics that are most frequently needed in physics. This book will serve several purposes: to provide an introduction for graduate students not previously acquainted with the material, to serve as a reference for mathematical physicists already working in the field, and to provide an introduction to various advanced topics which are difficult to understand in the literature.

Not all the techniques and application are treated in the same depth. In general, we give a very thorough discussion of the mathematical techniques and applications in quantum mechanics, but provide only an introduction to the problems arising in quantum field theory, classical mechanics, and partial differential equations.

This book is for physics students interested in the mathematics they use and for mathematics students interested in seeing how some of the ideas of their discipline find realization in an applied setting. The presentation tries to strike a balance between formalism and application, between abstract and concrete. The interconnections among the various topics are clarified both by the use of vector spaces as a central unifying theme, recurring throughout the book, and by putting ideas into their historical context. Enough of the essential formalism is included to make the presentation self-contained. This book features the applications of essential concepts as well as the coverage of topics in the this field.

Hilary D. Brewster

Contents

Chapter 1

Mathematical Basics

Before we begin our study of mathematical physics, we should review some mathematical basics. It is assumed that you know Calculus and are comfortable with differentiation and integration.

CALCULUS IS IMPORTANT

There are two main topics in calculus: derivatives and integrals. You learned that derivatives are useful in providing rates of change in either time or space. Integrals provide areas under curves, but also are useful in providing other types of sums over continuous bodies, such as lengths, areas, volumes, moments of inertia, or flux integrals. In physics, one can look at graphs of position versus time and the slope (derivative) of such a function gives the velocity.

Then plotting velocity versus time you can either look at the derivative to obtain acceleration, or you could look at the area under the curve and get the displacement:

$$x = \int_{to}^{t} v\,dt.$$

Of course, you need to know how to differentiate and integrate given functions. Even before getting into differentiation and integration, you need to have a bag of functions useful in physics. Common functions are the polynomial and rational functions. Polynomial functions take the general form

$$f(x) = a_n x^n + a_{n-1}\, xn^{n-1} + \ldots + a_1 x + a_0,$$

where $an \neq 0$.: This is the form of a polynomial of degree n. Rational functions consist of ratios of polynomials.

Their graphs can exhibit asymptotes. Next are the exponential and logarithmic functions. The most common are the natural exponential and the natural logarithm.

The natural exponential is given by $f(x) = e^x$, where $e \approx 2.718281828....$ The natural logarithm is the inverse to the exponential, denoted by ln x. The properties of the exponential function follow from our basic properties for exponents. Namely, we have:

$$e^0 = 1,$$
$$e^{-a} = \frac{1}{e^a}$$
$$e^a e^b = e^{a+b},$$
$$(e^{-a})^b = e^{ab}.$$

The relation between the natural logarithm and natural exponential is given by

$$y = ex \Leftrightarrow x = \ln y.$$

Some common logarithmic properties are

$$\ln 1 = 0,$$
$$\ln \frac{1}{a} = -\ln a,$$
$$\ln(ab) = \ln a + \ln b,$$
$$\ln \frac{a}{b} = \ln a - \ln b,$$
$$\ln \frac{1}{b} = -\ln b.$$

TRIGONOMETRIC FUNCTIONS

The trigonometric functions also called circular functions are functions of an angle. They are important in the study of triangles and modeling periodic phenomena, among many other applications. Trigonometric functions are commonly defined as ratios of two sides of a right triangle containing the angle, and can equivalently be defined as the lengths of various line segments from a unit circle.

More modern definitions express them as infinite series or as solutions of certain differential equations, allowing their extension to arbitrary positive and negative values and even to complex numbers.

In modern usage, there are six basic trigonometric functions, which are tabulated here along with equations relating them to one another. Especially in the case of the last four, these relations are often taken as the definitions of those functions, but one can define them equally well geometrically or by other means and then derive these relations.

They have their origins as far back as the building of the pyramids. Typical applications in your introductory math classes probably have included finding the heights of trees, flag poles, or buildings. It was recognized a long time ago that similar right triangles have fixed ratios of any pair of sides of the two similar triangles.

These ratios only change when the non-right angles change. Thus, the ratio of two sides of a right triangle only depends upon the angle. Since there are six possible ratios, then there are six possible functions. These are designated as sine, cosine, tangent and their reciprocals (cosecant, secant and cotangent). In your introductory physics class, you really only needed the first three.

Table: Table of Trigonometric Values

θ	$\cos\theta$	$\sin\theta$	$\tan\theta$
0	1	0	0
$\frac{\pi}{6}$	$\frac{\sqrt{3}}{2}$	$\frac{1}{2}$	$\frac{\sqrt{3}}{2}$
$\frac{\pi}{3}$	$\frac{1}{2}$	$\frac{\sqrt{3}}{2}$	$\sqrt{3}$
$\frac{\pi}{4}$	$\frac{\sqrt{2}}{2}$	$\frac{\sqrt{2}}{2}$	1
$\frac{\pi}{2}$	0	1	undefined

You also learned that they are represented as the ratios of the opposite to hypotenuse, adjacent to hypotenuse, etc. Hopefully, you have this down by now.

You should also know the exact values for the special angles $\theta = 0, \frac{\pi}{6}, \frac{\pi}{3}, \frac{\pi}{4}, \frac{\pi}{2}$, and their corresponding angles in the second, third and fourth quadrants. This becomes internalized after much use, but we provide these values in Table just in case you need a reminder.

We will have many an occasion to do so in this class as well. What is an identity?

It is a relation that holds true all of the time. For example, the most common identity for trigonometric functions is

$$\sin^2\theta + \cos^2\theta = 1.$$

This hold true for every angle θ! An even simpler identity is

$$\tan\theta = \frac{\sin\theta}{\cos\theta}.$$

Other simple identities can be derive from this one. Dividing the equation by $\sin^2\theta + \cos^2\theta$ yields

$$\tan^2\theta + 1 = \sec^2\theta,$$
$$1 + \cot^2\theta = \text{cosec}^2\,\theta.$$

Other useful identities stem from the use of the sine and cosine of the sum and difference of two angles. Namely, we have that

$$\sin(A \pm B) = \sin A \cos B \pm \sin B \cos A,$$
$$\cos(A \pm B) = \cos A \cos B \mp \sin B \cos A,$$

Note that the upper (lower) signs are taken together.

The double angle formulae are found by setting $A = B$:

$$\sin(2A) = 2\sin A \cos B,$$

$$\cos(2A) = \cos^2 A - \sin^2 A.$$

Using Equation, we can rewrite as

$$\cos(2A) = 2\cos^2 A - 1,$$

$$= 1 - 2\sin^2 A.$$

These, in turn, lead to the half angle formulae. Using $A = 2\alpha$, we find that

$$\sin^2\alpha = \frac{1-\cos 2\alpha}{2}$$

$$\cos^2\alpha = \frac{1+\cos 2\alpha}{2}.$$

Finally, another useful set of identities are the product identities. For example, if we add the identities for $\sin(A + B) +$ and $\sin(A - B)$, the second terms cancel and we have $\sin(A + B + \sin(A - B) = 2\sin A \cos B$.

Thus, we have that

$$\sin A \cos B = \frac{1}{2}(\sin(A+B)+\sin(A-B)).$$

Similarly, we have

$$\cos A \cos B = \frac{1}{2}(\sin(A+B)+\cos(A-B)).$$

and
$$\sin A \sin B = \frac{1}{2}(\sin(A-B)-\cos(A+B)).$$

These are the most common trigonometric identities. They appear often and should just roll off of your tongue. We will also need to understand the behaviors of trigonometric functions.

In particular, we know that the sine and cosine functions are periodic. They are not the only periodic functions. However, they are the most common periodic functions.

A periodic function $f(x)$ satisfies the relation

$$f(x + p) = f(x), \quad \text{for all } x$$

for some constant p. If p is the smallest such number, then p is called the period. Both the sine and cosine functions have period 2π. This means that the graph repeats its form every 2π units. Similarly $\sin bx$ and $\cos bx$, have the common period $p = \frac{2\pi}{b}$.

OTHER ELEMENTARY FUNCTIONS

So, are there any other functions that are useful in physics? Actually, there are many more. However, you have probably not see many of them to date. There are many important functions that arise as solutions of some fairly generic, but important, physics problems.

In calculus you have also seen that some relations are represented in parametric form.

However, there is at least one other set of elementary functions, which you should know about. These are the hyperbolic functions. Such functions are useful in representing hanging cables, unbounded orbits, and special traveling waves called solitons. They also play a role in special and general relativity.

Hyperbolic functions are actually related to the trigonometric functions. For now, we just want to recall a few definitions and an identity. Just as all of the trigonometric functions can be built from the sine and the cosine, the hyperbolic functions can be defined in terms of the hyperbolic sine and hyperbolic cosine:

$$\sinh x = \frac{e^x - e^{-x}}{2},$$

$$\cosh x = \frac{e^x + e^{-x}}{2}$$

There are four other hyperbolic functions. These are defined in terms of the above functions similar to the relations between the trigonometric functions.

Table: Table of Derivatives

Function	*Derivative*
a	0
x^n	nx^{n-1}
e^{ax}	ae^{ax}
$\ln ax$	$\frac{1}{x}$
$\sin ax$	$a \cos ax$
$\cos ax$	$-a \sin ax$
$\tan ax$	$a \sec^2 ax$
$\text{cosec } ax$	$-a \text{ cosec } ax \cot ax$
$\sec ax$	$a \sec ax \tan ax$
$\cot ax$	$-a \text{ cosec}^2 ax$
$\sinh ax$	$a \cosh ax$
$\cosh ax$	$a \sinh ax$
$\tanh ax$	$a \text{ sech}^2 ax$
$\text{cosech } ax$	$-a \text{ cosech } ax \coth ax$
$\text{sech } ax$	$-a \text{ sech} \tanh ax$
$\coth ax$	$-a \text{ cosech}^2 ax$

For example, we have

$$\tanh x = \frac{\sinh x}{\cosh x} = \frac{e^x - e^{-x}}{e^x + e^{-x}}.$$

There are also a whole set of identities, similar to those for the trigonometric functions. Some of these are given by the following:

$$\cosh^2 x - \sinh^2 x = 1,$$

$$\cosh(A \pm B) = \cosh A \cosh B \pm \sinh A \sinh B$$

$$\sinh(A \pm B) = \sinh A \cosh B \pm \sinh A \cosh A.$$

Others can be derived from these.

DERIVATIVES

Now that we know our elementary functions, we can seek their derivatives. We will not spend time exploring the appropriate limits in any rigorous way. We are only interested in the results.

We expect that you know the meaning of the derivative and all of the usual rules, such as the product and quotient rules.

Also, you should be familiar with the Chain Rule. Recall that this rule tells us that if we have a composition of functions, such as the elementary functions above, then we can compute the derivative of the composite function. Namely, if $h(x) = f(g(x))$, then

$$\frac{dh}{dx} = \frac{d}{dx}(f(g(x))) = \frac{df}{dg} = | g(x) \frac{dg}{dx} = f'(g(x)g'(x)$$

For example, let $H(x) = 5\cos(\pi \tanh 2x^2)$. This is a composition of three functions, $H(x) = f(g(h(x)))$, where $f(x) = 5 \cos x$, $g(x) = \pi \tanh x$. Then the derivative becomes

$$H'(x) = 5\left(-\sin\left(\pi \tanh 2x^2\right)\right)\frac{d}{dx}\left(\left(\pi \tanh 2x^2\right)\right)$$

$$= -5\pi \sin\left(\pi \tanh 2x^2\right)\operatorname{sech}^2 2x^2 \frac{d}{dx}(2x^2)$$

$$= -20\pi x \sin\left(\pi \tanh 2x^2\right)\operatorname{sech}^2 2x^2.$$

INTEGRALS

Integration is typically a bit harder. Imagine being given the result in equation and having to figure out the integral. As you may recall from the Fundamental Theorem of Calculus, the integral is the inverse operation to differentiation:

$$\int \frac{df}{dx} dx = f(x) + C.$$

However, it is not always easy to determine a given integral. In fact some integrals are not even doable! However, you learned in calculus that there are some methods that might yield an answer. While you might be happier using a computer with a computer algebra systems, such as Maple, you should know a few basic integrals and know how to use tables for some of the more complicated ones. In fact, it can be exhilarating when you can do a given integral without reference to a computer or a Table of Integrals. However, you should be prepared to do some integrals using what you have been taught in calculus. We will review a few of these methods and some of the standard integrals in this section.

Table: Table of Integrals

Function	*Indefinite Integral*
a	ax
x^n	$\frac{x^{n+1}}{n+1}$
e^{ax}	$\frac{1}{a}e^{ax}$
$\frac{1}{x}$	$\ln x$
$\sin ax$	$-\frac{1}{a}\cos ax$
$\cos ax$	$\frac{1}{a}\sin ax$
$\sec^2 ax$	$\frac{1}{a}\tan ax$
$\sinh ax$	$\frac{1}{a}\cosh ax$
$\cosh ax$	$\frac{1}{a}\sinh ax$
$\operatorname{sech}^2 ax$	$\frac{1}{a}\tanh ax$
$\frac{1}{a+bx}$	$\frac{1}{b}\ln(a+bx)$
$\frac{1}{a^2+x^2}$	$\frac{1}{a}\tan^{-1} ax$
$\frac{1}{\sqrt{a^2-x^2}}$	$\frac{1}{a}\sin^{-1} ax$
$\frac{1}{x^2-a^2}$	$\frac{1}{a}\tan^{-1} ax$

First of all, there are some integrals you should be expected to know without any work. These integrals appear often and are just an application of the Fundamental Theorem of Calculus. These are not the only integrals you should be able to do. However, we can expand the list by recalling a few of the techniques that you learned in calculus.

There are just a few: The Method of Substitution, Integration by Parts, Integration Using Partial Fraction Decomposition, and Trigonometric Integrals.

Example: When confronted with an integral, you should first ask if a simple substitution would reduce the integral to one you know how to do. So, as an example, consider the following integral

The ugly part of this integral is the $x^2 + 1$ under the square root. So, we let $u = x^2 + 1$. Noting that when $u = f(x)$, we have $du = f'(x)\, dx$. For our example, $du = 2x\, dx$. Looking at the integral, part of the integrand can be written as $x\, dx = \frac{1}{2} u du$: Then, our integral becomes:

$$\int \frac{x}{\sqrt{x^2+1}} dx = \frac{1}{2} \int \frac{du}{\sqrt{u}}.$$

The substitution has converted our integral into an integral over u. Also, this integral is doable! It is one of the integrals we should know. Namely, we can write it as

$$\frac{1}{2} \int \frac{du}{\sqrt{u}} = \frac{1}{2} \int u^{-1/2} du.$$

This is now easily finished after integrating and using our substitution variable to give

Note that we have added the required integration constant and that the derivative of the result easily gives the original integrand (after employing the Chain Rule).

Often we are faced with definite integrals, in which we integrate between two limits. There are several ways to use these limits. However, students often forget that a change of variables generally means that the limits have to change.

Example: Consider the above example with limits added.

$$\int_0^2 \frac{x}{\sqrt{x^2+1}} dx.$$

We proceed as before. We let $u = x^2 + 1$. As x goes from 0 to 2, u takes values from 1 to 5. So, our substitution gives

$$\int_0^2 \frac{x}{\sqrt{x^2+1}} dx = \frac{1}{2} \int_1^5 \frac{du}{\sqrt{u}} = \sqrt{u}\,\Big|_1^5 = \sqrt{5} - 1.$$

When the Method of substitution fails, there are other methods you can try. One of the most used is the Method of Integration by Parts.

$$\int u\,dv = uv - \int v\,du.$$

The idea is that you are given the integral on the left and you can relate it to an integral on the right. Hopefully, the new integral is one you can do, or at least it is an easier integral than the one you are trying to evaluate.

However, you are not usually given the functions u and v. You have to determine them. The integral form that you really have is a function of another variable, say x. Another form of the formula can be given as

$$\int f(x)g'(x)\,dx = f(x)g(x) - \int g(x)f'(x)\,dx.$$

This form is a bit more complicated in appearance, though it is clearer what is happening. The derivative has been moved from one function to the other. Recall that this formula was derived by integrating the product rule for differentiation.

The two formulae are related by using the relations

$$u\,f(x) \rightarrow du = f'(x)\,dx,$$
$$u\,g(x) \rightarrow dv = g'(x)\,dx.$$

This also gives a method for applying the Integration by Parts Formula.

Example: Consider the integral $\int x \sin 2x\,dx$. We choose $u = x$ and $dv = \sin 2x\,dx$. This gives the correct left side of the formula. We next determine v and du:

$$du = \frac{du}{dx}dx = dx,$$

$$v = \int dv = \int \sin 2x\,dx = -\frac{1}{2}\cos 2x.$$

We note that one usually does not need the integration constant. Inserting these expressions into the Integration by Parts Formula, we have

$$\int x \sin 2x\,dx = -\frac{1}{2}x\cos 2x + \frac{1}{2}\int \cos 2x\,dx.$$

We see that the new integral is easier to do than the original integral. Had we picked $u = \sin 2x$ and $dv = x dx$, then the formula still works, but the resulting integral is not easier.

For completeness, we can finish the integration. The result is

u sin $2x$ and $dv = x\,dx$,

$$\int x \sin 2x\,dx = -\frac{1}{4}x\cos 2x + \frac{1}{4}\sin 2x + C.$$

As always, you can check your answer by differentiating the result, a step students often forget to do. Namely,

$$\frac{d}{dx}\left(-\frac{1}{2}x\cos 2x+\frac{1}{4}\sin 2x+C\right) = -\frac{1}{2}\cos 2x+x\sin 2x+\frac{1}{4}(2\cos 2x)$$
$$= x \sin 2x.$$

So, we do get back the integrand in the original integral.

We can also perform integration by parts on definite integrals. The general formula is written as

$$\int_a^b f(x)g'(x)\,dx = f(x)g(x)\Big|_a^b - \int_a^b g(x)f'(x)\,dx\,.$$

Example: Consider the integral $\int_0^{\pi} x^2 \cos x\, dx$.

This will require two integrations by parts. First, we let $u = x^2$ and $dv = \cos x$. Then,

$$du = 2x\, dx. \qquad v = \sin x.$$

Inserting into the Integration by Parts Formula, we have

$$\int_0^{\pi} x^2 \cos x\, dx = x^2 \sin x\Big|_0^{\pi} - 2\int_0^{\pi} x \sin x\, dx$$

$$= -2\int_0^{\pi} x\sin x\, dx\,.$$

We note that the resulting integral is easier that the given integral, but we still cannot do the integral off the top of our head (unless we look at Example 3!). So, we need to integrate by parts again.

Note: In your calculus class you may recall that there is a tabular method for carrying out multiple applications of the formula. However, we will leave that to the reader and proceed with the brute force computation.

We apply integration by parts by letting $U = x$ and $dV = \sin x\, dx$. This gives that $dU = dx$ and $V = -\cos x$. Therefore, we have

$$\int_0^{\pi} x\sin x\, dx = -x\cos x\Big|_0^{\pi} + \int_0^{\pi} \cos x\, dx$$

$$= \pi + \sin x\Big|_0^{\pi}$$

$$= \pi.$$

The final result is

$$\int_0^{\pi} x^2 \cos x\, dx = -2\pi.$$

Other types of integrals that you will see often are trigonometric integrals. In particular, integrals involving powers of sines and cosines. For odd powers, a simple substitution will turn the integrals into simple powers.

Example: Consider $\int \cos^3 x\, dx$.

This can be rewritten as

$$\int \cos^3 x\,dx = \int \cos^2 x \cos x\,dx$$

Let $u = \sin x$. Then $du = \cos x\,dx$. Since $\cos^2 x = 1 - \sin^2 x$, we have

$$\begin{aligned}\int \cos^3 x\,dx &= \int \cos^2 x \cos x\,dx \\ &= \int (1-u^2)\,du \\ &= u - \frac{1}{3}u^3 + C \\ &= \sin x - \frac{1}{3}\sin^3 x + C.\end{aligned}$$

A quick check confirms the answer:

$$\begin{aligned}\frac{d}{dx}\left(\sin x - \frac{1}{3}\sin^3 x + C\right) &= \cos x - \sin^2 x \cos x \\ &= \cos x\,(1 - \sin^2 x) = \cos^3 x.\end{aligned}$$

Even powers of sines and cosines are a little more complicated, but doable. In these cases we need the half angle formulae:

$$\sin^2 \alpha = \frac{1-\cos 2\alpha}{2},$$

$$\cos^2 \alpha = \frac{1-\cos 2\alpha}{2}.$$

Example: We will compute $\int_0^{2\pi} \cos^2 x\,dx$.

Substituting the half angle formula for $\cos^2 x$,

$$\begin{aligned}\int_0^{2\pi} \cos^2 x\,dx &= \frac{1}{2}\int_0^{2\pi} (1+\cos 2x)\,dx \\ &= \frac{1}{2}\left(x - \frac{1}{2}\sin 2x\right)_0^{2\pi} \\ &= \pi.\end{aligned}$$

We note that this result appears often in physics. When looking at root mean square averages of sinusoidal waves, one needs the average of the square of sines and cosines.

The average of a function on interval $[a; b]$ is given as

$$\text{fave} = \frac{1}{b-a}\int_a^b f(x)\,dx.$$

So, the average of $\cos^2 x$ over one period is

$$\frac{1}{2\pi}\int_0^{2\pi}\cos^2 x\,dx = \frac{1}{2}.$$

The root mean square is then $\frac{1}{\sqrt{2}}$.

TECHNOLOGY AND TABLES

Many of you know that some of the tedium can be alleviated by using computers, or even looking up what you need in tables. However, you also need to be comfortable in doing many computations by hand. This is necessary, especially in your early studies, for several reasons.

For example, you should try to evaluate integrals by hand when asked to do them. This reinforces the techniques, as outlined earlier. It exercises your brain in much the same way that you might jog daily to exercise your body. Who knows, keeping your brain active this way might even postpone Alzheimer's.

The more comfortable you are with derivations and evaluations. You can always use a computer algebra system, or a Table of Integrals, to check on your work.

Problems can arise when depending purely on the output of computers, or other "black boxes". Once you have a firm grasp on the techniques and a feeling as to what answers should look like, then you can feel comfortable with what the computer gives you. Sometimes, programs like Maple can give you strange looking answers, and sometimes wrong answers. Also, Maple cannot do every integral, or solve every differential equation, that you ask it to do.

Even some of the simplest looking expressions can cause computer algebra systems problems. Other times you might even provide wrong input, leading to erroneous results.

BACK OF THE ENVELOPE COMPUTATIONS

Dimensional analysis is useful for recalling particular relationships between variables by looking at the units involved, independent of the system of units employed. Though most of the time you have used SI, or MKS, units in most of your physics problems.

There are certain basic units - length, mass and time. By the second course, you found out that you could add charge to the list. We can represent these as [L], [M], [T] and [C]. Other quantities typically have units that can be expressed in terms of the basic units.

These are called derived units. So, we have that the units of acceleration are $[L]/[T]^2$ and units of mass density are $[M]/[L]^3$. Similarly, units of magnetic field can be found, though with a little more effort. $F = qvB \sin\theta$ for a charge q moving with speed v through a magnetic field B at an angle of θ $\sin \mu$ has no units. So,

$$[B] = \frac{[F]}{[q][v]}$$

$$= \frac{\frac{[M][L]}{[T]^2}}{[C]\frac{[L]}{[T]}}$$

$$= \frac{[M]}{[C][T]}.$$

Now, assume that you do not know how B depended on F, q and v, but you knew the units of all of the quantities. Can you figure out the relationship between them? We could write and solve for the exponents by inserting the dimensions.

$$[B] = [F]^{\alpha}[q]^{\beta}[v]^{\gamma}$$

Thus, we have

$$[M][C]^{-1}[T]^{-1} = \left([M][L][T]^{-2}\right)^{\alpha}[C]^{\beta}\left([L][T]^{-1}\right)^{\gamma}.$$

Right away we can see that $\alpha = 1$ and $\beta = -1$ by looking at the powers of [M] and [C], respectively. Thus,

$$[M][C]^{-1}[T]^{-1} = [M][L][T]^{-2}[C]^{-1}\ \left([L][T]^{-1}\right)^{\gamma}$$

$$= [M][C]^{-1}[L]^{-1+\gamma}[T]^{-2-\gamma}.$$

We see that picking $\gamma = -1$ balances the exponents and gives the correct relation. An important theorem at the heart of dimensional analysis is the Buckingham Theorem.

In essence, this theorem tells us that physically meaningful equations in n variables can be written as an equation involving n–m dimensionless quantities, where m is the number of dimensions used. The importance of this theorem is that one can actually compute useful quantities without even knowing the exact form of the equation!

The Buckingham Theorem was introduced by E. Buckingham in 1914. Let qi be n physical variables that are related by

$$f(q_1, q_2, ..., q_n) = 0$$

Assuming that m dimensions are involved, we let πi be $k = n - m$ dimensionless variables. Then the equation can be rewritten as a function of these dimensionless variables as $F(\pi_1, \pi_2, ... \pi_k) = 0$, where the π_i's can be written in terms of the physical variables as

$$\pi_i = q_1^{k_1} q_2^{k_2} \cdots q_n^{k_n}, \quad i = 1, \ldots, k.$$

Well, this is our first new concept and it is probably a mystery as to its importance. It also seems a bit abstract. However, this is the basis for some of the proverbial "back of the envelope calculations" which you might have heard about. So, let's see how it can be used.

Example: Let's consider the period of a simple pendulum; e.g., a point mass hanging on a massless string.

The period, T, of the pendulum's swing could depend upon the the string length, ℓ the mass of the "pendulum bob", m, and gravity in the form of the acceleration due to gravity, g. These are the q_i's in the theorem. We have four physical variables. The only units involved are length, mass and time. So, $m = 3$. This means that there are $k = n - m = 1$ dimensionless variables, call it π. So, there must be an equation of the form

$$F(\pi) = 0$$

in terms of the dimensionless variable

$$\pi = \ell^{k_1} m^{k_2} T^{k_3} g^{k_4}.$$

We just need to find the k_i's. This could be done by inspection, or we could write out the dimensions of each factor and determine how π can be dimensionless. Thus,

$$\begin{aligned}
[\pi] &= [\ell]^{k_1} [m]^{k_2} [T]^{k_3} [g]^{k_4} \\
&= [L]^{k_1} [M]^{k_2} [T]^{k_3} \left(\frac{[L]}{[T]^2}\right)^{k_4} \\
&= [L]^{k_1+k_4} [M]^{k_2} [T]^{k_3-2k_4}.
\end{aligned}$$

π will be dimensionless when

$$\begin{aligned}
k_1 + k_4 &= 0, \\
k_2 &= 0, \\
k_3 - 2k_4 &= 0.
\end{aligned}$$

This is a linear homogeneous system of three equations and four unknowns. We can satisfy these equations by setting $k_1 = -k_4$, $k_2 = 0$, and $k_3 = 2k_4$. Therefore, we have

$$\pi = \ell^{-k_4} T^{2k_4} g^{k_4} = \left(\ell^{-1} T^2 g\right)^{k_4}.$$

k_4 is arbitrary, so we can pick the simplest value, $k_4 = 1$ Then,

$$F\left(\frac{T^2 g}{\ell}\right) = 0.$$

Assuming that this equation has one zero, z, which has to be verified by other means, we have that

$$\frac{gT^2}{\ell} = z = \text{cost...}$$

Thus, we have determined that the period is independent of the mass and proportional to the square root of the length. The constant can be determined by experiment as $z = 4\pi^2$.

Example: A more interesting example was provided by Sir Geoffrey Taylor in 1941 for determining the energy release of an atomic bomb. Let's assume that the energy is released in all directions from a single point. Possible physical variables are the time since the blast, t, the energy, E, the distance from the blast, r, the atmospheric density ρ and the atmospheric pressure, p. We have five physical variables and only three units. So, there should be two dimensionless quantities. Let's determine these.

We set

$$\pi = E^{k_1} t^{k_2} r^{k_3} p^{k_4} \rho^{k_5}.$$

Inserting the respective units, we find that

$$\begin{aligned}[\pi] &= [E]^{k_1} [t]^{k_2} [r]^{k_3} [p]^{k_4} [\rho]^{k_5} \\ &= \left([M][L]^2 [T]^{-2}\right)^{k_1} [T]^{k_2} [L]^{k_3} \left([M][L]^{-1} [T]^{-2}\right)^{k_4} \left([M][L]^{-3}\right)^{k_5} \\ &= [M]^{k_1+k_4+k_5} [L]^{2k_1+k_3-k_4-3k_5} [T]^{-2k_1+k_2-2k_4}.\end{aligned}$$

Note: You should verify the units used. For example, the units of force can be found using $F = ma$ and work (energy) is force times distance. Similarly, you need to know that pressure is force per area.

For π to be dimensionless, we have to solve the system:

$$\begin{aligned} k_1 + k_4 + k_5 &= 0, \\ 2k_1 + k_3 - k_4 - 3k_5 &= 0, \\ -2k_1 + k_2 - 2k_4 &= 0. \end{aligned}$$

This is a set of three equations and five unknowns. The only way to solve this system is to solve for three unknowns in term of the remaining two. (In linear algebra one learns how to solve this using matrix methods.) Let's solve for k_1, k_2, and k_5 in terms of k_3 and k_4. The system can be written as

$$\begin{aligned} k_1 + k_5 &= 0, \\ 2k_1 - 3k_5 &= k_3, \\ 2k_1 - k_2 &= -2k_4. \end{aligned}$$

These can be solved by solving for k_1 and k_4 using the first two equations and then finding k_2 from the last one. Solving this system yields:

$$k_1 = -\frac{1}{5}(2k_4 + k_3) \quad k_2 = \frac{2}{5}(3k_4 - k_3) \quad k_5 = \frac{1}{5}(k_3 - 3k_4).$$

We have the freedom to pick values for k_3 and k_4. Two independent set of simple values would be to pick one variable as zero and the other as one. This will give our two dimensionless variables:

Case I. $k_3 = 1$ and $k_4 = 0$.

In this case we then have $k_3 = -\frac{1}{5}$, $k_2 = -\frac{2}{5}$, $k_5 = \frac{1}{5}$. This gives

$$\pi_1 = E^{-1/5}t^{-2/5}r\rho^{1/5} = r\left(\frac{\rho}{Et^2}\right)^{1/5}.$$

Case II. $k_3 = 0$ and $k_4 = 1$.

In this case we then have $k_1 = -\frac{2}{5}$, $k_2 = \frac{6}{5}$, and $k_5 = -\frac{3}{5}$.

$$\pi_2 = E^{-2/5}t^{6/5}p\rho^{-3/5} = p\left(\frac{t^6}{\rho^3 E^2}\right)^{1/5}.$$

Thus, we have that the relation between the energy and the other variables is of the form

$$F\left(r\left(\frac{\rho}{Et^2}\right)^{1/5}, p\left(\frac{t^6}{\rho^3 E^2}\right)^{1/5}\right) = 0$$

Of course, this is not enough to determine the explicit equation. However, Taylor was able to use this information to get an energy estimate.

Note that π_1 is dimensionless. It can be represented as a function of the dimensionless variable π_2. So, assuming that $\pi_1 = h(\pi_2)$, we have that

$$h(\pi_2) = r\left(\frac{\rho}{Et^2}\right)^{1/5}.$$

Note that for $t = 1$ second, the energy is expected to be huge, so $\pi_2 \approx 0$. Thus,

$$r\left(\frac{\rho}{Et^2}\right)^{1/5} \approx h(0)$$

Simple experiments suggest that $h(0)$ is of order one, so

$$r \approx \left(\frac{Et^2}{\rho}\right)^{1/5}.$$

In 1947 Taylor applied his earlier analysis to movies of the first atomic bomb test in 1945 and his results were close to the actual values. How can one do this? We can rewrite the above result to get the energy estimate:

$$E \approx \frac{r^5\rho}{t^2}.$$

TENSORS OF FIRST ORDER (VECTORIAL QUANTITIES)

The complete presentation of a vectorial quantity requires the indication of the amount, the direction and the unit. Power, velocity, momentum, angular momentum etc. are examples for vectorial quantities. Graphically vectors are represented by arrows, whose length indicates the amount and the position of the arrowhead, indicates the direction. The derivable analytical description of vectorial quantities makes use of the indication of a vector component projected onto the axis of a Cartesian coordinate system, and the indication of the direction is shown by the signs of the resulting vector components. To represent the velocity vector {Ui} in a Cartesian coordinate system the components $Ui(i = 1, 2, 3)$.

$$\vec{U} = \{U_i\} = \begin{Bmatrix} U_1 \\ U_2 \\ U_3 \end{Bmatrix} = |U| \begin{Bmatrix} \cos\alpha_1 \\ \cos\alpha_2 \\ \cos\alpha_3 \end{Bmatrix} \left[\frac{m}{s}\right] \rightsquigarrow U_i = \underbrace{\pm}_{\text{Direction}} \underbrace{|U| \cdot |\cos\alpha_i|}_{\text{Amount}} \underbrace{\left[\frac{m}{s}\right]}_{\text{Unit}}$$

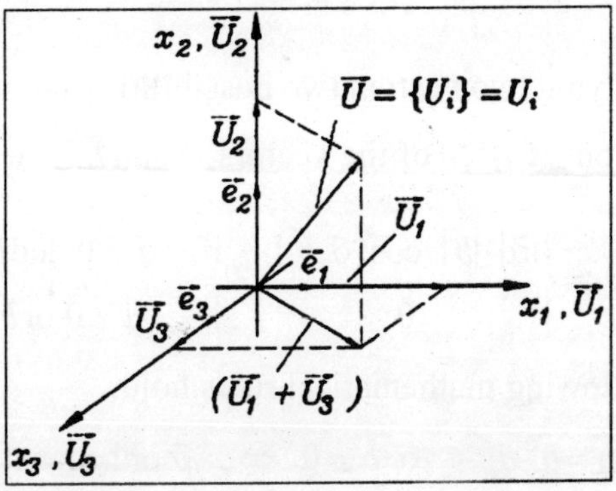

It holds: $\vec{U}_1 = U_1 \cdot \vec{e}_1, \vec{U}_2 = U_2 \cdot \vec{e}_2, \vec{U}_3 = U_3 \cdot \vec{e}_3$ where the unit vectors $\vec{e}_1, \vec{e}_2, \vec{e}_3$ in the coordinate directions x_1, x_2 and x_3 are employed. This is shown in Fig. $\alpha_1$designates the angle between $\breve{U}$ and the unit vector $\vec{e}_i$. Vectors can also be represented in other coordinate systems, through this the vector does not change in itself but its mathematical representation changes. Vector quantities which have the same unit can be added or subtracted vectorially. Laws are applied here that result in addition or subtraction of the components on the axes of a Cartesian coordinate system.

$$\vec{a} \pm \vec{b} = \{\alpha_i\} \pm \{b_i\} = \{(\alpha_i \pm b_i)\} = \{(\alpha_1 \pm b_1), (\alpha_2 \pm b_2), (\alpha_3 \pm b_3)\}$$

addition or subtraction of the components

Vectorial quantities with different units must not be added or subtracted vectorially. For the addition and subtraction of vectorial constants (having the same units) the following rules of addition hold:

$$\vec{\alpha}+\vec{0}=\{\alpha_i\}+\{0\}=\vec{\alpha} \quad \text{(neutral elements } \vec{0})$$

$$\vec{\alpha}+\{-\vec{\alpha}\}=\{\alpha_i\}+\{-\alpha_i\}=\vec{0} \quad (\vec{\alpha} \text{ element inverse to } -\vec{\alpha})$$

$$\vec{\alpha}+\vec{b}=\vec{b}+\vec{\alpha}, \text{ d.h.} \{\alpha_i\}+\{b_i\}=\{b_i\}+\{\alpha_i\}$$
$$=\{(\alpha_i+b_i)\} \quad \text{(commutative law)}$$

$$\vec{\alpha}+(\vec{b}+\vec{c})=(\vec{\alpha}+\vec{b})+\vec{c}, \text{ d.h.} \{\alpha_i\}+\{(b_i+c_i)\}$$
$$=\{(\alpha_i+b_i)\}+\{c_i\} \quad \text{(associative law)}$$

With $(\alpha\cdot\vec{\alpha})$ results a scalar multiple of $\vec{\alpha}$, if $\alpha>0$ have no unit of their own, i.e. $(\alpha\cdot\vec{\alpha})$ designates the vector that has the same direction as $\vec{\alpha}$ but α-times the amount.

In the case $\alpha<0$ one puts $(\alpha\cdot\vec{\alpha}):=-(|\alpha|\cdot\vec{\alpha})$. For $\alpha=0$ results the Zero vector $\vec{0}:0\cdot\vec{\alpha}=\vec{0}$.

When multiplying two vectors two possibilities should be distinguished:

The scalar product $\vec{\alpha}\cdot b$ of the vectors $\vec{\alpha}$ and $\vec{b}$ is defined as

$$\vec{\alpha}\cdot\vec{b}:=\begin{cases}|\vec{\alpha}|\cdot|\vec{b}|\cdot\cos(\vec{\alpha},\vec{b}) & \text{, if } \vec{\alpha}\neq\vec{0} \text{ and } \vec{b}\neq\vec{0}\\ 0 & \text{, if } \vec{\alpha}\neq\vec{0} \text{ or } \vec{b}\neq\vec{0}\end{cases}$$

where the following mathematical rules hold:

$$\vec{\alpha}\cdot\vec{b}=\vec{b}\cdot\vec{\alpha} \qquad \vec{\alpha}\cdot\vec{b}=0 \Leftrightarrow \vec{\alpha} \text{ orthogonal to } \vec{b}$$

$$(\alpha\vec{\alpha})\cdot\vec{b}\cdot\vec{\alpha}\cdot(\alpha\vec{b})=\alpha(\vec{\alpha}\cdot\vec{b}) \qquad |\vec{\alpha}| \underset{=}{\text{def}} \sqrt{\vec{\alpha}\cdot\vec{\alpha}}$$

$$(\vec{\alpha}+\vec{b})\cdot\vec{c}=\vec{\alpha}\cdot\vec{c}+\vec{b}\,\vec{c}$$

when vectors $\vec{\alpha}$ and $\vec{b}$ are represented in the Cartesian co-ordinates the following. Simple rules arise for the scalar product $(\vec{\alpha}\cdot\vec{b})$ and for $\cos(\vec{\alpha}\cdot\vec{b})$:

$$\vec{\alpha}\cdot\vec{b}=\alpha_1 b_1+\alpha_2 b_2+\alpha_3 b_3, |\alpha|=\sqrt{\alpha_1^2+\alpha_2^2+\alpha_3^2},$$

$$\cos(\vec{\alpha},\vec{b})=\frac{\vec{\alpha}\cdot\vec{b}}{|\vec{\alpha}||\vec{b}|}=\frac{\alpha_1 b_1+\alpha_2 b_2+\alpha_3 b_3}{\sqrt{\alpha_1^2+\alpha_2^2+\alpha_3^2}\sqrt{b_1^2+b_2^2+b_3^2}}$$

The above formulae hold for $\vec{\alpha},\vec{b}\neq 0$. Especially the directional cosine is calculated as

$$\cos(\vec{\alpha},\vec{e}_i)=\frac{|\alpha_i|}{\sqrt{\alpha_1^2+\alpha_2^2+\alpha_3^2}} \qquad i=1,2,3$$

i.e. as the angles between the vector $\vec{\alpha}$ and the base vectors

$$\vec{e}_1=\begin{Bmatrix}1\\0\\0\end{Bmatrix},\vec{e}_2=\begin{Bmatrix}0\\1\\0\end{Bmatrix},\vec{e}_3=\begin{Bmatrix}0\\0\\1\end{Bmatrix}.$$

The vector product $\vec{\alpha}\times\vec{b}$ Of the vectors $\vec{\alpha}$ and $\vec{b}$ has the below mentioned properties:

$\vec{\alpha}\times\vec{b}$ is a vector $\neq\vec{0}$, if $\vec{\alpha}\neq\vec{0}$ and $\vec{b}\neq\vec{0}$ and $\vec{\alpha}$ is not parallel to $\vec{b}$.

$|\vec{\alpha}\times\vec{b}|=|\vec{\alpha}|\cdot|\vec{b}|\sin(\vec{\alpha},\vec{b})$ (area of the parallelogram set up by $\vec{\alpha}$ and $\vec{b}$).

$\vec{\alpha}\times\vec{b}$ is a vector standing perpendicular to $\vec{\alpha}$ and $\vec{b}$ and can be presented with $(\vec{\alpha},\vec{b},\vec{\alpha}\times\vec{b})$ a right-handed system. It can easily be seen that $\vec{\alpha}\times\vec{b}=\vec{0}$, if $\vec{\alpha}=\vec{0}$ or $\vec{b}=\vec{0}$ or $\vec{\alpha}$ is parallel to $\vec{b}$ .One should take into consideration that for the vector product the associative law does not hold in general.

$$\vec{\alpha}\times(\vec{b}\times\vec{c})\neq(\vec{\alpha}\times\vec{b})\times\vec{c}$$

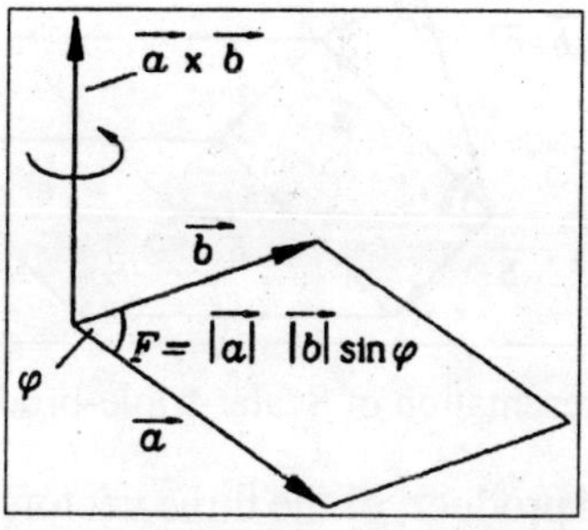

Fig. Graphical Representation of a Nector Product

$$\vec{\alpha}\times\vec{\alpha}=0,\ \vec{\alpha}\times\vec{b}=-(-\vec{b}\times\vec{\alpha}),$$

$$\alpha(\vec{\alpha}\times\vec{b})=(\alpha\vec{\alpha})\times\vec{b}=\vec{\alpha}\times(\alpha\vec{b})\ \ (\text{for } \alpha\in\mathrm{R}),$$

$$\vec{\alpha}\times(\vec{b}+\vec{c})=\vec{\alpha}\times\vec{b}+\vec{\alpha}\times\vec{c}$$

$$(\vec{\alpha}+\vec{b})\times\vec{c}=\vec{\alpha}\times\vec{c}+\vec{b}\times\vec{c} \qquad \text{(distributive laws)}$$

$$\vec{\alpha}\times\vec{b}=0 \Leftrightarrow \vec{\alpha}=0 \text{ or } \vec{b}=0 \text{ or } \vec{\alpha},\vec{b} \text{ parallel (parallelism test)},$$

$$|\vec{\alpha}\times\vec{b}|^2=|\vec{\alpha}|^2\cdot|\vec{b}|^2-(\vec{\alpha}\cdot\vec{b})^2.$$

If one represent the vectors $\vec{\alpha}$ and $\vec{b}$ in a Cartesian co-ordinate system $\vec{e}_i$ the following computation rules result:

$$\begin{Bmatrix}\alpha_1\\ \alpha_2\\ \alpha_3\end{Bmatrix}\times\begin{Bmatrix}b_1\\ b_2\\ b_3\end{Bmatrix}=\begin{vmatrix}\vec{e}_1 & \alpha_1 & b_1\\ \vec{e}_2 & \alpha_2 & b_2\\ \vec{e}_3 & \alpha_3 & b_3\end{vmatrix}=\begin{Bmatrix}\alpha_2 b_3 & -\alpha_3 b_2\\ \alpha_3 b_1 & -\alpha_1 b_3\\ \alpha_1 b_2 & -\alpha_2 b_1\end{Bmatrix}$$

The tensor of third order $\in_{ijk} := \vec{e}_i \cdot \left(\vec{e}_j \times \vec{e}_k\right)$ that permits moreover a computation according to $\{\alpha_i\}\times\{b_j\} := \{(\alpha\times b)k\} = \in_{ijk}\ \alpha_i b_j$

A combination of the scalar product and the vector product leads to the scalar-triple-product (STP) formed of three vectors

$$\left[\vec{\alpha},\vec{b},\vec{c}\right]=\vec{\alpha}\cdot\left(\vec{b}\times\vec{c}\right)$$

The properties of this product from three vectors can be seen from the sketch shown below. The scalar-triple-product (STP) of the vectors $\vec{\alpha},\vec{b},\vec{c}$ leads to the six times the volume of parallelopiped (ppd), Vppd defined by the vectors $\vec{\alpha},\vec{b}$ and $\vec{c}$.

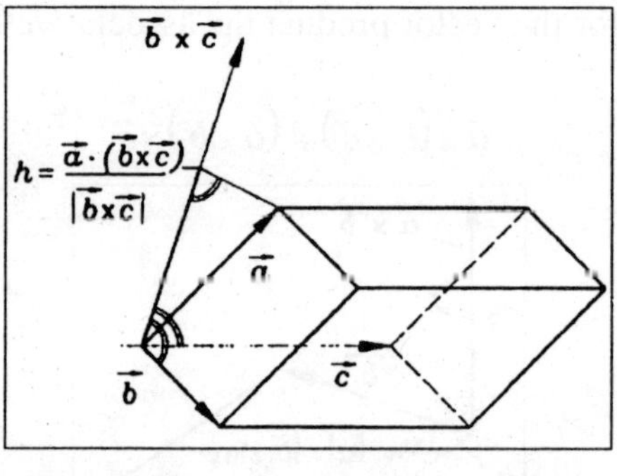

Fig. Graphical Representation of Scalar-triple-product by Three Vectors

The 'parallelopiped product' of the three vectors $\vec{\alpha},\vec{b},\vec{c}$ is calculated from the value of a triple-row determinant

$$\left[\vec{\alpha},\vec{b},\vec{c}\right]=\begin{vmatrix}\alpha_1 & b_1 & c_1\\ \alpha_2 & b_2 & c_2\\ \alpha_3 & b_3 & c_3\end{vmatrix}$$

$$V_{\text{ppd}}=\frac{1}{6}V_{\text{STP}}=\frac{1}{6}\left[\vec{\alpha},\vec{b},\vec{c}\right]$$

It is easy to show that for the scalar-triple-product

$$\vec{\alpha}\cdot\left(\vec{b}\times\vec{c}\right)=\vec{b}\cdot\left(\vec{\alpha}\times\vec{c}\right)=\vec{c}\cdot\left(\vec{\alpha}\times\vec{b}\right)$$

For the triple vector product $\vec{\alpha} \times \vec{b} \times \vec{c}$ the following relations hold:

$$\vec{\alpha} \cdot (\vec{b} \times \vec{c}) = (\vec{\alpha} \cdot \vec{c})\vec{b} - (\vec{\alpha} \times \vec{b})\vec{c}$$

TENSORS OF SECOND ORDER

In the preceding two sections tensors of zero order (scalar quantities) and tensors of first order (vectorial quantities) were introduced. In this chapter, a short summary concerning tensors of second order is included which can be formulated as matrices with nine elements:

$$\{\alpha_{ij}\} = \begin{Bmatrix} \alpha_{11} & \alpha_{12} & \alpha_{13} \\ \alpha_{21} & \alpha_{22} & \alpha_{23} \\ \alpha_{31} & \alpha_{32} & \alpha_{33} \end{Bmatrix} = \alpha_{ij}$$

In the matrix element, α_{ij} ; the index i represents the number of the row and j represents the number of the column, and the elements designating with i = j are referred to as the diagonal elements of the matrix. A tensor of second order is called symmetrical when aij = aji holds. The unit second order tensor is expressed by the Kronecker symbol:

$$\{\delta_{ij}\} = \begin{Bmatrix} 1 & 0 & 0 \\ 0 & 1 & 0 \\ 0 & 0 & 1 \end{Bmatrix}, i.e. \delta_{ij} = \begin{cases} +1 & \text{if } i = j \\ 0 & \text{if } i = j \end{cases}$$

The transposed tensor of $\{a_{ij}\}$ is formed by exchanging the rows and columns of the tensor: $\{a_{ij}\}^{T} = \{a_{ji}\}$.

When doing so, it is apparent that the transposed unit tensor of second order is again the unit tensor, i.e. $\delta_{ij}^{T} = \delta_{ij}$ As the sum or difference of two tensors of second order is defined that tensor of second order whose elements are formed from the sum or difference of the corresponding ij-elements of the initial tensors:

$$\{a_{ij} \pm b_{ij}\} = \begin{Bmatrix} a_{11} \pm b_{11} a_{12} \pm b_{12} a_{13} \pm b_{13} \\ a_{21} \pm b_{21} a_{22} \pm b_{22} a_{23} \pm b_{23} \\ a_{31} \pm b_{31} a_{32} \pm b_{32} a_{33} \pm b_{33} \end{Bmatrix}$$

In the case of the following presentation of tensor products often the so called Einstein's summation convention is applied. By this one understands the summation over the same indices in a product. When forming a product from tensors, one distinguishes the outer product and the inner product.

The outer product is again a tensor, where each element of the first tensor multiplied with each element of the second tensor results in an element of the new tensor. Thus the product of a scalar and a tensor of second order forms a tensor of second order, where each element results from the initial tensor of second order by scalar multiplication:

$$\alpha \cdot \{a_{ij}\} = \{\alpha \cdot a_{ij}\} = \begin{Bmatrix} \alpha \cdot a_{11} \alpha \cdot \alpha_{12} \alpha \cdot a_{13} \\ \alpha \cdot a_{21} \alpha \cdot a_{22} \alpha \cdot a_{23} \\ \alpha \cdot a_{31} \alpha \cdot \alpha_{32} \alpha \cdot a_{33} \end{Bmatrix}$$

The outer product of a vector (tensor of 1st order) and a tensor of 2nd order results in a tensor of 3rd order with altogether 27 elements. The inner product of tensors, however, can result in a contraction of the order. As examples are cited the products $a_{ij} \cdot b_j$:

$$\{a_{ij}\} \cdot \{b_j\} = \begin{Bmatrix} a_{11} & a_{12} & a_{13} \\ a_{21} & a_{22} & a_{23} \\ a_{31} & a_{32} & a_{33} \end{Bmatrix} \begin{Bmatrix} b_1 \\ b_2 \\ b_3 \end{Bmatrix} = \begin{Bmatrix} a_{11}b_1 + a_{12}b_2 + a_{13}b_3 \\ a_{21}b_1 + a_{22}b_2 + a_{23}b_3 \\ a_{31}b_1 + a_{32}b_2 + a_{33}b_3 \end{Bmatrix}$$

and

$$\{b_i\}^T \cdot \{a_{ij}\} = \{b_1, b_2, b_3\} \cdot \begin{Bmatrix} a_{11} & a_{12} & a_{13} \\ a_{21} & a_{22} & a_{23} \\ a_{31} & a_{32} & a_{33} \end{Bmatrix} = \begin{Bmatrix} b_1a_{11} + b_2a_{21} + b_3a_{31} \\ b_1a_{12} + b_2a_{22} + b_3a_{32} \\ b_1a_{13} + b_2a_{23} + b_3a_{33} \end{Bmatrix}^T$$

in summary this can be written as:

$\{a_{ij}\} \cdot \{b_j\} = \{(a_{ij} b_j)\} = \{(ab)_i\}$ respectively.

$$\{b_j\} \cdot \{a_{ij}\} = \{(b_i a_{ij})\} = \{(ab)_j\}$$

If one takes into account the above product laws

$$\{\delta_{ij}\} \cdot \{b_j\} = \{b_i\} \text{ and } \{b_{ij}\}^T \cdot \{\delta_{ij}\} = \{b_j\}^T$$

The multiplication of a tensor of second order by the unit tensor of second order, i.e. the 'Kronecker Delta', yields the initial tensor of second order

$$\{\delta_{ij}\} \cdot \{\alpha_{ij}\} = \begin{Bmatrix} 1 & 0 & 0 \\ 0 & 1 & 0 \\ 0 & 0 & 1 \end{Bmatrix} \cdot \begin{Bmatrix} \alpha_{11} & \alpha_{12} & \alpha_{13} \\ \alpha_{21} & \alpha_{22} & \alpha_{23} \\ \alpha_{31} & \alpha_{32} & \alpha_{33} \end{Bmatrix} = \{\alpha_{ij}\}$$

Further products can be formulated, as for example cross products between vectors and tensors of second order

$$\{\alpha_i\} \cdot \{b_{jk}\} = \epsilon_{ikl} \cdot \alpha_i \cdot b_{jk}$$

but these are not of special importance for the laws in fluid mechanics.

FIELD VARIABLES AND MATHEMATICAL OPERATIONS

In fluid mechanics it is usual to present thermodynamic state quantities of fluids, like density, pressure P, temperature T, internal energy e etc as a function of space and time, a Cartesian coordinate system being applied here generally.

To each point $p(x_1,x_2,x_3)=p(x_i)$ a value $\rho(x_i,t),P(x_i,t),$ $T(x_i,t),e(x_i,t)$ etc. is assigned, i.e. the entire fluid properties are presented as field variables and are thus functions of space and time. It is assumed that in each space the thermodynamics connections between the state quantities hold, as for example the state equations that can be formulated for thermodynamically ideal fluids as follows ρ = const (state equation of the thermodynamically ideal liquids) $P/\rho=RT$ (state equation of the thermodynamically ideal gases) In an analogous way the properties of the flows can be described by introducing the velocity vectors and their components as functions of space and time, i.e. as vector fields $u_j(x_i,z)$.

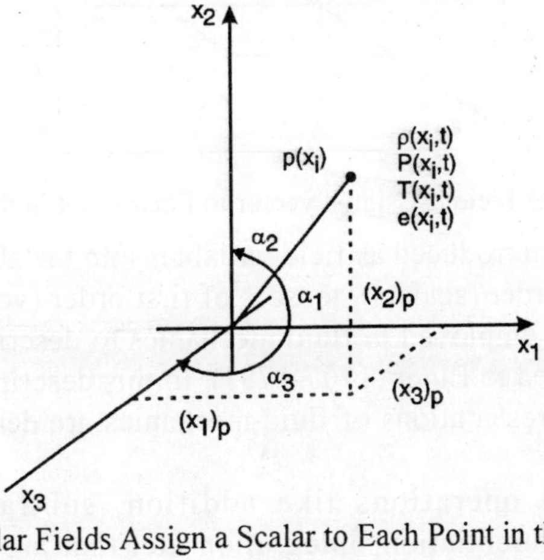

Fig. Scalar Fields Assign a Scalar to Each Point in the Space

Furthermore, the local rotation $\vec{w}=w_j(x_i,t)$ of the flow field as a field variable can be introduced into considerations of a flow field as well as the mass forces and mass accelerations, reacting locally on the fluid. Entirely analogous to this, the properties of the flows can be described by introducing the velocity vectors and their components as functions of space and time, i.e. as vector fields.

Furthermore the local rotation of the flow field can be included as a field quantity in considerations taken with respect to a flow, as well as the mass forces and mass acceleration acting locally on the fluid.

Thus the velocity $\vec{U}_j=U_j(x_i,t)$, the rotation $\vec{w}_j=w_j(x_i,t)$, the force

$\vec{K}_j = K_j\left(x_i,t\right)$ and the acceleration $\vec{g}_j\left(x_i,t\right)$ can be stated as field quantities and be employed as such quantities in the following considerations.

Analogously tensors of second and higher order can also be introduced as field variables.

For example $\tau_{ij}\left(x_i,t\right)$ which is the molecule caused momentum transport existing in the space i.e. at the point $p\left(x_i\right)$ at time t for the velocity components U_j acting in the direction.

Further represents the fluid-element deformation depending on the gradients of the velocity field at the location $p\left(x_i\right)$ at the time t.

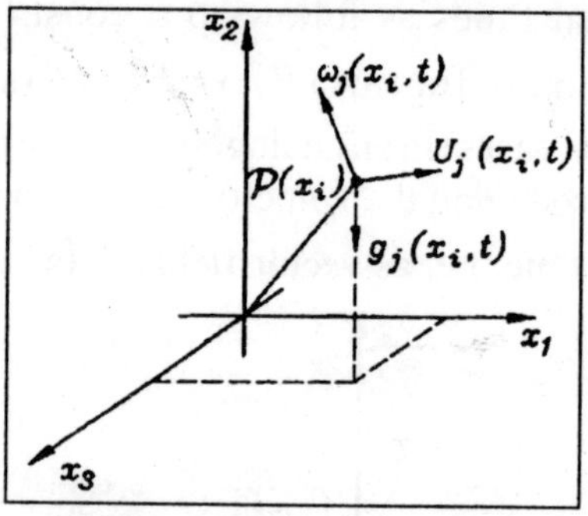

Fig. Vector Fields Assign a vector to Each Point in the Space

The properties introduced as field variables into the above presentations of tensors of zero order (scalars), tensors of first order (vectors) and tensors of second order are employed in fluid mechanics to describe the fluid flows and usually attributed to Euler (1707-1783). In this description all quantities considered in the presentations of fluid-mechanics are dealt as functions of space and time.

Mathematical operations like addition, subtraction, division, multiplication, differentiation, integration etc. that are applied to these quantities, are subject to the known laws of mathematics. For the differentiation of a scalar fields, for example density ρ, gives:

$$\frac{d\rho}{dt} = \frac{\partial\rho}{\partial t} + \frac{\partial\rho}{\partial x_i}\left(\frac{dx_1}{dt}\right) + \frac{\partial\rho}{\partial x_2}\left(\frac{dx_2}{dt}\right) + \frac{\partial\rho}{\partial x_3}\left(\frac{dx_3}{dt}\right)$$

$$= \frac{\partial\rho}{\partial t} + \sum_{i=1}^{3}\left(\frac{\partial\rho}{\partial x_i}\right)\left(\frac{dx_i}{dt}\right) = \frac{\partial\rho}{\partial t} + \left(\frac{\partial\rho}{\partial x_i}\right)\left(\frac{dx_i}{dt}\right)$$

In the last term, the summation symbol $\sum_{i=1}^{3}$ was omitted and the

"Einstein's summation convention" was employed, according to which the double index *i*: in $\left(\frac{\partial\rho}{\partial x_4}\right)\left(\frac{dx_4}{dt}\right)$ prescribes a summation over three terms

$i = 1,2,3$ i.e., $\sum_{i=1}^{3}\left(\frac{\partial\rho}{\partial x_i}\right)\left(\frac{dx_i}{dt}\right) = \frac{\partial\rho}{\partial x_i}\frac{dx_i}{dt}$

The differentiation of vectors yield is given by the following expressions:

$$\frac{d\vec{U}}{dt} = \left\{\frac{dU_1}{dt}, \frac{dU_2}{dt}, \frac{dU_3}{dt}\right\}^T \Rightarrow \frac{dU_i}{dt}, \qquad i = 1,2,3$$

i.e. each component of the vector is included in the differentiation. As the considered velocity vector depends on space xi and the time t, the following differentiation law holds:

$$\frac{d\vec{U}_j}{dt} = \frac{\partial U_j}{\partial t} + \frac{\partial U_j}{\partial x_i}\left(\frac{dx_i}{dt}\right)$$

When one applies the Nabla operator (or) Del operator

$$\nabla = \left\{\frac{\partial}{\partial x_1}, \frac{\partial}{\partial x_2}, \frac{\partial}{\partial x_3}\right\}^T = \left\{\frac{\partial}{\partial x_i}\right\}, \qquad i = 1,2,3$$

on a scalar field quantity a vector results

$$\nabla\alpha = \left\{\frac{\partial\alpha}{\partial x_1}, \frac{\partial\alpha}{\partial x_2}, \frac{\partial\alpha}{\partial x_3}\right\}^T = \text{grad } \alpha = \left\{\frac{\partial\alpha}{\partial x_i}\right\}, \qquad i = 1,2,3$$

This shows that the Nabla or Del operator ∇ results in a vector field, the gradient field. The different components of the resulting vector are formed from the prevailing partial derivations of the scalar field in the directions xi. The scalar product of the ∇ operator with a vector yields scalar variables:

$$\nabla\cdot\vec{\alpha} = \frac{\partial\alpha_1}{\partial x_1} + \frac{\partial\alpha_2}{\partial x_2} + \frac{\partial\alpha_3}{\partial x_3} = \text{div } \alpha = \frac{\partial\alpha_i}{\partial x_i}$$

Here in $\partial\alpha_i / \partial x_i$ the double index i indicates again the summation over all three terms, i.e.

$\sum_{i=1}^{3}\frac{\partial\alpha_i}{\partial x_i} \Rightarrow \frac{\partial\alpha_i}{\partial x_i}$ (Einstein's summation convention)

The vector product of the ∇ operator with the vector $\vec{\alpha}$: yields correspondingly:

$$\nabla\times\vec{\alpha}=\begin{vmatrix}e_1 & \partial/\partial x_1 & \alpha_1\\ e_2 & \partial/\partial x_2 & \alpha_2\\ e_3 & \partial/\partial x_3 & \alpha_3\end{vmatrix}=\begin{Bmatrix}\partial\alpha_3/\partial x_2 & -\partial\alpha_2/\partial x_3\\ \partial\alpha_1/\partial x_3 & -\partial\alpha_3/\partial x_1\\ \partial\alpha_2/\partial x_1 & -\partial\alpha_1/\partial x_2\end{Bmatrix}=\text{rot}\,\vec{\alpha}$$

or

$$\nabla\times\vec{\alpha}=\text{rot}\,\vec{\alpha}=-\in_{ijk}\frac{\partial\alpha_i}{\partial x_j}e_k=\in_{ijk}\frac{\partial\alpha_j}{\partial x_i}e_k$$

the Levi-Civita symbol called alternating unit tensor $\in_{ijk}$ is defined as follows:

$$\in_{ijk}=\begin{cases}0: \text{ if two of the three indices are equal}\\ +1: \text{if} \quad ijk=123, 231 \text{ or } 312\\ -1: \text{if} \quad ijk=132, 213 \text{ or } 321\end{cases}$$

Concerning the above-mentioned products of the ∇ operator the distributive law holds, but not the commutative and associative laws. If one applies the ∇ operator to the gradient field of a scalar function, the Laplace operator ∇^2 (alternative way of notation ∇),that when applied to a can be written as follows:

$$\nabla^2\alpha=(\nabla\cdot\nabla)\alpha=\frac{\partial^2\alpha}{\partial x_1^2}+\frac{\partial^2\alpha}{\partial x_2^2}+\frac{\partial^2\alpha}{\partial x_3^2}=\frac{\partial^2}{\partial x_i\partial x_i}$$

The Laplace operator can also be applied to vector fields of the components.

$$\nabla^2\vec{U}=\begin{pmatrix}\nabla^2U_1\\ \nabla^2U_2\\ \nabla^2U3\end{pmatrix}=\begin{pmatrix}(\partial^2U_1/\partial x_1^2)+(\partial^2U_1/\partial x_2^2)+(\partial^2U_1/\partial x_3^2)\\ (\partial^2U_2/\partial x_1^2)+(\partial^2U_2/\partial x_2^2)+(\partial^2U_2/\partial x_3^2)\\ (\partial^2U_3/\partial x_1^2)+(\partial^2U_3/\partial x_2^2)+(\partial^2U_3/\partial x_3^2)\end{pmatrix}$$

FUNCTIONAL ANALYSIS

METRIC SPACES

Definition: A metric space is a set X and a mapping $d: X \times X \to \mathbb{R}$, called a metric, which satisfies:

- $d(x, y) \geq 0$
- $d(x, y) = 0 \Leftrightarrow x = y$
- $d(x, y) = d(y, x)$
- $d(x, y) \leq d(x, z) + d(z, y)$

Definition: A sequence $\{x_n\}_{n=1}^{\infty}$, of elements of a metric space $(X; d)$ is said to converge to an element $x \in X$ if $d(x; x_n) \xrightarrow{n\to\infty} 0$.

Definition: Let (X, d) be a metric space.

- The set $B(y, r) = \{x \in X \mid d(x, y) < r\}$ is called the open ball of radius r about y; the set $B(y, r) = \{x \in X| \, d(x, y) \leq r|$ is calle the closed ball of radius r about y; the set $S(y, r) = |x \in X| \, d(x, y) = r\}$ is called the sphere of radius r about y;
- A set $O \subset X$ is called open if $\forall y \in O \, \exists r > 0$: $B(y, r) \leq O$;
- A set $N \subset X$ is called a neighborhood of $y \in N$ if $\exists r > 0$: $B(y, r) \in N$;
- A point $x \in X$ is a limit point of a set $E \subset X$ if $\forall r > 0$: $B(x; r) \cap (E \backslash \{x\}) = \emptyset$, i.e. if E contains points other than x arbitrarily close to x;
- A set $F \subset X$ is called closed if F contains all its limit points;
- $x \in G \leq X$ is called an interior point of G if G is a neighborhood of x.
- The intersection S of all closed sets containing $S \leq X$ is called the closure of S. The closure of $S \leq X$ is the smallest set containg S.
- The interior of S, S°, is the set of interior points. It is the largest open set contained in S.
- The boundary of S is the set $\partial S = \overline{S} \backslash S^\circ$.

Theorem: Let (X, d) be a metric space.

- A set O is open iff $X \backslash O$ is closed
- $x_n \xrightarrow{d} x$ iff $\forall$ neighborhood N of $x \, \exists m$: $n \geq m$ implies $x_n \in N$;
- The set of interior points of a set is open;
- The union of a set and all its limit points is closed;
- A set is open iff it is a neighborhood of each of its points.
- The union of any number of open sets is open.
- The intersection of finite number of open sets is open.
- The union of finite number of closed sets is closed.
- The intersection of any number of closed sets is closed.
- The empty set and the whole space are both open and closed.

Theorem: *A* subset S of a metric space X is closed iff every convergent sequence in S has its limit in S, i.e.

$$\{x_n\}_{n=1}^{\infty}, \quad x_n \in S, x_n \to x \quad \Rightarrow \quad x \in S$$

Theorem: The closure of a subset S of a metric space X is the set of limits of all convergent sequences in S, i.e.

$$\overline{S} = \{x \in X| \, \exists x_n \in S: x_n \to x\}.$$

Definition: A subset, $Y \subset X$, of a metric space (X, d) is called dense if

$$\forall x \in X \, \exists \{y_n\}_{n=1}^{\infty}, y_n \in Y,: y_n \xrightarrow{d} x.$$

Theorem: Let S be a subset in a metric space X. Then the following conditions are equivalent:

- S is dense in X,

- $\overline{S} = X$,
- Every non-empty subset of X contains an element of S.

Definition: A metric space X is called separable if it has a countable dense set.

Definition: A subset $S \subset X$ of a metric space X is called compact if every sequence $\{x_n\}$ in S contains a convergent subsequence whose limit belongs to. S.

Theorem: Compact sets are closed and bounded.

Definition: A sequence $\{x_n\}_{n=1}^{\infty}$, of elements of a metric space (X, d) is called a Cauchy sequence if $\forall \in > 0\ \exists N$: n; $m \geq N$ implies $d(x_n, x_m) < \in$.

Proposition: Any convergent sequence is Cauchy.

Definition: A metric space in which all Cauchy sequences converge is called complete.

Definition: A mapping f: $X \to Y$ from a metric space (X, d) to a $\leq$ metric space $(Y; \rho)$ is called continuous at x if $f(x_n) \xrightarrow{\rho} (x) \forall \{x_n\}_{n=1}^{\infty}$,

$x_n \in \xrightarrow{d} x$, i.e. the image of a convergent sequence converges to the image of the limit.

Definition: A bijection (one-to-one onto mapping) h: $X \to Y$ from (X, d) to (Y, ρ) is called isometry if it preserves the metric, i.e.

$$\rho(h(x), h(y)) = d(x, y) \qquad \forall x; y \in X$$

Proposition: Any isometry is continuous.

Theorem: If (X, d) is an incomplete metric space, it is possible to$\leq$ nd a complete metric space $(\overline{X}, \overline{d})$ so that X is isometric to a dense subset of $\overline{X}$.

VECTOR SPACES

Definition: A complex vector space is a nonempty set V with two operations: $+$: $V \times V \to V$ and $\cdot$: $\mathbb{C} \leq V \to V$ that satisfy the following conditions:

$$\forall x, y, z \in V$$

- $x + y = y + x$
- $(x + y) + z = x + (y + z)$
- $\exists 0 \in V$: $\forall x \in V$: $x + 0 = x$
- $\forall x \in V\ \exists(-x)\ 2\ V$: $x + (-x) = 0$
- $\forall \alpha, \beta \in \mathbb{C}, \forall x, y \in V$
- $\alpha(\beta x) = (\alpha\beta)x$
- $(\alpha + \beta)x = \alpha x + \beta x$
- $\alpha(x + y) = \alpha x + \alpha y$

- $1 \cdot x = x$

A real vector space is defined similarly.

Examples: (Function Spaces). Let $\Omega \subset \mathbb{R}^n$ be an open subset of $\mathbb{R}^n$.

- $P(\Omega)$ is the space of all polynomials of n variables as functions on Ω.
- $C(\Omega)$ is the space of all continuous complex valued functions on Ω.
- $C^k(\Omega)$ is the space of all complex valued functions with continuous partial derivatives of order k on Ω.
- $C^\infty(\Omega)$ is the space of all infinitely differentiable complex valued (smooth) functions on Ω.

Example: (Sequence Spaces (l^p-Spaces)). Let $p \geq 1$. l^p is the space of all infinite sequences $\{z_n\}_{n=1}^{\infty}$ of complex numbers such that $\left(\sum_{n=1}^{\infty} |z_n|^p\right)^{\frac{1}{p}} < \infty$.

Definition: Let V be a complex vector space and let $x_1,...x_k \in V$ and $\alpha_1,...,\alpha_k \in C$. A vector $x = \alpha_1 x_1 + \cdots \alpha_k x_k$ is called a linear combination of $x_1,... x_k$.

Definition: *A* finite collection of vectors $x_1,..., x_k$ is called linearly independent if

$$\sum_{i=1}^{k} \alpha_i x_i = 0 \quad \Leftrightarrow \quad \alpha_i = 0, \quad i = 1, 2, \ldots, k.$$

An arbitrary colection of vectors $B = \{x_n\}_{n=1}^{\infty}$ is called linearly independent if every finite subcollection is linearly independent. A collection of vectors which is not linearly independent is called linearly dependent.

Definition: Let $B \subset V$ be a subset of a vector space V. Then span B is the set of all finite linear combinations of vectors from B

$$\text{span } B = \left\{ \sum_{i=1}^{k} \alpha_i x_i \,\middle|\, x_i \in B, \alpha_i \in \mathbb{C}, k \in \mathbb{N} \right\}$$

Proposition: Let $B \subset V$ be a subset of a vector space V. Then span B is a subspace of V.

Definition: A set of vectors $B \subset V$ is called a basis of V (or a base of V) if B is linearly independent and span $B = V$. If $\exists$ a finite basis in V, then V is called finite dimensional vector space. Otherwise V is called in finite dimensional vector space.

Proposition: The number of vectors in any basis of a finite dimensional vector space is the same.

Definition: The number of vectors in a basis of a finite dimensional vector space is called the dimension of V, denoted by dim V.

NORMED LINEAR SPACES

Definition: A normed linear space is a vector space, V, over $\mathbb{C}$ (or R) and a mapping $\|\cdot\|: V \to R$, called a norm, that satisfies:

- $\|v\| \geq 0 \; \forall v \in V$
- $\|v\| = 0 \Leftrightarrow v = 0$
- $\|\alpha v\| = |\leq| \, \|v\| \quad \forall v \in V, \forall \alpha \in \mathbb{C}$
- $\|v + w\| \leq \|v\| + \|w\| \quad \forall v, w \in V$

Examples:

- Norms in $\mathbb{R}^n$:

$$\|x\|_2 = \left(\sum_{i=1}^{k} x_i^2 \right)^{\frac{1}{2}}$$

$$\|x\|_1 = \sum_{i=1}^{n} |x_i|$$

$$\|x\|_\infty = \max_{1 \leq i \leq n} \{|x_i|\}$$

- A norm in $\mathbb{C}^n$

$$\|z\| = \left(\sum_{i=1}^{k} |z_i|^2 \right)^{\frac{1}{2}}$$

- Let $\overline{\Omega}$ $\mathbb{R}^n$ be a closed bounded subset of $\mathbb{R}^n$ and $dx = dx_1 \ldots dx_n$ be a measure in $\mathbb{R}^n$. Norms in $C(\Omega)$ can be defined by

$$\|f\|_\infty = \sup_{x \in \Omega} |f(x)|$$

$$\|f\|_p = \left(\int_\Omega |f(x)|^p \, dx \right)^{\frac{1}{p}}$$

$$\|f\|_1 = \int_\Omega |f(x)| \, dx$$

- A norm in l^p

$$\|z\| = \left(\sum_{i=1}^{k} |z_i|^p \right)^{\frac{1}{p}}$$

- A norm in l^∞

$$\| z \| = \sup_{x \in \mathbb{N}} |z_n|$$

Proposition: A normed linear space $(V, \| \cdot \|)$ is a metric space (V, d) with the induced metric $d(v, w) = \|v - w\|$.

Convergence, open and closed sets, compact sets, dense sets, completeness, in a normed linear space are defined as in a metric space in the induced metric.

Definition: A normed linear space is complete if it is complete as a metric space in the induced metric.

Definition: A complete normed linear space is called the Banach space.

Definition: A bounded linear transformation from a normed linear space $(V, \|\cdot\|_V)$ to a normed linear space $(W, \|\cdot\|_W)$ is a mapping $T: V \to W$ that satisfies:

- $T(\alpha v + \beta w) = \alpha T(v) + \beta T(w), \quad \forall v, w \in V, \forall \alpha, \beta \in \mathbb{C}$,
- $\|T(v)\|_V \leq C\|v\|_W$ for some $C \leq 0$.
- The number

$$\|T\| = \sup_{v \in V\, v \neq 0} \frac{\| T(v) \|_W}{\| v \|_V}$$

is called the norm of T.

Theorem: Any bounded linear tranformation between two normed linear spaces is continuous.

Theorem: A bounded linear transformation, $T: V \to W$, from a normed linear space $(V, \| \cdot \|_V)$ to a complete normed linear space $(W, \| \cdot \|_W)$ can be uniquely extended to a bounded linear transformation, T, from the completion V of V to $(W, \| \leq \|_W)$. The extension of T preserves the norm $\|T\| = \|T\|$.

NOTES ON LEBESGUE INTEGRAL

Definition: Characteristic function of a set $A \subset X$ is a mapping $\chi_A: X \to \{0, 1\}$ defined by

$$\chi_A(x) = \begin{cases} 1, & \text{if } x \in A \\ 0, & \text{if } x \notin A \end{cases}$$

Definition: For a non-zero function $f: \mathbb{R}^n \to \mathbb{R}$, the set, supp f, of all points $x \in \mathbb{R}^n$ for which $f(x) \neq 0$ is called the support of f, i.e.

$$\text{supp} f = \{x \in \mathbb{R}^n | f(x) \neq 0\}.$$

Clearly, supp $\chi_A = A$.

Definition: Let I be a semi-open interval in $\mathbb{R}^n$ defined by

$$I = \{x \in \mathbb{R}^n \mid a_k \leq x_k < b_k,\ k = 1, ..., n\}$$

for some $a_k < b_k$. The measure of the set I is defined to be

$$\mu(I) = (b_1 - a_1) .... (b_n - a_n):$$

The Lebesgue integral of a characteristic function of the set I is defined

$$\int \chi_I \, dx = \mu(I)$$

Definition: A fiinite linear combination of characteristic functions of semiopen intervals

$$f = \sum_{k=1}^{N} \alpha_k \chi I_k$$

is called a step function.

Definition: The Lebesgue integral of a step function is defined by linearity

$$\int \sum_{k=1}^{N} \alpha_k \chi I_k \, dx = \sum_{k=1}^{N} \alpha_k \mu(I_k)$$

Definition: A function $f: \mathbb{R}^n \to \mathbb{R}$ is Lebesgue integrable if $\exists$ a sequence of step functions $\{f_k\}$ such that

$$f = \sum_{k=1}^{\infty} f_k,$$

which means that two conditions are satisfied

- $\sum_{k=1}^{\infty} \int |f_k| dx < \infty$
- $f(x) = \sum_{k=1}^{\infty} f_k(x) \quad \forall x \in \mathbb{R}^n$ such that $\sum_{k=1}^{\infty} |f_k(x)| < \infty$

The Lebesgue integral of f is then defined by

$$\int f \, dx = \sum_{k=1}^{\infty} \int f_k \, dx$$

Proposition: The space, $L^1(\mathbb{R}^n)$, of all Lebesgue integrable functions on Rn is a vector space and $\int$ is a linear functional on it.

Theorem: If $f, g \in L^1(\mathbb{R}^n)$ and $f \leq g$, then $\int f dx \leq \int g dx$.

If $f \in L^1(\mathbb{R}^n)$, then $|f| \in L^1(\mathbb{R}^n)$ and $|\int f dx| \leq \int f dx \leq \int g dx$.

Theorem: If $\{f_k\}$ is a sequence of integrable functions and

$$f = \sum_{k=1}^{\infty} f_k,$$

then $$\int f = \sum_{k=1}^{\infty} \int f_k ,$$

Definition: The L^1-norm in $L^1(\mathbb{R}^n)$ is defined by

$$\|f\| = \int |f|\,dx$$

Definition: A function f is called a null function is it is integrable and $\|f\| = 0$. Two functions f and g are said to be equivalent if $f - g$ is a null function.

Definition: The equivalence class of $f \in L^1(\mathbb{R}^n)$, denoted by $[f]$, is the set of all functions equivalent to f.

Remark: Strictly speaking, to make $L^1(\mathbb{R}^n)$ a normed space one has to consider instead of functions the classes of equivalent functions.

Definition: A set $X \subset \mathbb{R}^n$ is called a null set (or a set of measure zero) if its characteristic function is a null function.

Theorem:

- Every countable set is a null set.
- A countable union of null sets is a null set.
- Every subset of a null set is a null set.

Definition: Two integrable functions, $f, g \in L^1(\mathbb{R}^n)$, are said to be equal almost everywhere, $f = g$ a.e., if the set of all $x \in R^n$ for which $f(x) \neq g(x)$ is a null set.

Theorem:

$$f = g \text{ a.e} \Leftrightarrow \| f - g \| = \int |f - g| = 0$$

Theorem: The space $L^1(\mathbb{R}^n)$ is complete.

HILBERT SPACES

GEOMETRY OF HILBERT SPACE

Definition: A complex vector space V is called an inner product space (or a pre-Hilbert space if there is a mapping (.,.): $V \leq V \to \mathbb{C}$, called an inner product, that satisfies: $\forall x, y, z \in V$, $\forall \alpha \in \mathbb{C}$:

- $(x, x) \geq 0$
- $(x, x) = 0 \Leftrightarrow x = 0$
- $(x, y + z) = (x, y) + (x, z)$
- (x, ≤y) = ≤(x, y)
- $(x, y) = (y, x)^*$

Definition: Let V be an inner product pace.

- Two vectors $x, y \in V$ are said to be orthogonal if $(x, y) = 0$,
- A collection, $\{x_i\}_{i=1}^{N}$, of vectors in V is called an orthonormal set if $(x_i, x_j) = \delta_{ij}$, i.e. $(x_i, x_j) = 1$ if $i = j$ and $(x_i, x_j) = 0$ if $i \neq j$.

Theorem: Every inner product space is a normed linear space with the norm $\|x\| = \sqrt{(x,x)}$ and a metric space with the metric $d(x, y) = \sqrt{(x-y, x-y)}$.

Theorem: (Pythagorean Theorem) Let V be an inner product space and be an orthonormal set in V. Then $\forall x \in V$

$$\|x\|^2 = \sum_{n=1}^{N} |(x, x_n)|^2 + \left\| x - \sum_{n=1}^{N} (x, x_n) x_n \right\|^2$$

Theorem: (Bessel inequality) Let V be an inner product space and $\{x_n\}_{n=1}^{N}$ be an orthonormal set in V. Then $\forall x \in V$

$$\|x\|^2 \geq \sum_{n=1}^{N} |(x, x_n)|^2$$

Theorem: (Schwarz inequality) Let V be an inner product space. Then $\forall x, y \in V$

$$|(x, y)| \leq \|x\| \, \|y\|.$$

Theorem: (Parallelogram Law) Let V be an inner product space. Then $\forall x, y \in V$

$$\|x + y\|^2 + \|x - y\|^2 = 2\|x\|^2 + 2\|y\|^2.$$

Definition: A sequence $\{x_n\}$ of vectors in an inner product space V is called strongly convergent to $x \in V$, denoted by $x_n \xrightarrow{w} x$, if

$$\|x_n\| \xrightarrow{b \to 0} 0$$

and weakly convergent to $x \in V$, denoted by $x_n \xrightarrow{w} x$, if

$$(x_n - x, y) \xrightarrow{n \to 0} 0 \quad \forall y \in V.$$

Theorem:

- $x_n \to x \Rightarrow x_n \xrightarrow{w} x$
- $x_n \xrightarrow{w} x$ and $\|x_n\| \to \|x\|$

$$x_n \to x$$

Definition: A complete inner product space is called a Hilbert space.

Definition: A linear transformation U: $H_1 \to H_2$ from a Hilbert space H_1 onto the Hilbert space H_2 is called unitary if if it preserves the inner product, i.e. $\forall x, y \in H_1$

$$(Ux, Uy)_{H2} = (x, y)H_1.$$

Definition: Two Hilbert spaces H_1 and H_2 are said to be isomorphic if there is a unitary linear transformation U from H_1 onto H_2.

Definition: Let H_1 and H_2 be Hilbert spaces. The direct sum $H_1 \oplus H_2$ of Hilbert spaces H_1 and H_2 is the set of ordered pairs $z = (x, y)$ with $x \in H_1$ and $y \in H_2$ with inner product

$$(z_1, z_2)_{H1\oplus H2} = (x_1, x_2)_{H1} + (y_1, y_2)_{H2}$$

EXAMPLES OF HILBERT SPACES

Finite Dimensional Vectors. $\mathbb{C}^N$ is the space of N-tuples $x = (x_1, ..., x_N)$ of complex numbers. It is a Hilbert space with the inner product

$$(x, y) = \sum_{n=1}^{N} x_n^* y_n.$$

Square Summable Sequences of Complex Numbers. l^2 is the space of sequences of complex numbers $x = \{x_n\}_{n=1}^{\infty}$ such that

$$\sum_{n=1}^{\infty} |x_n|^2 < \infty.$$

It is a Hilbert space with the inner product

$$(x, y) = \sum_{n=1}^{\infty} x_n^* y_n$$

Square Integrable Functions on $\mathbb{R}$. $L^2(\mathbb{R})$ is the space of complex valued functions such that

$$\int_{\mathbb{R}} \left| f(x)^2 \right| dx < \infty$$

It is a Hilbert space with the inner product

$$(f, g) = \int_{\mathbb{R}} f^* (x) g(x) dx$$

Square Integrable Functions on R^n. Let be an open set in R^n (in particular, can be the whole $\mathbb{R}^n$). The space $L^2(\Omega)$ is the set of complex valued functions such that

$$\int_{\Omega} |f(x)|^2 dx < \infty$$

where $x = (x_1,..., x_n) \in$ and $dx = dx_1... dx_n$. It is a Hilbert space with the inner product

$$(f, g) = \int_{\Omega} f^*(x) g(x) dx$$

Square Integrable Vector Valued Functions. Let be an open set in $\mathbb{R}^n$ (in particular, Ω can be the whole $\mathbb{R}^n$) and V be a finite-dimensional vector space. The space $L^2(V, \Omega)$ is the set of vector valued functions $f = (f_1,..., f_N)$ on Ω such that

$$\sum_{i=1}^{N} \int_{\Omega} |f_i(x)|^2 dx < \infty.$$

It is a Hiibert space with the inner product

$$(f, g) = \sum_{i=1}^{N} \int_{\Omega} f_i^*(x) g_i(x)\, dx$$

Sobolev Spaces: Let Ω be an open set in $\mathbb{R}^n$ (in particular, Ω can be the whole $\mathbb{R}^n$) and V a finite-dimensional complex vector space. Let $C^m(V, \Omega)$ be the space of complex vector valued functions that have partial derivatives of all orders less or equal to m.

Let $\alpha = (\alpha_1, ..., \alpha_n)$, $\alpha \in N$, be a multiindex of nonnegative integers, $\alpha_i \geq 0$, and let $|\leq| = \alpha_1 + ... + \alpha_n$. Define

$$D^\alpha f = \frac{\partial^{|\alpha|}}{\partial x_1^{\alpha_1} \cdots \partial x_n^{\alpha n}} f.$$

Then $f \in C^m(V, \Omega)$ iff

$$\left|D^\alpha f_i(x)\right| < \infty \quad \forall \alpha, |\alpha| \leq m, \forall i = 1, \ldots, N, \forall x \in \Omega.$$

The space $H^m(V, \Omega)$ is the space of complex vector valued functions such that $D^\alpha f \in L^2(V, \Omega)$ $\forall \alpha$, $|\alpha| \leq m$, i.e. such that

$$\sum_{i=1}^{N} \int_{\Omega} \left|D^\alpha f_i(x)\right|^2 dx < \infty \qquad \forall \alpha, |\alpha| \leq m.$$

It is a Hilbert space with the inner product

$$(f, g) = \sum_{\alpha, |a| \leq m} \sum_{i=1}^{N} \int_{\Omega} (D^\alpha f_i(x))^* D^\alpha g_i(x)\, dx$$

Remark. More precisely, the Sobolev space $H^m(V, \Omega)$ is the completion of the space defined above.

PROJECTION THEOREM

Definition: Let M be a closed subspace of a Hilbert space H. The set, $M^\perp$, of vectors in H which are orthogonal to M is called the othogonal complement of M.

Theorem: A closed subspace of a Hilbert space and its orthogonal complement are Hilbert spaces.

Theorem: Let M be a closed subspace of a Hilbert space H. Then $\forall x \in H \, \exists$ a unique element $z \in M$ closest to x.

Theorem: (Projection Theorem) Let M be a closed subspace of a Hilbert space H. Then $\forall x \in H \; \exists \; z \in M$ and $\exists w \in M^\perp$ such that $x = z + w$. That is

$$H = M \oplus M^\perp$$

Remark: The set, $L(H, H')$, of linear transformations from a Hilbert space H to H' is a Banach space under the norm

$$\|T\| = \sup_{\|x\|_{H=1}} \|T_x\|_{H'}$$

Definition: The space $H^* = L(H, \mathbb{C})$ of linear transformations from a Hilbert space H to $\mathbb{C}$ is called the dual space of H. The elements of H^* are called continuous linear functionals.

Theorem: (Riesz Lemma) Let H be a Hilbert pace.

Then $\forall T \in H^* \exists y_T \in H$ such that $\forall x \in H$

$$T(x) = (y_T, x), \quad \text{and} \quad \|T\|_{H^*} = \|y_T\|_H$$

ORTHONORMAL BASES

Definition: Let S be an orthonormal set in a Hilbert pace H. If there is no other orthonormal set that contains S as a proper subset, then S is called orthonormal basis (or complete orthonormal system) for H.

Theorem: Every Hilbert space has an othonormal basis.

Theorem: Let $S = \sum_{\alpha \in A} (x_\alpha, y) x_\alpha$ be an orthonormal basis for a Hilbert space H. Then $\forall y \in H$

$$y = \sum_{\alpha \in A} (x_\alpha, y) x_\alpha$$

$$\|y\|^2 = \sum_{\alpha \in A} |(x_\alpha, y)|^2$$

Definition: Let $S = \{x_\alpha\}_{\alpha \in A}$ be an orthonormal basis for a Hilbert space H. The coefficients (x_α, y) are called the Fourier coefficients of $y \in H$ with respect to the basis S.

Definition: A metric space which has a countable dense subset is said to be separable.

Theorem: A Hilbert space H is separable iff it has a countable orthonormal basis S. If S contains finite number, N, of elements, then H is isomorphic to $\mathbb{C}^N$. If S contains countably many elements, then H is isomorphic to l^2.

TENSOR PRODUCTS OF HILBERT SPACES

Let H_1 and H_2 *b* Hilbert spaces. For each $\varphi_1 \in H_1$ and $\varphi_2 \in H_2$ let $\varphi_1 \otimes \varphi_2$ denote the conjugate bilinear form on H_1 X H_2 defined by

$$(\varphi_1 \otimes \varphi_2)(\psi_1, \psi_2) = (\psi_1, \varphi_1)_{H_1} (\psi_2, \varphi_2)_{H_2}$$

where $\psi_1 \in H_1$ and $\psi_2 \in H_2$. Let E be the set of finite linear combinations of such bilinear forms. An inner product on E can be defined by

$$(\varphi_1 \otimes \psi, \eta \otimes \mu)_E = (\varphi, \eta)_{H_1}(\psi, \mu)_{H_2}$$

(with $\varphi, \eta \in H_1$ and, $\psi, \mu \in H_2$) and extending by linearity on E.

Definition: The tensor product $H_1 \otimes H_2$ of the Hilbert paces H_1 and H_2 is defined to be the completion of E under the inner product defined above.

Theorem: Let H_1 and H_2 be Hilbert spaces. If $\{\varphi_k\}$ and $\{\psi_l\}$ are or-thonormal bases for H_1 and H_2 respectively, then $\{\varphi_k \otimes \psi_l\}$ basis for the tensor product $H_1 \otimes H_2$.

Fock Spaces. Let H be a Hilbert space and let

$$H^0 = \mathbb{C}$$

$$H^n = \underbrace{H \otimes \cdots \otimes H}_{n}$$

denote the n-fold tensor product of H. The space

$$F(H) = \oplus_{n=0}^{\infty} H^n$$

is called the Fock space over H. Fock space F (H) is separable if H is separable. For example, if $H = L_2(\mathbb{R})$, then an element ψ $F(H)$ is a sequence of functions

$$\psi = \{\psi_0, \psi_1(x), | \psi(x_1, x_2), \psi_3(x_1, x_2, x_3), ...\}$$

so that

$$||\psi|| = |\psi_0|^2 + \sum_{n=1}^{\infty} \int |\psi_n(x_1, \cdots, x_n)|^2 dx_1 ... dx_n < \infty.$$

Let P_n be the permutation group of n elements and let $\{\psi_k\}$ be a basis for H. Each $\sigma \in P_n$ defines a permutation

$$\sigma(\varphi k_1 \otimes \cdots \otimes \varphi k_n\} = \varphi k_{\sigma(1)} \otimes \cdots \otimes \varphi k_{\sigma(n)}.$$

By linearity this can be extended to a bounded operator on H^n, so one can define

$$S_n = \frac{1}{n!} \sum_{\sigma \in P_n} \sigma$$

$$A_n = \frac{1}{n!} \sum_{\sigma \in p_n} \varepsilon(\sigma)\sigma$$

where

$$\varepsilon(\sigma) = \begin{cases} 1, & \text{if } \sigma \text{ is even} \\ -1 & \text{if } \sigma \text{ is odd} \end{cases}$$

Finally, the Boson (symmetric) Fock space is defined by

$$F_s(H) = \oplus_{n=0}^{\infty} S_n H^n$$

and the Fermion (antisymmetric) Fock space is defined by

$$F_a(H) = \oplus_{n=0}^{\infty} A_n H^n$$

In the case $H = L^2(\mathbb{R})$, $\psi_n \in S_n H^n$ is a function of n variables symmetric under any permutations of variables, and $\psi_n \in A_n H^n$ is a function of n variables that is odd function under interchanges of any two variables.

ASYMPTOTIC EXPANSIONS

ASYMPTOTIC ESTIMATES

Let M be a set of real or complex numbers with a limit point a.
Let f, g: $M \to \mathbb{R}$ (or f, g: $M \to \mathbb{C}$) be some functions on M.
Definition: The following are asymptotic estimates

- $f(x) \sim g(x)$ $(x \to a, x \in M)$

  If $\lim_{x \to a,\, x \in M} \dfrac{f(x)}{g(x)} = 1.$

- $f(x) = o(g(x))$ $(x \to a, x \in M)$

  If $\lim_{x \to a,\, x \in M} \dfrac{f(x)}{g(x)} = 0.$

- $f(x) = O(g(x))$ $(x \in M)$
  If $\exists C$: $|f(x)| \le C|g(x)|$ $\forall\, x \in M$
- $f(x) = O(g(x))$ $(x \to a, x \in M)$
  If $\exists$C and a neighborhood U of a such that:
  $|f(x)| \le C|g(x)|$ $\forall x \in M \cap U$

ASYMPTOTIC SEQUENCES

Definition: Let φ_n: $M \to \mathbb{R}$, $n \in \mathbb{N}$, and a be a limit point of M.
Let $\varphi_n(x) \neq 0$ in a neighborhood U_n of a. The the sequence $\{\varphi_n\}$ is called asymptotic sequence at $x \to a$, $x \in M$ if $\forall n \in \mathbb{N}$

$$\varphi_{n+1}(x) = o(\varphi_n(x)) \quad (x \to a, x \in M)$$

Examples:

1. Power asymptotic sequences
   (a) $\{(x - a)^n\}$, $x \to a$.
   (b) $\{x^{-n}\}$, $x \to \infty$.
2. Let $\{\lambda_n\}$ be a decreasing sequence of real numbers, i.e. $\lambda_n < \lambda_{n+1}$, and let $0 < \varepsilon \le \pi/2$. Then the sequence

$$\left\{e^{\lambda_n z}\right\},\ z \to \infty, \quad |\arg z| \le \frac{\pi}{2} - \varepsilon$$

   is an asymptotic sequence.

Properties of Asymptotic Sequences

1. Any subsequence of an asymptotic sequence is an asymptotic sequence.
2. Let $f(x) \neq 0$ for $x \in M$ in some neighborhood of a and $\{\varphi_n\}$ be an asymptotic sequence at $x \to a$, $x \in M$. Then the sequence $\{f(x)\ \varphi_n(x)\}$ is an asymptotic sequence as $x \to a$, $x \in M$.
3. Let $\{\varphi_n(x)\}, \{\psi_n(x)\}$ be asymptotic sequences as $x \to$ a, $x \in M$. Then the sequence $\{\varphi_n(x)\}, \{\psi_n(x)\}$ is an asymptotic sequence as $x \to$ a, $x \in$ M.

Asymptotic Series

Let $f: M \to \mathbb{R}$ and a be a limit point of M.

Definition: Let $\{\varphi_n\}$ be an asymptotic sequence as $x \to$ a, $x \in M$. We say that the function f is expanded in an asymptotic series

$$f(x) \sim \sum_{n=0}^{\infty} a_n \varphi_n(x), \quad (x \to a, x \in M),$$

where a_n are constants, if $\forall N \geq 0$

$$R_N(x) \equiv f(x) - \sum_{n=0}^{\infty} a_n \varphi_n(x) = o(\varphi_N(x)), \quad (x \to a, x \in M)$$

This series is called asymptotic expansion of the function f with respect to the asympotic sequence $\{\varphi_n\}$. $RN(x)$ is called the rest term of the asymptotic series.

Remarks:

1. The condition $R_N(x) = o(\varphi_N(x))$ means, in particular, that
$$\lim_{x \to a} R_N(x) = 0 \quad \text{for any fixed } N$$
2. Asymptotic series could diverge. This happens if
$$\lim_{N \to \infty} R_N(x) \neq 0 \quad \text{for some fixed } x$$
3. There are three possibilities:
   (a) Asymptotic series converges to $f(x)$,
   (b) Asymptotic series converges to a function $g(x) \neq f(x)$,
   (c) Asymptotic series diverges.

Theorem: Asymptotic expansion of a function with respect to an asymptotic sequence is unique.

Remark: Two different functions can have the same asymptotic expansion.

For example, $f(x) = e^x$ and $g(x) = e^x + e^{-1/x}$ have the same asymptotic expansion with respect to the asymptotic sequence $\{x^n\}$:

$$e^x \sim e^x + e^{-1/x} \sim \sum_{n=0}^{\infty} \frac{x^n}{n!}, \quad x \to 0+$$

Theorem: Asymptotic series can be added and multiplied by numbers, but, cannot be multipied by asymptotic series.

Theorem: One can multiply and divide power asymptotic series.

Definition: Let f: M X $S \to \mathbb{R}$ be a function of two variables and a be a limit point of M and $\{\varphi_n\}$ be an asymptotic sequence as $x \to$ a.

Let for any xed $y \in S$ the function f is expanded in an asymptotic series

$$f(x, y) \sim \sum_{n=0}^{\infty} a_n(y)\varphi_n(x), \quad (x \to a, x \in M).$$

This asymptotic expansion is called uniform with respect to the parameter $y \in S$, if the relation

$$R_N(x, y) \equiv f(x, y) - \sum_{n=0}^{\infty} a_n(y)\varphi_n(x) = 0(\varphi_N(x)), \quad (x \to a, x \in M)$$

is valid uniformly with respect to $y \in S$.

Theorem: A uniform asymptotic expansion can be integrated with respect to the parameter term by term.

Remark: One cannot, in general, differentiate asymptotic series, neither with respect to x nor with respect to a parameter.

Asymptotics of Integrals: Weak Singularities

Let us consider the integrals of the form

$$F(\varepsilon) = \int_0^a f(x, \varepsilon)\, dx$$

where $a > 0$ and $\varepsilon > 0$ is a small positive parameter. Here $f \in C^\infty([0, a] \times (0, \varepsilon_0])$ is a smooth function for $0 \le x \le a$, $0 < \varepsilon \le \varepsilon_0$, with some ε_0. Then the integral converges for $\varepsilon > 0$.

Let f have a singularity when $\varepsilon = 0$, i.e. $g(x) = f(x, 0)$ has a singularity at some $0 \le x \le a$. If this singularity is of power or logarithmic type then we say that the integral $F(\varepsilon)$ has a weak singularity.

This Definition is obviously extended for unbounded intervals. Let

$$F(\varepsilon) = \int_a^\infty f(x, \varepsilon)\, dx$$

where $a > 0$ and $\varepsilon > 0$ is a small parameter. Here $f \in C^\infty([a, 1) \times (0, \varepsilon_0])$ is a smooth function for $a \le x \le 1$, $0 < \varepsilon \le \varepsilon_0$, with some ε_0. Let the integral converge for $\varepsilon > 0$ and diverge for $\varepsilon = 0$. If the function $g(x) = f(x, 0)$ is of power or logarithmic order at $x \to \infty$, then we say that the integral $F(\varepsilon)$ has a weak singularity.

Power Singularity on a Bounded Interval

Let $a, \alpha, \beta \in \mathbb{R}$, $a, \beta > 0$, be some real numbers and $\varepsilon > 0$ be a small positive prameter. Let $\varphi \in C^{\infty}[0, a]$ be a smooth function on $[0, a]$. We will study the asymptotics as $\varepsilon \to 0+$ of the integrals of the form

$$F(\varepsilon) = \int_0^a t^{\beta-1}(t+\varepsilon)^{\alpha}\,\varphi(t)\,dt.$$

Remarks. The function F is holomorphic in complex plane ε with a cut along the negative half-axis. At the point $\varepsilon = 0$ this function has a singularity (if $\alpha > 0$ is not integer).

The type of this singularity is determined by the behaviour of the function φ at small $t \le 0$.

Standard Integral. One needs the following result. Let α and β be two complex numbers such that Re $\beta > 0$ Re $\alpha < 0$ and Re $(\beta + \alpha) < 0$. Then

$$\int_0^{\infty} t^{\beta-1}(t+1)^{\alpha}\,dt = \frac{\Gamma(\beta)\Gamma(-\alpha-\beta)}{\Gamma(-\alpha)}, \quad (0 < \mathrm{Re}\,\beta < -\mathrm{Re}\,\alpha).$$

Theorem: Let $\varphi \in C^{\infty}[0, a]$. Let $r, \delta > 0$ and $S_{\delta} = \{\varepsilon \in \mathbb{C} | 0 < |\varepsilon\} \le$ r, $|\arg \varepsilon| \le \pi \quad - \delta\}$ be a sector in the complex plane of ε.

1. If $\alpha + \beta$ is not integer, then

$$F(\varepsilon) \sim \sum_{n=0}^{\infty} \frac{\Gamma(\beta+n)\Gamma(-\alpha-\beta-n)\varphi^{(n)}(0)}{\Gamma(-\alpha)}\,\varepsilon^{\alpha+\beta+n} + \sum_{n=0}^{\infty} a_n \varepsilon^n$$

$$(\varepsilon \to 0, \quad \varepsilon \in S_{\delta}).$$

2. If $\alpha + \beta = N$ is an integer, then

$$F(\varepsilon) \sim \sum_{n \ge \max\{0,-N\}}^{\infty} \frac{\Gamma(N+n)}{\Gamma(\alpha)\Gamma(N+n-\alpha)} \frac{\varphi^{(n)}(0)}{n!}\,\varepsilon^{n+N} \ln \varepsilon + \sum_{n=0}^{\infty} b_n \varepsilon^n$$

$$(\varepsilon \to 0, \quad \varepsilon \in S_{\delta}).$$

The coefficients a_n and b_n depend on the values $\varphi(t)$ for $0 \le t \le a$. The branch for the functions ε and ln ε is choosen in such a way that $\varepsilon^{\gamma} > 0$ and ln ε is real for $\varepsilon > 0$.

Examples: In all examples $\varphi \in C^{\infty}([0, a])$ is a smooth function bounded with all its derivatives.

1. Let $0 < a < 1$ and

$$F(\varepsilon) = \int_0^a \frac{\varphi(t)}{t+\varepsilon}\,dt.$$

Then

$$F(\varepsilon) = -\varphi(0)\ln\varepsilon + O(1), \quad (\varepsilon \to 0^+)$$

2. Let $0 < a < 1$ and

$$F(\varepsilon) = \int_0^a \frac{\varphi(t)}{t^2+\varepsilon^2}\,dt.$$

By using

$$\frac{1}{t^2+\varepsilon^2} = \frac{1}{2i\varepsilon}\left(\frac{1}{t-i\varepsilon} - \frac{1}{t+i\varepsilon}\right)$$

we obtain from the previous example

$$F(\varepsilon) = \varphi(0)\frac{\pi}{2\varepsilon} + O(1) \quad (\varepsilon \to 0^+)$$

3. Let $\alpha > 1/2$ and

$$F(\varepsilon) = \int_0^a \frac{\varphi(t)}{(t^2+\varepsilon^2)^\alpha}\,dt.$$

Then

$$F(\varepsilon) = \varphi(0)\frac{\sqrt{\pi}}{2}\frac{\Gamma(\alpha-1/2)}{\Gamma(\alpha)}\varepsilon^{1-2\alpha} + O(\varepsilon^{3-2\alpha}) + O(1), \quad (\varepsilon \to 0^+)$$

POWER SINGULARITY ON UNBOUNDED INTERVAL

Standard Integral

To compute the following asymptotics one needs the following standard integral.

Let $\operatorname{Re}\beta > 0$ and $\operatorname{Re}\alpha > -1$. Then

$$\int_0^\infty t^\alpha e^{-t^\beta}\,dt = \frac{1}{\beta}\Gamma\left(\frac{\alpha+1}{\beta}\right), \quad (\operatorname{Re}\beta > 0, \operatorname{Re}\alpha > -1).$$

Examples: Let $\varphi \in C^\infty([a, \infty))$ be a smooth function on $[a, \infty)$ that has asymptotic expansion as $x \to \infty$.

$$\varphi(x) \sim \sum_{k=0}^{\infty} a_k x^{-k}.$$

1. Let $a, \beta > 0$, and

$$F(\varepsilon) = \int_a^\infty \varphi(x)x^\alpha e^{-\varepsilon x^\beta}\,dx$$

If $\alpha < -1$, then the integral is not singular as $\varepsilon \to 0^+$. Its asymptotic expansion can be obtained either by integration by parts or by a change of variables. So, let now $\alpha + 1 > 0$ and let $N = [\alpha + 1]$ $\varepsilon \to 0^+$ be the integer part of $\alpha + 1$. Let us single out the first $N + 1$ terms of the asymptotic expansion in φ, i.e.

$$\varphi(x) = \sum_{k=0}^{N} a_k x^{-k} + R_N(x)$$

Since $R_N(x) = O(x^{-(N+1)})$ as $x \to \infty$, we have

$$F(\varepsilon) = \sum_{k=0}^{N} a_k \int_a^{\infty} x^{a-k} e^{-\varepsilon x^{\beta}} dx + O(1).$$

Now by changing the variables and extending the interval to $[0, \infty)$ we obtain

$$F(\varepsilon) = \sum_{k=0}^{N} \frac{a_k}{\beta} \varepsilon^{-\frac{\alpha-k+1}{\beta}} \left[\Gamma\left(\frac{\alpha - k + 1}{\beta}\right) + O(1) \right] + O(1)$$

$$= \frac{a_0}{\beta} \varepsilon^{-\frac{\alpha+1}{\beta}} \Gamma\left(\frac{\alpha + 1}{\beta}\right) + O(\varepsilon^{-\frac{\alpha}{\beta}})$$

If $\alpha = -1$, then

$$F(\varepsilon) = -\frac{a_0}{\beta} \ln \varepsilon + O(1), \quad (\varepsilon \to 0^+)$$

2. Let a > 0 and let
$$P(x) = x^n + \ldots + a_1 x \quad n \geq 1.$$
Consider the integral

$$F(\varepsilon) = \int_a^{\infty} \varphi(x) x^{\alpha} e^{-\varepsilon P(x)} dx.$$

If $\alpha > -1$, then the main term of the asymptotics is

$$F(\varepsilon) = \frac{a_0}{n} \Gamma\left(\frac{a+1}{n}\right) \varepsilon^{\frac{-\alpha+1}{n}} + O\left(e^{-\frac{\alpha}{n}}\right), \quad \left(\varepsilon \to 0^+\right)$$

If $\alpha = -1$ then

$$F(\varepsilon) = -\frac{a_0}{n} \ln \varepsilon + O(1), \quad \left(\varepsilon \to 0^+\right)$$

If $\alpha < 1$, then the integral is not singular as $\varepsilon \to 0^+$. By integration by parts the integral can be reduced to the cases considered above.

Chapter 2

Laplace and Saddle Point Method

LAPLACE INTEGRALS IN ONE DIMENSION

Let $M = [a, b]$ be a closed bounded interval, $S: M \to \mathbb{R}$ be a real valued function, $\varphi: M \to \mathbb{C}$ a complex valued function and λ be a large positive parameter. Consider the integrals of the form

$$F(\lambda) = \int_a^b \varphi(x)\exp[\lambda S(x)]dx.$$

Such integrals are called Laplace integrals. We will study the asymptotics of the Laplace integrals as $\lambda \to \infty$.

Lemma: Let $\sup_{a<x<b} S(x) = L < \infty$ and the integral converges absolutely for some $\lambda_0 > 0$. Then

1. $|F(\lambda)| \le C|e^{\lambda L}|$ $(\mathrm{Re}\,\lambda \ge \lambda_0)$.
2. If f, $S \in C(a, b)$, then $F(\lambda)$ is holomorphic in the halfplane $\mathrm{Re}\,\lambda > \lambda_0$.

Watson Lemma

Lemma: (Watson) Let $0 < a < \infty$, $\alpha > 0$, $\beta > 0$ and let S_ε be the sector $S_\varepsilon = \{\lambda \in \mathbb{C} \,||\, \arg\lambda| \le \pi/2\ \varepsilon\}$ in the complex plane λ. Let $\varphi \in C^\infty([0, a])$ and let

$$\Phi(\lambda) = \int_0^a \varphi(x)^{x\beta-1}\exp(-\lambda x^\alpha)dx$$

Then, there is an asymptotic expansion as $\lambda \to \infty$, $\lambda \in S_\varepsilon$,

$$\Phi(\lambda) \sim \frac{1}{\alpha}\sum_{k=0}^{\infty}\lambda^{-(\beta+k)/\alpha}\Gamma\left(\frac{\beta+k}{\alpha}\right)\frac{\varphi^{(k)}(0)}{k!}$$

Laplace Transform

Let $\varphi \in C^\infty(\mathbb{R}_+)$ be a smooth function on the positive real axis such that its Laplace transform

$$\mathcal{L}(\varphi)(\lambda) = \int_0^\infty \varphi(x)e^{-\lambda x}dx$$

converges absolutely for some λ_0. Then

$$\mathcal{L}(\varphi)(\lambda) \sim \sum_{k=0}^{\infty} \lambda^{-k} \varphi^{(k)}(0) \qquad (|\lambda| \to \infty,\ \lambda \in S_\varepsilon)$$

Interior Nondegenerate Maximum Point

Let now S and φ be smooth functions and the function S have a maximum at an interior point x_0 of the interval $[a, b]$, i.e. $a < x_0 < b$.

Then $S'(x_0) = 0$. Assume, for simplicity, that $S''(x_0) < 0$. Then $S''(x_0) < 0$. In other words, in a neighborhood of x_0 the function S has the following Taylor expansion

$$S(x) = S(x_0) + S''(x_0)\frac{(x - x_0)^2}{2} + O((x - x_0)^3).$$

Such a point is called nondegenerate critical point. Then, as $\lambda \to \infty$ the main contribution to the integral comes from a small neighborhood of x_0.

In this neighborhood the function φ is almost constant and can be replaced by its value at x_0. The terms of order $(x - x_0)^3$ can be neglected in the exponent and the remaining integral can be extended to the whole real line. By using the standard Gaussian integral

$$\int_{-\infty}^{\infty} \exp\left(-\frac{\alpha}{2} y^2\right) dy = \sqrt{\frac{2\pi}{\alpha}}, \qquad (\operatorname{Re} \alpha > 0),$$

one obtains finally the main term of the asymptotics

$$F(\lambda) \sim \lambda^{-1/2} \sqrt{\frac{2\pi}{-S''(x_0)}} \varphi(x_0) e^{\lambda S(x_0)} \qquad (\lambda \to \infty)$$

One can prove the general theorem.

Theorem: Let $M = [a, b]$ and $\varphi, S \in C^\infty(M)$, S has a maximum only at one point x_0, $a < x_0 < b$ and $S''(x_0) \neq 0$. Then as $\lambda \to \infty$, $\lambda \in S_\varepsilon$ there is asymptotic expansion

$$\mathrm{F}(\lambda) \sim e^{\lambda S(x_0)} \sum_{k=0}^{\infty} c_k \lambda^{-1/2-k}.$$

The coefficients c_k are expressed in terms of derivatives of φ and S at x_0. The theorem can be proved as follows. First, we change the integration variable

$$x = x_0 + \lambda^{-1/2} y.$$

So, y is the scaled fluctuation from the maximum point x_0. The interval of integration should be changed accordingly, so that the maximum point is now $y = 0$. Then, we expand both functions S and φ in Taylor series at x_0 getting

$$\varphi S(x_0 + \lambda^{-1/2}y) = \lambda S(x_0) + \frac{1}{2}S''(x_0)y^2 + \sum_{n=3}^{\infty} \frac{S^{(n)}(x_0)}{n!} y^n \lambda^{-(n-2)/2}$$

$$\varphi(x_0 + \lambda^{-1/2}y) = \sum_{n=0}^{\infty} \frac{\varphi^{(n)}(x_0)}{n!} y^n \lambda^{-n/2}.$$

Since the quadratic terms are of order $O(1)$ we leave it in the exponent and expand the exponent of the rest in a power series. Next, we extend the integration interval to the whole real line and compute the standard Gaussian integrals of the form

$$\int_{-\infty}^{\infty} \exp\left(-\frac{\alpha}{2}y^2\right) y^{2k+1} dy = 0,$$

$$\int_{-\infty}^{\infty} \exp\left(-\frac{\alpha}{2}y^2\right) y^{2k} dy = \Gamma\left(k+\frac{1}{2}\right)\left(\frac{\alpha}{2}\right)^{-k-1},$$

where k is a nonnegative integer and α has a positive real part, Re $\alpha > 0$. Finally, we get a power series in inverse powers of λ. The coefficients c_k of the asymptotic expansion are polynomials in the higher derivatives $S^{(k)}(x_0)$, $k \geq 3$, and derivatives $\varphi^{(l)}(x_0)$, $l \geq 0$, and involve inverse powers of $S''(x_0)$.

Stirling Formula

$$\Gamma(x + 1) = \sqrt{2\pi} x^{x+1/2} e^{-x}[1 + O(x^{-1})], \qquad (x \to \infty)$$

is obtained by applying the Laplace method to the integral

$$\Gamma(x + 1) = x^{x+1} \int_0^{\infty} \exp[x(\ln t - t)]dt$$

Stieltjes Transform

Let $\varphi: \mathbb{R}_+ \to \mathbb{C}$ have finite moments

$$m_n(\varphi) = \int_0^{\infty} t^n \varphi(t)dt < \infty \qquad \forall_n \in \mathbb{N}$$

Then the Stieltjes transform of φ

$$S(\varphi)(x) = \int_0^{\infty} \frac{\varphi(t)}{t+x} dt$$

has asymptotic expansion as $x \to \infty$

$$S(\varphi)(x) \sim \sum_{k=0}^{\infty} (-1)^k m_k(\varphi) x^{-1-k}$$

Boundary Maximum Point

Let the function S have a maximum at a boundary point $x_0 = a$. Let, for simplicity, $S'(a) \neq 0$, i.e. $S'(a) < 0$, and $\varphi(a) \neq 0$. Then, as $\lambda \to \infty$, the main contribution to the integral comes from the interval $[a, a + \varepsilon]$, where

$$S(x) = S(a) + (x - a) S'(a) + O((x - a)^2).$$

Now, by replacing the function φ by its value at a and neglecting nonlinear terms in $S(x)$, we get

$$F(\lambda) \sim \lambda^{-1} \frac{\varphi(a)}{-S'(a)} e^{\lambda S(a)}, \quad (\lambda \to \infty)$$

In this way one can prove the following theorem.

Theorem: Let $M = [a, b]$, φ, $S \in C^\infty(M)$, S has a maximum only at the point $x = a$ and $S'(a) \neq 0$. Then, as $\lambda \to \infty$, $\lambda \in S_\varepsilon$, there is asymptotic expansion

$$F(\lambda) \sim e^{\lambda S(a)} \sum_{k=0}^{\infty} c_k \lambda^{-1-k}.$$

The coefficients c_k are expressed in terms of derivatives of φ and S at $x =$ a.

Error Function: The asymptotic expansion of the (complementary) error function as $x \to \infty$ has the form

$$\text{Erfc } x = \int_0^x t^{a-1} e^{-t} dt, \quad (0 < a < \infty, x > 0)$$

Incomplete Gamma Function: The incomplete gamma-function

$$\gamma(a,x) = \int_0^x t^{a-1} e^{-t} dt, \quad (0 < a < \infty, x > 0)$$

has the following asympotic expansion as $x \to \infty$

$$\gamma(a,x) = \Gamma(a) + e^{-x} x^{a-1} \sum_{k=0}^{\infty} \frac{\Gamma(a)}{\Gamma(a-k)} x^{-k}$$

BACKGROUND FROM ANALYSIS

1. Let x^j, $j = 1,\dots, n$ be real numbers and let x be the n-tuple $x = (x^1,\dots x^n)$. The set of all n-tuples of real numbers is denoted by R^n. A connected open subset Ω of $\mathbb{R}^n$ is called a domain. The set $\partial\Omega$ of boundary points of Ω is called the boundary of Ω. The union $[\Omega \cup \partial\Omega$ is called the closure of and is denoted by $\overline{\Omega}$.
2. Let $x, \chi \in \mathbb{R}^n$. Then the scalar product of x and ξ is defined by $x \cdot \xi = x^1\xi^1 + \dots + x^n\xi^n$.
3. On $\mathbb{R}^n$ there is standard Lebesgue measure $dx = dx^1 \dots dx^n$.
4. Let σ_j, $j = 1,\dots, n$, be non-negative integers, $\alpha_j \geq 0$. The n-tupl $\alpha =$ $(\alpha_1,\dots, \alpha_n)$ is called a multi-index. Further, let

$$|\alpha| = \alpha_1 + \dots + \alpha_n$$

$$\alpha! = \alpha_1! .... \alpha_n!$$

$$\partial^\alpha f(x) = \frac{\partial^{|\alpha|} f(x)}{\partial (x^1)^{\alpha_1} \cdots \partial (x^n)^{\alpha_n}}$$

$$D^\alpha f(x) = \left(\frac{1}{i}\frac{\partial}{\partial x^1}\right)^{\alpha_1} \cdots \left(\frac{1}{i}\frac{\partial}{\partial x^n}\right)^{\alpha_n} f(x)$$

Then the Taylor expansion of a smooth function j at x_0 can be written in the form

$$\varphi(x) = \sum_{|\alpha|=0}^{\infty} \frac{1}{a!}\left[\partial^\alpha \varphi(x_0)\right](x - x_0)^\alpha .$$

5. The boundary $\partial\Omega$ is said to be smooth, denoted $\partial\Omega \in C^\infty$, if in a neighborhood of any boundary point $x_0 \in \partial\Omega$ it can be locally defined by an equation $x_j = \varphi(x')$ with a smooth function φ.
6. The set of all continuous functions on $\overline{\Omega}$ is denoted by $C(\overline{\Omega})$.
7. The set of all functions with continuous partial derivatives up to order k on is denoted by $C^k(\Omega)$.
8. The set of all functions with continuous partial derivatives up to order k on $\overline{\Omega}$ is denoted by $C^k(\overline{\Omega})$.
9. The set of all functions with continuous partial derivatives up to order k on that vanish in a neighborhood of the boundary ∂ is denoted by $C^{k0}(\overline{\Omega})$.
10. The closure of the set where a function is not equal to zero is calle the support of the function, denoted by

$$\text{supp} f = \overline{\{x \in \Omega \mid f(x) \neq 0\}} .$$

11. A mapping φ: $\Omega \to \Omega$, is said to be of class C^k if $\varphi \in C^k(\Omega)$.
12. A one-to-one mapping φ: $\Omega \to \Omega$ of onto is called di≦eomorphism of class C^k if $\varphi \in C^k(\Omega)$ and $\varphi^{-1} \in C^k(\Omega)$.
13. Let φ^j, $j = 1,..., k$ be some scalar functions on $\mathbb{R}^n$ and φ be a k-tuple $\varphi = (\varphi^1,..., \varphi^k)$. In other words, φ: $\mathbb{R}^n \to \mathbb{R}^k$. The matrix

$$\partial_x \varphi(x) = \left(\frac{\partial \varphi^i(x)}{\partial x^i}\right), \; i = 1,..., k; j = 1,..., n$$

is called the Jacobi matrix.

Theorem: (Inverse Function Theorem) Let φ: $\mathbb{R}^n \to \mathbb{R}^n$ is of class C^k, $k \geq 1$, in a neighborhood of a point x_0 and det $\partial_x\varphi(x_0) \neq 0$. Then φ is local diffeomorphism of class C^k in a neighborhood of the point x_0.

Theorem: (Implicit Function Theorem) Let be a domain in $\mathbb{R}^{2n}$, let F:

F: $\Omega \to \mathbb{R}^n$ be a mapping of class $C^k\Omega)$ and let $(x_0, y_0) \in \Omega$ be a poin in Ω such that

$$F(x_0, y_0) = 0, \det \partial_y F(x_0, y_0) \neq 0.$$

Then in a neighborhood of the point x_0 there is a mapping $y = f(x)$ of class C^k such that $y_0 = f(x_0)$ and

$$F(x, f(x)) \equiv 0.$$

Morse Lemma

Let S: $\Omega \to \mathbb{R}$ be a real valued function of class C^k on a domain Ω in $\mathbb{R}^n$ with $k \geq 2$. Let

$$\partial_x^2 S(x) = \left(\frac{\partial^2 S(x)}{\partial x^i \partial x^j}\right), \quad i, j = 1, \ldots, n.$$

Definition:

1. The point x_0 is called a critical point of the function S if $\partial S(x_0) = 0$
2. A critical point x_0 is called non-degenerate if $\det \partial_x^2 S(x_0) \neq 0$.
3. The determinant $\det \partial_x^2 S(x_0)$ is called the Hessian of the function S at the point x_0.

Lemma: (Morse) Let S: $\mathbb{R}^n \to$ R and $x_0 \in \mathbb{R}^n$ be a non-degenerate critical point of the function S.

Let $S \in C^\infty$ in a neighborhood of the point x_0 and let $\mu_j \neq 0, j = 1, \ldots, n$ be the eigenvalues of the matrix $\partial_x^2 S(x_0)$. Then there are neighborhoods U and V of the points x_0 and 0 and a smooth local diffeomorphism φ: $V \to$ U of class C^∞ such that $\det \partial_y \varphi(0) = 1$ and

$$S(\varphi(y)) = S(x_0) + \frac{1}{2}\sum_{j=1}^{n} \mu_j (y^j)^2.$$

Remark: Nondegenerate critical points are isolated.

Gaussian Integrals

Proposition: Let $A = (a_{ij}$ be a complex symmetric nondegenerate n X n matrix with the eigenvalues $\mu_j(A), j = 1, \ldots, n$,. Let Re $A \geq 0$, which means that $x \cdot \text{Re}\, A\, x \geq 0\ \forall x \in \mathbb{R}^n, x \neq 0$, or Re $\mu_j(A) \geq 0, j = 1, \ldots, n$. Then for $\lambda > 0, \xi \in \mathbb{R}^n$ there holds

$$\int_{\mathbb{R}^n} \exp\left(-\frac{\lambda}{2} x \cdot Ax - i\xi \cdot x\right) dx = \left(\frac{\lambda}{2}\right)^{n/2} (\det A)^{-1/2} \exp\left(-\frac{1}{2\lambda}\xi \cdot A^{-1}\xi\right)$$

The branch of $\sqrt{\det A}$ is choosen as follows

$$(\det A)^{-1/2} = |\det A|^{-1/2} \exp(-i \operatorname{Ind} A),$$

where

$$\operatorname{Ind} A = \frac{1}{2}\sum_{j=1}^{1} \arg\mu_j(A), \quad |\arg\mu_j(A)| \le \frac{\pi}{2}$$

By expanding both sides of this equation in Taylor series in ≤ we obtain the following result.

Corollary:

$$\int_{\mathbb{R}^n} \exp\left(-\frac{\lambda}{2} x\cdot Ax\right) x^{i_1}\cdots x^{i_{2k+1}}\, dx = 0$$

$$\int_{\mathbb{R}^n} \exp\left(-\frac{\lambda}{2} x\cdot Ax\right) x^{i_1}\cdots x^{i_{2k}}\, dx =$$

$$\left(\frac{2\pi}{\lambda}\right)^{n/2} (\det A)^{1/2} (2\lambda)^{-k} \frac{(2k)!}{k!} G^{(i_1 i_2} \cdots G^{i_{2k-1} i_{2k})}$$

Here k is a non-negative integer, $G = A^{-1}$, and the round brackets denote complete symmetrization over all indices included.

An important particular case of the previous formula is when the matrix A is real.

Proposition: Let A be a real symmetric nondegenerate $n \times n$ matrix. Let $\nu_+(A)$ and $\nu_-(A)$ be the numbers of positive and negative eigenvalues of A and

$$\operatorname{sgn} A = \nu_+ - \nu_-$$

be the signature of the matrix A. Then for $\lambda > 0$, $\xi \in \mathbb{R}^n$ there holds

$$\int_{\mathbb{R}^n} \exp\left(i\frac{\lambda}{2} x\cdot Ax - i\xi\cdot x\right) dx$$

$$= \left(\frac{2\pi}{\lambda}\right)^{n/2} |\det A|^{1/2} \left(-\frac{i}{2\lambda}\xi\cdot A^{-1} + \frac{i\pi}{4}\operatorname{sgn}(A)\right)$$

LAPLACE INTEGRALS IN MANY DIMENSIONS

Interior Maximum Point

Let be a bounded domain in $\mathbb{R}^n$, $S: \to \mathbb{R}$, $f: \Omega \to \mathbb{C}$ are some functions on Ω and $\lambda > 0$ be a large positive parameter. We will study the asymptotics as $\lambda \to \infty$ of the multidimensional Laplace integrals

$$F(\le) = \int_{\Omega} f(x) \exp[\lambda S(x)]\, dx.$$

Let S and f be smooth functions and the function S have a maximum only at onc interior nondegenerate critical point $x_0 \in \Omega$. Then $\partial_x S(x_0) = 0$ and $[\det \partial_x^2 S(x_0)] < 0$. Then in a neighborhood of x_0 the function S has the following Taylor expansion

$$S(x) = S(x_0) + \frac{1}{2}(x - x_0)\cdot\left[\partial_x^2 S(x_0)\right](x - x_0) + O((x - x_0)^3)$$

One could also use the Morse Lemma to replace the function S by a quadratic form. Then as $\lambda \to \infty$ the main contribution to the integral comes from a small neghborhood of x_0. In this neighborhood the terms of the third order in the Taylor expansion of S can be neglected. Also, since the function f is continuous at x_0, it can be replaced by its value at x_0. Then the region of integration can be extended to the whole $\mathbb{R}^n$. By using the formula for the standard Gaussian integral one gets then the leading asymptotics of the integral $F(\lambda)$ as $\lambda \to \infty$

$$F(\lambda) \sim \exp\left[\lambda S(x_0)\right]\left(\frac{2\pi}{\lambda}\right)^{n/2}\left[-\det\partial_x^2 S(x_0)\right]^{-1/2} f(x_0)$$

One can prove the general theorem.

Theorem: Let $f, S \in C^\infty(\Omega)$ and let x_0 be a nondegenerate critical point of the function S where it has the only maximum in Ω. Let $0 < \varepsilon < \pi/2$. Then there is asymptotic expansion as $\lambda \to \infty$ in the sector $S_\varepsilon = \{\lambda \in \mathbb{C} \| \arg\lambda| \le \pi/2 - \varepsilon\}$

$$F(\lambda) \sim \exp\left[\lambda S(x_0)\right]\lambda^{-n/2}\sum_{k=0}^{\infty} a_k \lambda^{-k}.$$

The coefficients a_k are expressed in terms of the derivatives of the functions f and S at the point x_0.

The idea of the proof is the same as in the one-dimensional case and goes as follows. First, we change the integration variables

$$x_i = x_0^i + \lambda^{-1/2} y^i.$$

So, y is the scaled uctuation from the maximum point x_0. The interval of integration should be changed accordingly, so that the maximum point is now $y = 0$. Then, we expand both functions S and φ in Taylor series at x_0 getting

$$\lambda S(x_0 + \lambda^{-1/2} y) = \lambda S(x_0) + \frac{1}{2} y\cdot[\partial_x^2 S(x_0)]y + \sum_{|\alpha|}^{\infty} \frac{\lambda^{-(|\alpha|-2)/2}}{\alpha!}[\partial^\alpha S(x_0)]y^\alpha$$

$$\varphi(x_0 + \lambda^{1/2} y) = \sum_{|\alpha|=0}^{\infty} \frac{\lambda^{-|\alpha|/2}}{\alpha!}\partial^\alpha \varphi(x_0) y^\alpha$$

Since the quadratic terms are of order $O(1)$ we leave them in the exponent and expand the exponent of the rest in a power series. Next, we extend the integration domain to the whole $\mathbb{R}^n$ and compute the standard Gaussian integrals.

Finally, we get a power series in inverse powers of λ. The coefficients

a_k of the asymptotic expansion are polynomials in the higher derivatives $\partial^\alpha S(x_0)$, $|\alpha| \geq 3$, and derivatives $\partial^\alpha \varphi(x_0)$, $|\alpha| \geq 0$, and involve inverse matrices $G = [\partial_x^2 S(x_0)]^{-1}$.

Remark: If x_0 is a degenerate maximum point of the function S, then the asymptotic expansion as $\lambda \to \infty$ has the form

$$F(\lambda) \sim \exp[\lambda S(x_0)]\lambda^{-n/2} \sum_{k=0}^{\infty}\sum_{l=0}^{N} a_{kl}\lambda^{-rk} (\ln \lambda)^l,$$

where N is some positive integer and $\{r_k\}$, $r_k \geq n/2$, $k \in \mathbb{N}$, is a increasing sequence of nonnegative rational numbers.

The coefficients a_k (and a_{kl}) of the asymptotic expansion of the integral $F(\lambda)$ are invariants under smooth local diffieomorphisms in a neighborhood of x_0 and play very important role in various applications.

Boundary Maximum Point

Let now S has maximum at the boundary point $x_0 \in \partial\Omega$. We assume, for simplicity, that both the boundary and the function S are smooth, i.e. $S \in C^\infty$ and $\partial\Omega \in C^\infty$.

Since the boundary is smooth we can smoothly parametrize it in a neighborhood of x_0 by $(n-1)$ parameters $\xi = (\xi^a)$, $(a = 1,\ldots, n-1)$. Let the the parametric equations of the boundary be

$$x^i = x^i(\xi), \; i = 1,\ldots, n.$$

Then

$$T_a = \left(T_a^i\right) = \left(\frac{\partial x^i}{\partial \xi^a}\right)$$

are tangent vectors to the boundary.

Let $r = r(x)$ be the normal distance to the boundary. Then the equation of the boundary can be written as

$$r(x) = 0.$$

a and for $x \in \Omega$ we obtain that the vector we have $r > 0$. Obviously, $r(x(\xi)) \equiv 0$. From this equation

$$N = (N_i) = \left(\frac{\partial r}{\partial x^i}\right)$$

is orthogonal to all tangent vectors and is therefore normal to the boundary. It can be certainly normalized, since it is nowhere zero. We choose it to be the inward normal. The normal and tangential derivatives are defined as usual

$$\partial_r = \sum_{i=1}^{n} \frac{\partial x^i}{\partial r}\frac{\partial}{\partial x^i}, \quad \frac{\partial}{\partial \xi^a} = \sum_{i=1}^{n} \frac{\partial x^i}{\partial r}\frac{\partial}{\partial x^i}$$

The point x_0 is not, in general, a critical point of S, since the normal derivative of S at x_0 does not have to be equal to zero.

Definition: The point x_0 is said to be a nondegenerate boundary maximum point of S if

$$\partial_r S(x_0) \neq 0$$

and the $(n-1) \times (n-1)$ matrix $\partial_\xi^2 S(x(\xi))$ is negative definite.

In a neighborhood of a nondegenerate boundary maximum point the function S has the following Taylor expansion

$$S(x) = S(x_0) + [\partial_r S(x_0)]r + \frac{1}{2}[\partial_r^2 S(x_0)]r^2$$

$$+ [\partial_r \partial_\xi S(x_0)] \cdot (\xi - \xi_0)r + \frac{1}{2}(\xi - \xi_0) \cdot [\partial_r^2 S(x_0)](\xi - \xi_0)$$

$$+ \ldots,$$

up to third order terms in r and $(\xi - \xi_0)$.

Now we replace the integral $F(\lambda)$ by an integral over a small neighborhood of x_0. We change the variables of integration from x^i, $i = 1,\ldots, n$, to (ξ^a, r), $a = 1,\ldots, n-1$, and neglect the terms of third order in the Taylor series. We also replace the function f by its value at the point x_0. In the remaining integral we extend the integration to the whole space $\mathbb{R}_+ \times \mathbb{R}^{n-1}$, i.e. we integrate over r from 0 to ∞ and integrate over the whole tangent plane at x_0. These integrals are standard Gaussian integrals and we obtain the leading asymptotics as $\leq \to \infty$

$$F(\lambda) \sim -\lambda^{-(n+1)/2}(2\pi)^{(n-1)/2}\exp[\lambda S(x_0)]$$

$$\times [\partial_r S(x_0)]^{-1}[-\det \partial_\xi^2 S(x_0)]^{-1/2} J(x_0) f(x_0)$$

where $J(x_0)$ is the Jacobian of change of variables.

The general form of the asymptotic expansion is given by the following theorem.

Theorem: Let $f, S \in C^\infty(\Omega)$ and let S have a maximum only at a non-degenerate boundary maximum point $x_0 \in \partial\Omega$. Then as $\lambda \to \infty$, $\lambda \in S_\varepsilon$,

$$F(\lambda) \sim \lambda^{-(n+1)/2}\exp[\lambda S(x_0)]\sum_{k=0}^{\infty} a_k \lambda^{-k}$$

Integral Operators with Singular Kernels

Let be a bounded domain in $\mathbb{R}^n$ including the origin, $0 \in \Omega$. Let S be a real valued non-positive function on Ω of class C^2 that has maximum equal to zero, $S(0) = 0$, only at a nondegenerate maximum critical point $x_0 = 0$. Let K_λ: $C^\infty(\Omega)$ be a linear integral operator defined by

$$(K_\lambda f)(x) = \left(\frac{\lambda}{2\pi}\right)^{n/2} \int_\Omega \exp[\lambda S(x-y)] f(y)\,dy$$

Let M be a compact subset of Ω. Then

$$\lim_{\lambda\to\infty} (K_\lambda f)(x) = \left[-\det \partial_x^2 S(0)\right]^{-1/2} f(x)$$

uniformly for $x \in M$. Formally

$$\left(\frac{\lambda}{2\pi}\right)^{n/2} \exp[\lambda S(x-y)] \to [-\det \partial_x^2 S(0)]^{-1/2} \partial(x-y)$$

STATIONARY PHASE METHOD

STATIONARY PHASE METHOD IN ONE DIMENSION

Fourier Integrals

Let $M = [a, b]$ be a closed bounded interval, $S: M \to \mathbb{R}$ be a real valued nonconstant function, $f: M \to \mathbb{C}$ be a complex valued nonzero function and $\leq$ be a large positive parameter. Consider the integrals of the form

$$F(\lambda) = \int_a^b f(x) \exp[i\lambda S(x)]\,dx.$$

The function S is called phase function and such integrals are called Fourier Integrals.

We will study the asymptotics of such integrals.

As $\lambda \to \infty$ the integral $F(\lambda)$ is small due to rapid oscillations of $\exp(i\lambda S)$.

Lemma: (Riemann-Lebesgue) Let f be an integrable function on the real line, i.e. $f \in L^1(\mathbb{R})$. Then

$$\int_{\mathbb{R}} f(x) e^{i\lambda x}\,dx = o(1), \quad (\lambda \to \infty).$$

Definition: 1. A point x_0 is called the regular point of the Fourier integral $F(\lambda)$ if the functions f and S are smooth in a neighborhood of x_0 and $S'(x_0) \neq 0$.

2. A point x_0 is called the critical point of the integral $F(\lambda)$ if it is not a regular point.
3. A critical point x_0 is called isolated critical point if there is a neighborhood of x_0 that does not contain any other critical points.
4. An interior isolated critical point is called stationary point.
5. The integral over a neighborhood of an isolated critical point that does not contain other critical points will be called the contribution of the critical point to the integral.

Clearly the main contribution comes from the critical points since close to these points the oscillations slow down. As always, we will assume that functions S and f are smooth, i.e. of class $C^\infty(M)$. Otherwise, the singularities of the functions S and f and their derivatives would contribute significantly to $F(\lambda)$.

Localization Principle

Lemma: Let $S\in C^\infty(\mathbb{R})$ be smooth function and $f\in C_0^\infty(\mathbb{R})$ be a smooth function of compact support. Then as $\lambda\to\infty$

$$\int_{\mathbb{R}} f(x)\exp[i\lambda S(x)]dx = O(\lambda^{-\infty})$$

Remarks:

1. Since the function f has compact support, the integral is, in fact, over a finite interval.
2. This is the main technical lemma for deriving the (power) asympotics of the Fourier integrals. It means that such integrals can be neglected in a power asymptotic expansion.
3. The Fourier integrals are in general much more subtle object than the Laplace integrals. Instead of exponentially decreasing integrand one has a rapidly oscillating one. This requires much finer estimates and also much stronger conditions on the phase function S and the inte-grand f.

Theorem: Let the Fourier integral $F(\lambda)$ have finite number of isolated critical points.

Then as $\lambda\to\infty$ the integral $F(\lambda)$ is equal to the sum of the contributions of all critical points up to $O(\lambda^{-\infty})$.

Thus, the problem reduces to computing the asymptotics of the contributions of critical points.In a neighborhood of a critical point we can replace the functions S and f by more simple functions and then compute some standard integrals.

Boundary Points

If the phase function does not have any stationary points, then by integration by parts one can easily obtain the asymptotic expansion.

Theorem: Let $S'(x)\neq 0\ \forall x\in M$. Then as $\lambda\to\infty$

$$F(\lambda)\sim\sum_{k=0}^{\infty}(i\lambda)^{-k-1}\left(\frac{1}{-S'(x)}\frac{\partial}{\partial x}\right)^k\left(\frac{f(x)}{S'(x)}\right)e^{i\lambda S(x)}\Bigg|_a^b.$$

The leading asymptotics is

$$F(\lambda) = (i\lambda)^{-1}\{f(b)\exp[i\lambda S(b)] - f(a)\exp[i\lambda S(a)]\} + O(\lambda^{-2})$$

The same technique, i.e. integration by parts, applies to the integrals over an unbounded interval, say,

$$F(\lambda) = \int_0^\infty f(x)\exp[i\lambda S(x)]dx$$

with some additional conditions that guarantee the converges at ∞ as well as to the integrals of the form

$$F(x) = \int_x^\infty f(t)\exp[iS(t)]dt$$

as $x \to \infty$.

Standard Integrals

Consider the integral

$$\Phi(\lambda) = \int_0^a f(x)x^{\beta-1}e^{i\lambda x^\alpha}.$$

Lemma: (Erdeyi) Let $\alpha \geq 1$, $\beta > 0$ and f is a smooth function on a closed bounded interval $[0, a]$, $f \in C^\infty([0, a])$, that vanish at $x = a$ with all its derivatives. Then as $\lambda \to \infty$

$$\Phi(\lambda) = \int_0^a f(x)x^{\beta-1}e^{i\lambda x^\alpha} \sim \sum_{k=0}^{\infty} a_k \lambda^{-(\beta+k)/\alpha}.$$

where

$$a_k = \frac{f^{(k)}(0)}{k!}\frac{1}{\alpha}\Gamma\left(\frac{\beta+k}{\alpha}\right)\exp\left[i\pi\frac{\beta+k}{\alpha}\right]$$

This lemma plays he same role in the stationary phase method as Watson lemma in the Laplace method.

Stationary Point

Theorem: Let $M = [a, b]$ be a closed bounded interval, $S \in C^\infty(M)$ be a smooth real valued nonconstant function, $f \in C_0^\infty(M)$ be a complex valued function with compact support in M. Let S have a single isolated nondegenerate critical point x_0 in M, i.e. $S'(x_0) = 0$ and $S''(x_0) \neq 0$. Then as $\lambda \to \infty$ there is asymptotic expansion of the Fourier integral

$$F(\lambda) = \int_a^b f(x)\exp[i\lambda S(x)]dx$$

$$\sim \exp\left[i\lambda S(x_0) + [\operatorname{sgn} S''(x_0)]i\frac{\pi}{4}\right]\lambda^{-1/2}\sum_{k=0}^{\infty}\lambda^{-k}.$$

The coefficients a_k are determined in terms of the derivatives of the functions S and f at x_0.

The leading asymptotics as $\lambda \to \infty$ is

$$F(\lambda) = \sqrt{\frac{2\pi}{|S''(x_0)|}}\lambda^{-1/2}\exp\left[i\lambda S(x_0) + [\operatorname{sgn} S''(x_0)]i\frac{\pi}{4}\right](f(x_0) + O(\lambda^{-1}))$$

To prove this theorem one does a change of variables in a sufficiently small neghborhood of x_0.

Principal Values of Integrals

Let f be a smooth function and consider the integral

$$\int_a^b \frac{f(x)}{x} dx$$

This integral diverges, in general, at $x = 0$. One can regularize it by cutting out a symmetric neighborhood of the singular point

$$I(\varepsilon) = \int_a^{-\varepsilon} \frac{f(x)}{x} dx + \int_\varepsilon^b \frac{f(x)}{x} dx .$$

Definition: If the limit of $I(\varepsilon)$ as $\varepsilon \to 0^+$ exists, then it is called the principal value of the integral I

$$\mathcal{P}\int_a^b \frac{f(x)}{x} dx = \lim_{\varepsilon \to 0^+} \left(\int_a^{-\varepsilon} \frac{f(x)}{x} dx + \int_\varepsilon^b \frac{f(x)}{x} dx \right)$$

In this section we consider the asymptotics of the integrals of the form

$$F(\lambda) = \mathcal{P}\int_{\mathbb{R}} e^{\pm i\lambda S(x)} f(x) \frac{dx}{x}.$$

as $\lambda \to \infty$.

Lemma: Let $f \in C_0^\infty(\mathbb{R})$ be a smooth function of compact support. Then as $\lambda \to \infty$

$$\mathcal{P}\int_{\mathbb{R}} e^{\pm i\lambda x} f(x) \frac{dx}{x} = \pm i\pi f(0) + O(\lambda^{-\infty}).$$

Theorem: Let $f \in C_0^\infty(\mathbb{R})$ be a smooth function of compact support, $S \in C^\infty(\mathbb{R})$ be a real valued smooth function and $S'(0) \neq 0$. Then as $\lambda \to \infty$

$$F(\lambda) = \mathcal{P}\int_{\mathbb{R}} e^{\pm i\lambda x} f(x) \frac{dx}{x} = [\operatorname{sgn} S'(0)] i\pi f(0) \exp[i\lambda S(0)] + O(\lambda^{-\infty}).$$

Theorem: Let $f \in C_0^\infty(\mathbb{R})$ be a smooth function of compact support, $S \in C^\infty(\mathbb{R})$ be a real valued smooth function. Let $x = 0$ be the only statinary point of the function S on supp f, and let it be nondegenerate, i.e. $S'(0) = 0$ and $S''(0) \neq 0$. Then as $\lambda \to \infty$ there is asymptotic expansion

$$F(\lambda) = \mathcal{P}\int_{\mathbb{R}} e^{\pm i\lambda S(x)} f(x) \frac{dx}{x}$$

$$\sim \exp[i\lambda S(0)] \lambda^{-1/2} \sum_{k=0}^{\infty} a_k \lambda^{-k}.$$

The leading asymptotics has the form

$$F(\lambda) = \exp\left[i\lambda S(0) + [\operatorname{sgn} S''(0) i\frac{\pi}{4}\right] \sqrt{\frac{2\pi}{|S''(0)|}}$$

$$\times \lambda^{-1/2}\left[-\frac{S'''(0)}{6S''(0)} f(0) + f'(0) + O(\lambda^{-1}) \right].$$

STATIONARY PHASE METHOD IN MANY DIMENSIONS

Let Ω be a domain in $\mathbb{R}^n$ and $f \in C_0^\infty(\Omega)$ be a smooth function of compact support, $S \in C^\infty(\Omega)$ be a real valued smooth function. In this section we study the asympotics as $\lambda \to \infty$ of the multi-dimensional Fourier integrals

$$F(\lambda) = \int_\Omega f(x) \exp[i\lambda S(x)]\, dx.$$

Nondegenerate Stationary Point

Localization Principle

Lemma: Let Ω be a domain in $\mathbb{R}^n$ and $f \in C_0^\infty(\Omega)$ be a smooth function of compact support, $S \in C^\infty(\Omega)$ be a real valued smooth function without stationary points in supp f, i.e. $\partial_x S(x) \neq 0$ for $x \in$ supp f. Then as $\lambda \to \infty$

$$F(\lambda) = O(\lambda^{-\infty})$$

This lemma is proved by integration by parts.

Definition: The set $S(\mathbb{R}^n)$ of all smooth functions on R^n that decrease at $|x| \to \infty$ together with all derivatives faster than any power of $|x|$ is called the Schwartz space.

For any integrable function $f \in L^1(\mathbb{R}^n)$ the Fourier transform is defined by

$$F(f)(\xi) = (2\pi)^{-n/2} \int_{\mathbb{R}^n} \exp(i x \cdot \xi) f(x)\, dx$$

Proposition: Fourier transform is a one-to-one onto map (bijection) $\mathcal{F}$: $\mathcal{S}(\mathbb{R}^n) \to S(\mathbb{R}^n)$, i.e. if $f \in \mathcal{S}(\mathbb{R}^n)$, then $\mathcal{F}(f) \in \mathcal{S}(\mathbb{R}^n)$.

The inverse Fourier transform is

$$\mathcal{F}^{-1}(f)(x) = (2\pi)^{-n/2} \int_{\mathbb{R}^n} \exp(i x \cdot \xi) f(\xi)\, d\xi$$

Theorem: Let Ω be a finite domain in $\mathbb{R}^n$, $f \in C^\infty{}_0(\Omega)$ be a smooth function with compact support in and $S \in C^\infty(\Omega)$ be a real valued smooth function. Let S have a single stationary point x_0 in and let it be non-degenerate. Then as $\lambda \to \infty$ there is asymptotic expansion:

$$F(\lambda) \sim \lambda^{-n/2} \exp[i\lambda S(x_0)] \sum_{k=0}^{\infty} a_k \lambda^{-k}.$$

The coefficients a_k are determined in terms of derivatives of the functions f and S at x_0.

The leading asymptotics is

$$F(\lambda) = \left(\frac{2\pi}{\lambda}\right)^{n/2} \exp\left[i\lambda S(x_0) + [\operatorname{sgn} \partial_x^2 S(x_0) i\frac{\pi}{4}\right]$$

$$\times \left|\det \partial_x^2 (S(x_0)\right|^{-1/2} \left[f(x_0) + O(\lambda^{-1})\right].$$

Recall that sgn A = $\nu_+(A) - \nu_-(A)$ denotes the signature of a real symmetric nondegenerate matrix A, where $\nu_\pm(A)$ are the number of positive and negative eigenvalues of A.

Integral Operators with Singular Kernels

Let be a bounded domain in $\mathbb{R}^n$ including the origin, $0 \in \Omega$. Let $f \in C$ be a smooth function with compact support and $S \in C^\infty(\mathbb{R}^n)$ be a real valued non-positive smooth function.

Let S have a single stationary point $x = 0$, and let it to a nondegenerate, i.e. let $S(0) = S'(0) = 0$ and $\partial_x^2 S(0) \neq 0$. Let $K_\lambda: C_0^\infty(\Omega) \to C^\infty(\Omega)$ be a linear integral operator defined by

$$(K_\lambda f)(x) = \left(\frac{\lambda}{2\pi}\right)^{n/2} \int_\Omega \exp[i\lambda S(x-y)] f(y) dy$$

Then as $\lambda \to \infty$

$$(K_\lambda f)(x) = \exp\left[\operatorname{sgn} \delta_x^2 S(0) i\frac{\pi}{4}\right] \left|\det \partial_x^2 S(0)\right|^{-1/2} \left[f(x) + O(\lambda^{-1})\right]$$

uniformly for $x \in \Omega$. On the other hand, if $x \in \Omega$, then as $\lambda \to \infty$

$$(K_\lambda f)(x) = O(\lambda^{-\infty})$$

SADDLE POINT METHOD FOR LAPLACE INTEGRALS

Let be a contour in the complex plane and the functions f and S are holomorphic in a neighborhood of this contour. In this section we will study the asymptotics as $\lambda \to \infty$ of the Laplace integrals

$$F(\lambda) = \int_\gamma f(z) \exp[\lambda S(z)] dz$$

Heuristic Ideas of the Saddle Point Method

The idea of the saddle point method (or method of steepest descent) is to deform the contour in such a way that the main contribution to the integral comes from a neighborhood of a single point. This is possible since the functions f and S are holormorphic.

First of all, let us find a bound for $|F(\lambda)|$. For a contour $\gamma = \gamma_0$ finite length $l(\gamma_0)$ we have, obviously,

$$|F(\lambda)| \leq l(\gamma_0) \max_{z \in \gamma_0} f(z) \exp[\lambda \operatorname{Re} S(z)].$$

Now, let Γ be the set of all contours obtained by smooth deformations of the contour γ_0 keeping the endpoints fixed. Then such an estimate is valid for any contour $\gamma \in \Gamma$, hence,

$$|F(\lambda)| \leq \inf_{\gamma \in \Gamma} \left\{ l(\gamma) \max_{z \in \gamma} f(z)[\lambda \operatorname{Re} S(z)] \right\}$$

Since we are interested in the limit $\lambda \to \infty$, we expect that the length of the contour does not affect the accuracy of the estimate. Also, intuitively it is clear that the behaviour of the function S is much more important that that of the function f (since S is in the exponent and its variations are scaled significantly by the large parameter λ). Thus we expect an estimate of the form

$$|F(\lambda)| \leq C(\gamma, f) \inf_{\gamma \in \Gamma} \left\{ \max_{z \in \gamma} \exp[\lambda \operatorname{Re} S(z)] \right\}$$

where $C(\gamma, f)$ is a constant that depends on the contour γ and the function f but does not depend on λ. So, we are looking for a point on a given contour where the maximum of Re $S(z)$ is attained. Then, we look for a contour γ^* where the minimum of this maximum is attained, i.e. we assume that there exists a contour γ_* where

$$\min_{\gamma \in \Gamma} \max_{z \in \gamma} \operatorname{Re} S(z)$$

is attained. Such a contour will be called a minimax contour.

Let $z_0 \in \lambda_*$ be the only point on the contour γ_* where the maximum of Re $S(z)$ is attained. Then, we have an estimate

$$|F(\lambda)| \leq C(\gamma_*, f) \exp[\lambda \operatorname{Re} S(z_0)].$$

By deforming the contour of integration to γ_* we obtain

$$F(\lambda) = \int_{\gamma_*} f(z) \exp[\lambda S(z)] dz.$$

The asymptotics of this integral can be computed by Laplace method.

1. Boundary Point. Let z_0 be an endpoint of γ_*, say, the initial point. Suppose that $S'(z_0) \neq 0$. Then one can replace the integral $F(\lambda)$ by an integral over a small arc with the initial point z_0. Finally, integrating by parts gives the leading asymptotics

$$F(l) \quad = \frac{1}{-S'(z_0)} \exp[\lambda S(z_0)] \lambda^{-1} [f(z_0) + O(\lambda^{-1})].$$

Interior Point. Let z_0 be an interior point of the contour γ_*. From the minimax property of the contour γ_* it follows that the point z_0 is the saddle point of the function Re $S(z)$. Let $z = x + iy$. Since the saddle point is a stationary point, then

$$\frac{\partial}{\partial x}\operatorname{Re} S(z_0) = \frac{\partial}{\partial y}\operatorname{Re} S(z_0).$$

Then from Cauchy-Riemann conditions it follows that $S'(z_0) = 0$.

Definition: *A* point $z_0 \in \mathbb{C}$ is called a saddle point of the complex valued function $S\colon \mathbb{C} \to \mathbb{C}$ if $S'(z_0) = 0$.

2. A saddle point z_0 is said to be of order n if
$$S'(z_0) = \cdots = S^{(n)}(z_0) = 0, \quad S^{(n+1)}(z_0) \neq 0.$$
3. A first order saddle point is called simple, i.e. for a simple saddle point $S''(z_0) \neq 0$.
4. The number Re $S(z_0)$ is called the height of the saddle point.

To compute the asymptotics at an interior saddle point, we replace the contour γ_* by a small arc γ_*^0 containing the point z_0. Then we expand the function S in the Taylor series in the neighborhood of z_0 and neglect the terms of third order and higher, i.e. we replace S by

$$S(z) = S(z_0) + \frac{1}{2}S''(z_0)(z - z_0)^2 + O((z - z_0)^3).$$

Finally, by changing the variables and evaluating the integral by Laplace method we obtain the asymptotics as $\lambda \to \infty$

$$F(\lambda) = \sqrt{\frac{2\pi}{-S''(z_0)}}\,\lambda^{-1/2}\exp[\lambda S(z_0)]\,f(z_0) + O(\lambda^{-1})].$$

The saddle point method consists of two parts: the toplogical part and the analytical part.

The topological part consists of the deformation of the contour to the minimax contour γ_* that is most suitable for asymptotical estimates. The analytical part contains then the evaluation of the asymptotics over the contour γ_*.

The analytical part is rather straightforward. Here one can apply the same methods and as in the Laplace method; in many cases one can even use the same formulas.

The topological part is usually much more complicated since it is a global problem.

It could happen, for example, that a contour γ_* where the minimax $\min_{\gamma\in\Gamma}\max_{z\in\gamma}$ Re $S(z)$ is attained does not exist at all! Next, strictly speaking we need to look for a contour where $\min_{\gamma\in\Gamma}\max_{z\in\gamma} f(z)\exp[\operatorname{Re} S(z)]$ is attained what makes the problem even more complicated.

Thus, if one can find the minimax contour, then one can compute the asymptotics of $F(\lambda)$ as $\lambda \to \infty$. Unfortunately, there is no simple algorithm that would always enable one toλ find the minimax contour. Nevertheless, under certain conditions one can prove that such a contour exists and, in fact, find one.

Level Curves of Harmonic Functions

Lemma: Let $S: \mathbb{C} \to \mathbb{C}$ be holomorphic at z_0 and $S'(z_0) \neq 0$. Then in a small neighborhood of the point z_0 the arcs of the level curves

$$\text{Re } S(z) = \text{Re } S(z_0), \quad \text{Im } S(z) = \text{Im } S(z_0),$$

are analytic curves. These curves are orthogonal at z_0.

Let $\varphi(z) = S(z)\, S(z_0)$. Since $S'(z_0) \neq 0$ the function $w = \varphi(z)$ is is a one-to-one holomorphic, in fact, conformal, mapping of a neghborhood of the point $z = z_0$ onto a neighborhood of the point $w = 0$. The inverse function $z = \varphi^{-1}(w)$ is holomorphic in a neighborhood of the origin $w = 0$. Let $w = u + iv$ and $(u, v) \equiv \varphi^{-1}(w)$.

The arc of the level curve Re $S(z)$ = Re $S(z_0)$ is mapped onto an open interval on the imaginary axis. It is defined by $z = (0, v)$ and is analytic. The same is true for the level curve Im $S(z)$ = Im $S(z_0)$. It is defined by $z = \psi(u, 0)$ and is analytic as well. The tangent vectors to the level curves at z_0 are determined by

$$\left.\frac{\partial\psi(u,0)}{\partial u}\right|_{u=0} = (\partial_w \varphi^{-1})(0), \left.\frac{\partial\psi(0,v)}{\partial v}\right|_{v=0} = i(\partial_w \varphi^{-1})(0)$$

and are obviously orthogonal. One could also conlude this from the fact that the map is conformal and, therefore, preserves the angles.

Lemma: Let z_0 be a simple saddle point of the function S. Then in a small neighborhood of the point z_0 the level curve Re $S(z)$ = Re $S(z_0)$ consists of two analytic curves that intersect orthogonally at the point z_0 and separate the neighborhood of z_0 in four sectors. The signs of the function Re $[S(z) - S(z_0)]$ in adjacent sectors are different.

In a neighborhood of a simple saddle point the there is a one-to-one holomorphic function $z = \psi(w)$ such that $\psi(0) = z_0$, $\psi'(0) \neq 0$, and $S(\psi(w)) = S(z_0) + w^2$.

In the complex plane of w the level curve Re $S(z)$ = Re $S(z_0)$ takes the form Re $w^2 = 0$. Its solution consists of two orthogonal lines $w_{\pm} = (1 \pm i)t$ with $-\varepsilon < t < \varepsilon$ that intersect at the point $w = 0$. The level curves $z_{\pm} = (w_{\pm})$ have listed properties.

Lemma: Let z_0 be a saddle point of the function S of order n. Then in a small neighborhood of the point z_0 the level curve Re $S(z)$ = Re $S(z_0)$ consists of $(n + 1)$ analytic curves that intersect at the point z_0 and separate the neighborhood of z_0 in $2(n + 1)$ sectors with angles $\pi(n + 1)$ at the vertex. The signs of the function Re $[S(z) - S(z_0)]$ in adjacent sectors are different.

Definition: Let S be a complex valued function and γ be a simple curve with the initial point z_0.

The curve γ is called curve of steepest descent of the function Re S if Im $S(z)$ = const and Re $S(z) <$ Re $S(z_0)$ for $z \in \gamma$, $z \neq z_0$. If Im $S(z)$ = const and Re $S(z) >$ Re $S(z_0)$ for $z \in \gamma$, $z \neq z_0$, then the curve γ is called curve of steepest ascent of the function Re S.

Lemma: 1. If z_0 is not a saddle point, then there is exactly one curve of steepest descent.

2. If z_0 is a simple saddle point, then there are 2 curves of steepest descent.
3. If z_0 is a saddle point of order n, then there are $(n + 1)$ curves of steepest descent.
4. In a neighborhood of a saddle point z_0 in each sector in which Re $[S(z) - S(z_0)] > 0$ there is exatly one curve of steepest descent.

This is proved by a change of variables in a neighborhood of z_0.

Remarks: Let S: $D \to \mathbb{C}$ be a nonconstant holomorphic function in a domain D. Let z = x+iy and $S(z) = u(x, y) + iv(x, y)$. Then both u: $D \to \mathbb{R}$ and v: $D \to \mathbb{R}$ are harmonic functions in D. Harmonic functions do not have maximum or minimum points in the interior of D. They are attained only at the boundary of the domain D.

All critical points of harmonic functions, i.e. the points where $\partial u = \partial v = 0$ are saddle points.

These are exactly the points where $S'(z) = 0$. That is why such points are called saddle points of the function S. In the simplest case $S = z^2$ the surface $u = x^2 - y^2$ is hyperbolic paraboloid (saddle).

Definition: Two contours γ_1 and γ_2 are called equivalent if

$$\int_{\gamma_1} f(z)\exp[\lambda S(z)]dz = \int_{\gamma_2} f(z)\exp[\lambda S(z)]dz$$

Lemma: Let S and f be holormorphic functions on a finite contour Let the points where $\max_{z\in\gamma}$ Re $S(z)$ is attained are neither saddle points nor the endpoints of the contour γ. Then there is a contour γ' equivalent to the contour γ and such that

$$\max_{z\in\gamma'} \operatorname{Re} S(z) < \max_{z\in\gamma} \operatorname{Re} S(z).$$

Theorem: Let $F(\lambda)$ be a Laplace integral if there exists a contour γ_* such that: *i)* it is equivalent to the contour γ and *ii)* the integral $F(\lambda)$ attains the minimax $\min_{\gamma\in\Gamma}\max_{z\in\gamma}$ Re $S(z)$ on it. Then among the points where $\max_{z\in\gamma^*}$ Re $S(z)$ is attained there are either endpoints of the contour or saddle points z_j such that in aneighborhood of z_j the contour γ_* goes through two different sectors where Re $S(z) <$ Re $S(z_j)$.

Analytic Part of Saddle Point Method

In this section we always assume that γ is a simple smooth (or piece-wise smooth) curve in the complex plane, which may be γ finite or infinite. The functions f and S are assumed to be holomorphic on γ. Also, we assume that the integral $F(\lambda)$ converges absolutely.

First of all, we have

Lemma: If $\max_{z\in\gamma} \operatorname{Re} S(z) \geq C$, then

$$F(\lambda) = O(e^{C}\lambda), \quad (\lambda \geq 1).$$

Theorem: Let z_0 be the initial endpoint of the curve γ. Let f and S are analytic at z_0, $\operatorname{Re} S(z_0) > \operatorname{Re} S(z)\ \forall z \in \gamma$, and $S'(z_0) \neq 0$. Then, as $\lambda \to \infty$ there is asympotic expansion

$$F(\lambda) \sim \exp[\lambda S(z_0)] \sum_{k=0}^{\infty} a_k \lambda^{-k-1}$$

where

$$a_k = -\left(-\frac{1}{S'(z)}\frac{\partial}{\partial z}\right)^k \left[\frac{f(z)}{S'(z)}\right]\Bigg|_{z=z_0}$$

The proof is by integration by parts.

Theorem: Let z_0 be an interior point of the curve γ. Let f and S are analytic at z_0, $\operatorname{Re} S(z_0) > \operatorname{Re} S(z)\ \forall z \in \gamma$. Let z_0 be a simple saddle point of S such that in a neighborhood of z_0 the contour γ goes through two different sectors where $\operatorname{Re} S(z) < \operatorname{Re} S(z_0)$. Then, as $\lambda \to \infty$ there is asympotic expansion

$$F(\lambda) \sim \exp[\lambda S(z_0)] \sum_{k=0}^{\infty} a_k \lambda^{-k-1/2}$$

The branch of the square root is choosen so that $\arg\sqrt{-S''(z_0}$ is equal to the angle between the positive direction of the tangent to the curve γ at the point z_0 and the positive direction of the real axis.

In a neighborhood of z_0 there is a mapping $z = \psi(w)$ such that

$$S(\psi(w)) = S(z_0) - \frac{w^2}{2}.$$

After this change of variables and deforming the contour to the steepest descent contour the integral becomes

$$F(\lambda) = \exp[\lambda S(z_0)] \int_{\varepsilon}^{\varepsilon} e^{-\lambda w^2/2} f(\psi(w))\psi'(w)\,dw + O(\lambda^{-\infty})$$

Since both f and ψ are holomorphic there is a Taylor expansion

$$f(\psi(w))\psi'(w) = \sum_{k=0}^{\infty} c_k w^k.$$

Then the coefficients a_k are easily computed in terms of c_k

$$ak = 2^{k+1/2}\Gamma\left(k+\frac{1}{2}\right)c_{2k}$$

Theorem: Let γ be a finite contour and f and S be holomorphic in a neighborhood of λ. Let $\max_{z\in\gamma} \operatorname{Re}S(z)$ be attained at the points z_j and these points are either the endpoints of the curve or saddle points such that in a neighborhood of a saddle points the contour goes through two different sectors where $\operatorname{Re} S(z) < \operatorname{Re} S(z_j)$.

Then as $\lambda \to \infty$ the integral $F(\lambda)$ is asymptotically equal to the sum of contributions of the points z_j.

Remark: The integral over a small arc containing a saddle point is called the contribution of the saddle point to the integral.

Proposition: Let f and S be holomorphic functions on and $\operatorname{Im} S(z) =$ const on γ. If has γ finite number of saddle points, then as |lambda| $\to \infty$, $|\arg \lambda| \leq \pi/2 - \epsilon < \pi/2$, the asymptotic expansion of the integral $F(\lambda)$ is equal to the sum of contributions of the saddles points and the endpoints of the contour.

Chapter 3

Free Fall and Harmonic Oscillators

THE SIMPLE HARMONIC OSCILLATOR

The next physical problem of interest is that of simple harmonic motion. Such motion comes up in many places in physics and provides a generic first approximation to models of oscillatory motion.

This is the beginning of a major thread running throughout our course. You have seen simple harmonic motion in your introductory physics class. We will review SHM (or SHO in some texts) by looking at springs and pendula (the plural of pendulum).

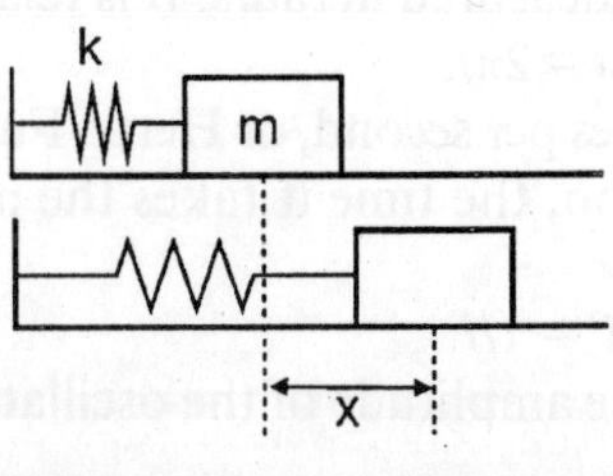

Fig. Spring-Mass System

Mass-Spring Systems

We begin with the case of a single block on a spring. The net force in this case is the restoring force of the spring given by Hooke's Law,

$$Fs = -kx,$$

where $k > 0$ is the spring constant. Here x is the elongation, or displacement, of the spring from equilibrium. When the displacement is positive, the spring force is negative and when the displacement is negative the spring force is positive. We have depicted a horizontal system sitting on a frictionless surface. A similar model can be provided for vertically oriented springs. However, you need to account for gravity to determine the location of equilibrium. Otherwise, the oscillatory motion about equilibrium is modelled the same.

From Newton's Second Law, $F = m\ddot{x}$, we obtain the equation for the motion of the mass on the spring.

$$m\ddot{x} + kx = 0.$$

For now we note that two solutions of this equation are given by

$$x(t) = A \cos \omega t$$

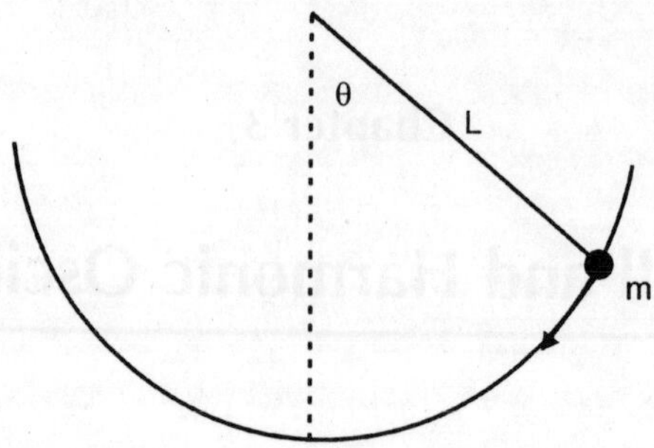

Fig. A Simple Pendulum Consists of a Point Mass m Attached to a String of Length L. It is Released from an Angle θ_0

$$x(t) = A \sin \omega t,$$

where

$$\omega = \sqrt{\frac{k}{m}}$$

is the angular frequency, measured in rad/s. It is related to the frequency by

$$\omega = 2\pi f,$$

where f is measured in cycles per second, or Hertz. Furthermore, this is related to the period of oscillation, the time it takes the mass to go through one cycle.

$$T = 1/f.$$

Finally, A is called the amplitude of the oscillation.

The Simple Pendulum

The simple pendulum consists of a point mass m hanging on a string of length L from some support. One pulls the mass back to some stating angle, θ_0, and releases it. The goal is to find the angular position as a function of time.

There are a couple of possible derivations. We could either use Newton's Second Law of Motion, $F = ma$, or its rotational analogue in terms of torque. We will use the former only to limit the amount of physics background needed.

There are two forces acting on the point mass. The first is gravity. This points downward and has a magnitude of mg, where g is the standard symbol for the acceleration due to gravity. The other force is the tension in the string. These forces and their sum are shown. The magnitude of the sum is easily found as $F = mg \sin \theta$ using the addition of these two vectors.

Now, Newton's Second Law of Motion tells us that the net force is the mass times the acceleration. So, we can write

$$m\ddot{x} = -mg \sin \theta.$$

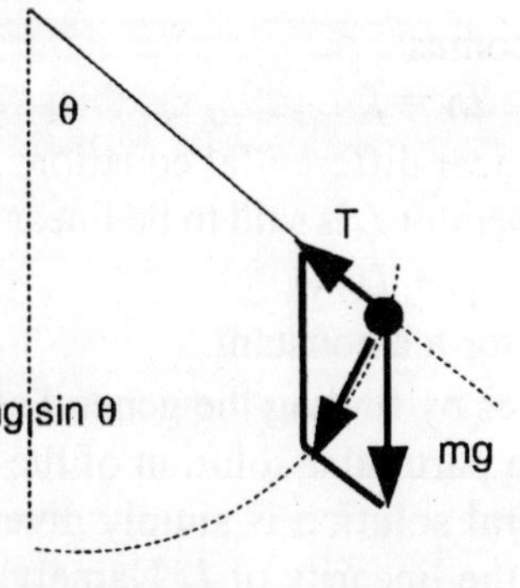

Fig. There are two Forces Acting on the Mass, the Weight mg and the Tension T. The net Force is found to be $F = mg \sin \theta$

Next, we need to relate x and θ. x is the distance traveled, which is the length of the arc traced out by our point mass. The arclength is related to the angle, provided the angle is measure in radians. Namely, $x = r\theta$ for $r = L$. Thus, we can write

$$mL\ddot{\theta} = -mg \sin \theta.$$

Cancelling the masses, this then gives us our nonlinear pendulum equation

$$L\ddot{\theta} + g \sin \theta = 0.$$

There are several variations of equation which will be used in this text. The first one is the linear pendulum. This is obtained by making a small angle approximation. For small angles we know that $\sin \theta \; \pi \; \theta$. Under this approximation becomes

$$L\ddot{\theta} + g\theta = 0.$$

We note that this equation is of the same form as the mass-spring system.

We define $\omega = \sqrt{g/L}$ and obtain the equation for simple harmonic motion,

$$\ddot{\theta} + \omega^2\theta = 0.$$

SECOND ORDER LINEAR DIFFERENTIAL EQUATIONS

In the last section we saw how second order differential equations naturally appear in the derivations for simple oscillating systems. In this section we will look at more general second order linear differential equations.

Second order differential equations are typically harder than first order. In most cases students are only exposed to second order linear differential equations. A general form is given by

$$a(x)y''(x) + b(x)y'(x) + c(x)y(x) = f(x).$$

One can rewrite this equation using operator terminology. Namely, we first define the differential operator $L = a(x)D^2 + b(x)D + c(x)$, where $D = \frac{d}{dx}$.

Then equation becomes

$$Ly = f.$$

The solutions of linear differential equations are found by making use of the linearity of L. An operator L is said to be linear if it satisfies two properties.

1. $L(y_1 + y_2) = L(y_1) + L(y_2)$.
2. $L(ay) = aL(y)$ for a a constant.

One typically solves by finding the general solution of the homogeneous problem, $Ly_h = 0$, and a particular solution of the nonhomogeneous problem, $Ly_p = f$. Then the general solution is simply given as $y = y + h + y_p$. This is found to be true using the linearity of L. Namely,

$$Ly = L(y_h + y_p) = Ly_h + Ly_p = 0 + f = f.$$

There are methods for finding a particular solution, $y_p(x)$, of the equation. These range from pure guessing to either using the Method of

Undetermined Coefficients or the Method of Variation of Parameters.

Determining solutions to the homogeneous problem is not laways so easy. However, others have studied a variety of second order linear equations and have saved us the trouble in the case of differential equations that keep reappearing in applications. Again, linearity is useful.

If y_1 and y_2 are solutions of the homogeneous equation, then the linear combination $c_1y_1 + c_2y_2$ is also a solution of the homogeneous equation. In fact, if y_1 and y_2 are linearly independent, namely,

$$c_1 y_1 + c_2y_2 = 0 \Leftrightarrow c_1 = c_2 = 0,$$

then $c_1y_1 + c_2y_2$ is the general solution of the homogeneous problem.

Linear independence is established if the Wronskian of the solutions in not zero.

$$W(y_1, y_2) = y_1(x)y'_2(x) - y'_1(x)y_2(x) \neq 0.$$

CONSTANT COEFFICIENT EQUATIONS

The simplest and most taught equations are those with constant coefficients. The general form for a homogeneous constant coefficient second order linear differential equation is given as

$$ay''(x) + by'(x) + cy(x) = 0.$$

Solutions are obtained by making a guess of $y(x) = e^{rx}$ and determining what possible values of r will yield a solution. Inserting this guess into leads to the characteristic equation

$$ar^2 + br + c = 0.$$

The roots of this equation lead to three types of solution depending upon the nature of the roots.

1. Real, distinct roots r_1, r_2. In this case the solutions corresponding to each root are linearly independent. Therefore, the general solution is simply $y(x) = c_1e^{r_1x} + c_2e^{r_2x}$.

2. Real, equal roots $r_1 = r_2 = r = -\frac{b}{2a}$. In this case the solutions corresponding to each root are linearly dependent. To find a second linearly independent solution, one uses what is called the Method of Reduction of Order. This gives the second solution as xe^{rx}. Therefore, the general solution is found as
3. Complex conjugate roots In this case the solutions corresponding to each root are linearly independent. Making use of Euler's identity, $e^{i\theta} = \cos(\theta) + i\sin(\theta)$, these complex exponentials can be rewritten in terms of trigonometric functions. Namely, one has that $e^{\alpha x}\cos(\beta x)$ and $e^{\alpha x}\sin(\beta x)$ are two linearly independent solutions. Therefore the general solution becomes $y(x) = e^{\alpha x}(c_1\cos(\beta x) + c_2\sin(\beta x))$.

The solution of constant coefficient equations now follows easily. One solves the characteristic equation and then determines which case applies. Then one simply writes down the general solution. We will demonstrate this with a couple of examples. In the last section of this chapter we review the class of equations called the Cauchy-Euler equations. These equations occur often and follow a similar procedure.

Example: $y'' - y' - 6y = 0$ $y(0) = 2$, $y'(0) = 0$.

The characteristic equation for this problem is $r^2 - r - 6 = 0$. The roots of this equation are found as $r = -2, 3$. Therefore, the general solution can be quickly written down.

$$y(x) = c_1e^{-2x} + c_2e^{3x}.$$

Note that there are two arbitrary constants in the general solution. Therefore, one needs two pieces of information to find a particular solution. Of course, we have them with the information from the initial conditions. One needs

$$y'(x) = -2c_1e^{-2x} + 3c_2e^{3x}$$

in order to attempt to satisfy the initial conditions. Evaluating y and y' at $x =$ 0 yields

$$2 = c_1 + c_2$$
$$0 = -2c_1 + 3c_2$$

These two equations in two unknowns can readily be solved to give $c_1 =$ 6/5 and $c_2 = 4/5$. Therefore, the solution of the initial value problem is $y(x) = \frac{6}{5}e^{-2x} + \frac{4}{5}e^{3x}$. You should verify that this is indeed a solution.

Example: $y'' + 6y' + 9y = 0$.

In this example we have $r^2 + 6r + 9 = 0$. There is only one root, $r = -3$. Again, the solution is found as $y(x) = (c_1 + c_2x)e^{-3x}$.

Example: $y'' + 4y = 0$.

The characteristic equation in this case is $r^2 + 4 = 0$. The roots are pure imaginary roots, $r = \pm 2i$ and the general solution consists purely of sinusoidal functions. $y(x) = c_1 \cos(2x) + c_2 \sin(2x)$.

Example: $y'' + 4y = \sin x$.

This is an example of a nonhomogeneous problem. The homogeneous problem was actually solved in the last example. According to the theory, we need only seek a particular solution to the nonhomogeneous problem and add it to the solution of the last example to get the general solution.

The particular solution can be obtained by purely guessing, making an educated guess, or using variation of parameters. We will not review all of these techniques at this time. Due to the simple form of the driving term, we will make an intelligent guess of $y_p(x) = A \sin x$ and determine what A needs to be. (Recall, this is the Method of

Undetermined Coefficients.) Inserting our guess in the equation gives $(-A + 4A)\sin x = \sin x$. So, we see that $A = 1/3$ works. The general solution of the nonhomogeneous problem is therefore

$$y(x) = c_1 \cos(2x) + c_2 \sin(2x) + \frac{1}{3} \sin x.$$

As we have seen, one of the most important applications of such equations is in the study of oscillations. Typical systems are a mass on a spring, or a simple pendulum. For a mass m on a spring with spring constant $k > 0$, one has from Hooke's law that the position as a function of time, $x(t)$, satisfies the equation

$$m\ddot{x} + kx = 0.$$

This constant coefficient equation has pure imaginary roots ($\alpha = 0$) and the solutions are pure sines and cosines. This is called simple harmonic motion. Adding a damping term and periodic forcing complicates the dynamics, but is nonetheless solvable.

LRC CIRCUITS

Another typical problem often encountered in a first year physics class is that of an LRC series circuit. The resistor is a circuit element satisfying Ohm's Law. The capacitor is a device that stores electrical energy and an inductor, or coil, store magnetic energy. The physics for this problem stems from Kirchoff's Rules for circuits. Namely, the sum of the drops in electric potential are set equal to the rises in electric potential. The potential drops across each circuit element are given by

1. Resistor: $V = IR$.
2. Capacitor: $V = \dfrac{q}{C}$.
3. Inductor: $V = L\dfrac{dI}{dt}$.

Furthermore, we need to define the current as $I = \frac{dq}{dt}$. where q is the charge in the circuit. Adding these potential drops, we set them equal to the voltage supplied by the voltage source, $V(t)$. Thus, we obtain

$$LR + \frac{q}{C} + L\frac{dI}{dt} = V(t).$$

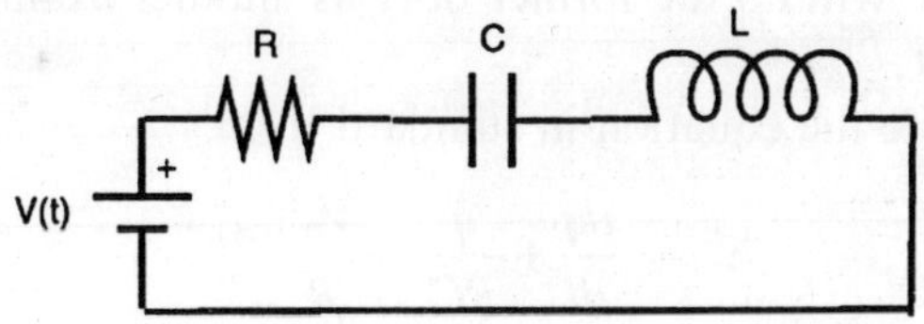

Fig. LRC Circuit.

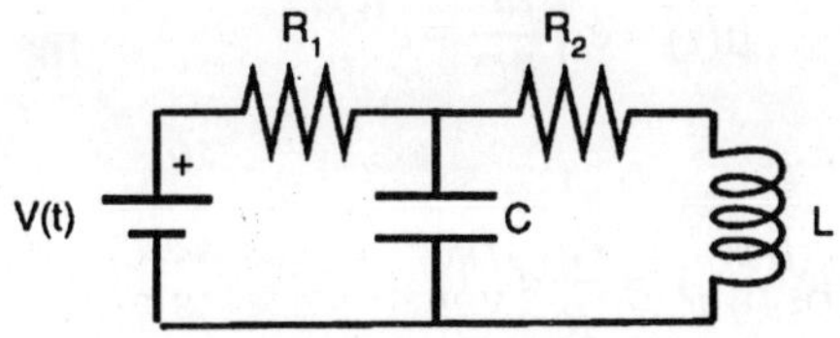

Fig. LRC Circuit.

Since both q and I are unknown, we can replace the current by its expression in terms of the charge to obtain

$$L\ddot{q} + R\dot{q} + \frac{1}{C}q = V(t)$$

This is a second order equation for $q(t)$.

More complicated circuits are possible by looking at parallel connections, or other combinations, of resistors, capacitors and inductors. This will result in several equations for each loop in the circuit, leading to larger systems of differential equations.

An example of another circuit setup. This is not a problem that can be covered in the first year physics course. One can set up a system of second order equations and proceed to solve them.

Special Cases

In this section we will look at special cases that arise for the series LRC circuit equation. These include *RC* circuits, solvable by first order methods and *LC* circuits, leading to oscillatory behaviour.

Case I. RC Circuits: We first consider the case of an RC circuit in which there is no inductor. Also, we will consider what happens when one charges a capacitor with a DC battery ($V(t) = V_0$) and when one discharges a charged capacitor ($V(t) = 0$).

For charging a capacitor, we have the initial value problem

$$R\frac{dq}{dt}+\frac{q}{C}=V_0,\quad q(0)=0.$$

This equation is an example of a linear first order equation for q(t). However, we can also rewrite it and solve it as a separable equation, since V_0 is a constant. We will do the former only as another example of finding the integrating factor.

We first write the equation in standard form.

$$\frac{dq}{dt}+\frac{q}{RC}=\frac{V_0}{R}.$$

The integrating factor is then

$$\mu(t)=e^{\int\frac{dt}{RC}}=e^{t/RC}.$$

Thus,

$$\frac{d}{dt}(qe^{t/RC})=\frac{V_0}{R}e^{t/RC}.$$

Integrating, we have

$$qe^{t/RC}=\frac{V_0}{R}\int e^{t/RC}=\frac{V_0}{C}\int e^{t/RC}+K.$$

Note that we introduced the integration constant, K. Now divide out the exponential to get the general solution:

$$q=\frac{V_0}{C}+Ke^{-t/RC}.$$

(If we had forgotten the K, we would not have gotten a correct solution for the differential equation.) Next, we use the initial condition to get our particular solution. Namely, setting $t = 0$, we have that

$$0=q(0)=\frac{V_0}{C}+K.$$

So, $K=-\frac{V_0}{C}$. Inserting this into our solution, we have

Now we can study the behaviour of this solution. For large times the second term goes to zero. Thus, the capacitor charges up, asymptotically, to the final value of $q_0=-\frac{V_0}{C}$. This is what we expect, because the current is no longer flowing over R and this just gives the relation between the potential difference across the capacitor plates when a charge of q_0 is established on the plates.

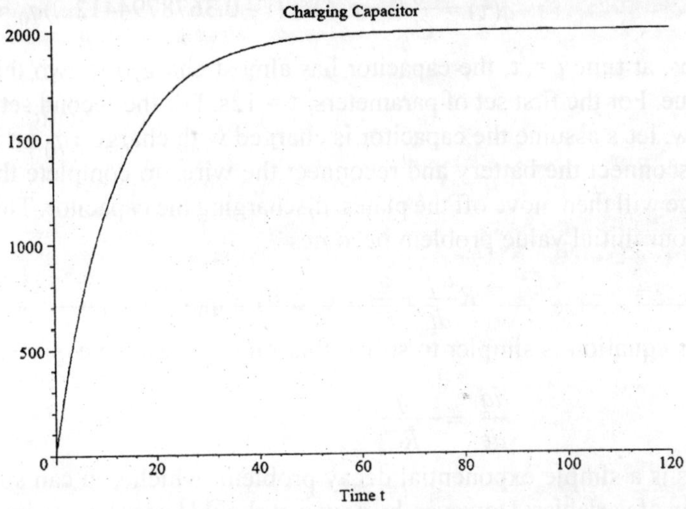

Fig. The Charge as a Function of time for a Charging Capacitor with R = 2.00 kΩ, $C = 6.00\ mF$, and $V_0 = 12\ V$

Let's put in some values for the parameters. We let $R = 2.00\ k\Omega$, $C = 6.00\ mF$, and $V_0 = 12\ V$. A plot of the solution is given. We see that the charge builds up to the value of $V_0 = C = 2000\ C$. If we use a smaller resistance, $R = 200\Omega$, that the capacitor charges to the same value, but much faster.

The rate at which a capacitor charges, or discharges, is governed by the time constant, $\tau = RC$. This is the constant factor in the exponential. The larger it is, the slower the exponential term decays.

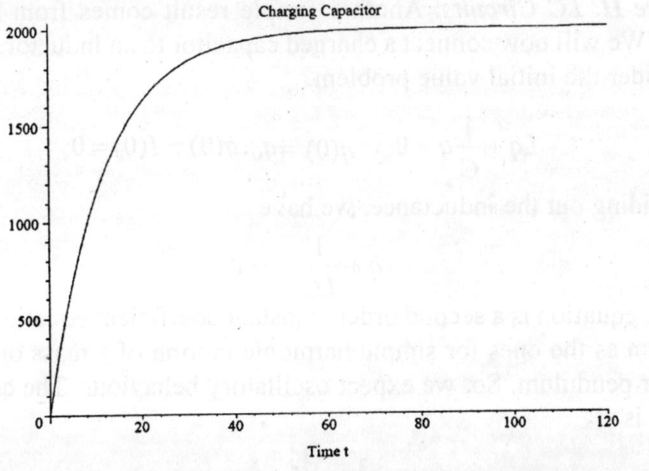

Fig. The Charge as a Function of time for a Charging Capacitor with $R = 200\Omega$, $C = 6.00\ mF$, and $V_0 = 12\ V$.

If we set $t = \tau$, we find that

$$q(\tau) = \frac{V_0}{C}(1 - e^{-1}) = (1 - 0.3678794412...)q_0 \approx 0.63q_0$$

Thus, at time $t = \tau$, the capacitor has almost charged to two thirds of its final value. For the first set of parameters, $\tau = 12$s. For the second set, $\tau = 1.2$s.

Now, let's assume the capacitor is charged with charge $\pm q_0$ on its plates. If we disconnect the battery and reconnect the wires to complete the circuit, the charge will then move off the plates, discharging the capacitor. The relevant form of our initial value problem becomes

$$R\frac{dq}{dt} + \frac{q}{C} = 0,\ q(0) = q_0.$$

This equation is simpler to solve. Rearranging, we have

$$\frac{dq}{dt} = -\frac{q}{RC}.$$

This is a simple exponential decay problem, which you can solve using separation of variables. However, by now you should know how to immediately write down the solution to such problems of the form $y' = ky$. The solution is

$$q(t) = q_0 e^{-t/\tau}, \quad \tau = RC.$$

We see that the charge decays exponentially. In principle, the capacitor never fully discharges. That is why you are often instructed to place a shunt across a discharged capacitor to fully discharge it.

In figure we show the discharging of our two previous RC circuits. Once again, $\tau = RC$ determines the behaviour. At $t = \tau$ we have

$$q(\tau) = q_0 e^{-1} = (0.3678794412...)q_0 \approx 0.37q_0.$$

So, at this time the capacitor only has about a third of its original value.

Case II. *LC Circuits*: Another simple result comes from studying *LC* circuits. We will now connect a charged capacitor to an inductor. In this case, we consider the initial value problem.

$$L\ddot{q} + \frac{1}{C}q = 0, \quad q(0) = q_0, \dot{q}(0) = I(0) = 0.$$

Dividing out the inductance, we have

$$\ddot{q} + \frac{1}{LC}q = 0.$$

This equation is a second order, constant coefficient equation. It is of the same form as the ones for simple harmonic motion of a mass on a spring or the linear pendulum. So, we expect oscillatory behaviour. The characteristic equation is

$$r^2 + \frac{1}{LC} = 0.$$

The solutions are

$$r_{1,2} = \pm \frac{i}{\sqrt{LC}}.$$

Thus, the solution of equation is of the form

$$q(t) = c_1 \cos(\omega t) + c_2 \sin(\omega t), \qquad \omega = (LC)^{-1/2}.$$

Inserting the initial conditions yields

$$q(t) = q_0 \cos(\omega t).$$

The oscillations that result are understandable. As the charge leaves the plates, the changing current induces a changing magnetic field in the inductor. The stored electrical energy in the capacitor changes to stored magnetic energy in the inductor.

However, the process continues until the plates are charged with opposite polarity and then the process begins in reverse. The charged capacitor then discharges and the capacitor eventually returns to its original state and the whole system repeats this over and over.

The frequency of this simple harmonic motion is easily found. It is given by

$$f = \frac{\omega}{2\pi} = \frac{1}{2\pi}\frac{1}{\sqrt{LC}}.$$

This is called the tuning frequency because of its role in tuning circuits.

This is an ideal situation. There is always resistance in the circuit, even if only a small amount from the wires. So, we really need to account for resistance, or even add a resistor. This leads to a slightly more complicated system in which damping will be present.

DAMPED OSCILLATIONS

As we have indicated, simple harmonic motion is an ideal situation. In real systems we often have to contend with some energy loss in the system. This leads to the damping of our oscillations. This energy loss could be in the spring, in the way a pendulum is attached to its support, or in the resistance to the flow of current in an LC circuit. The simplest models of resistance are the addition of a term in first derivative of the dependent variable. Thus, our three main examples with damping added look like.

$$m\ddot{x} + b\dot{x} + kx = 0.$$

$$L\ddot{\theta} + b\dot{\theta} + g\theta = 0.$$

$$L\ddot{q} + R\dot{q} + \frac{1}{C}q = 0.$$

These are all examples of the general constant coefficient equation

$$ay''(x) + by'(x) + cy(x) = 0.$$

We have seen that solutions are obtained by looking at the characteristic equation $ar^2 + br + c = 0$. This leads to three different behaviors depending on the discriminant in the quadratic formula:

$$r = \frac{-b \pm \sqrt{b^2 - 4ac}}{2a}.$$

We will consider the example of the damped spring. Then we have

$$r = \frac{-b \pm \sqrt{b^2 - 4mk}}{2m}.$$

For $b > 0$, there are three types of damping.

I. Overdamped, $b^2 > 4mk$

In this case we obtain two real root. Since this is Case I for constant coefficient equations, we have that

$$x(t) = c_1 e^{r}1\ t + c_2 e^{r2t}.$$

We note that $b^2 - 4mk < b^2$. Thus, the roots are both negative. So, both terms in the solution exponentially decay. The damping is so strong that there is no oscillation in the system.

II. Critically Damped, $b^2 = 4mk$

In this case we obtain one real root. This is Case II for constant coefficient equations and the solution is given by

$$x(t) = (c_1 + c_2 t)e^{rt},$$

where $r = -b/2m$. Once again, the solution decays exponentially. The damping is just strong enough to hinder any oscillation. If it were any weaker the discriminant would be negative and we would need the third case.

III. Underdamped, $b^2 < 4mk$

In this case we have complex conjugate roots. We can write $\alpha = -b/2m$ and $\beta = \sqrt{4mk - b^2 / 2m}$. Then the solution is

$$x(t) = e^{\alpha t}(c_1 \cos \beta t + c_2 \sin \beta t).$$

These solutions exhibit oscillations due to the trigonometric functions, but we see that the amplitude may decay in time due the the overall factor of $e^{\alpha t}$ when $\alpha < 0$. Consider the case that the initial conditions give $c_1 = A$ and $c_2 = 0$. Then, the solution, $x(t) = Ae^{\alpha t} \cos \beta t$, looks like the plot in Figure.

FORCED OSCILLATIONS

All of the systems presented at the beginning of the last section exhibit the same general behaviour when a damping term is present. An additional term can be added that can cause even more complicated behaviour. In the case of LRC circuits, we have seen that the voltage source makes the system nonhomogeneous.

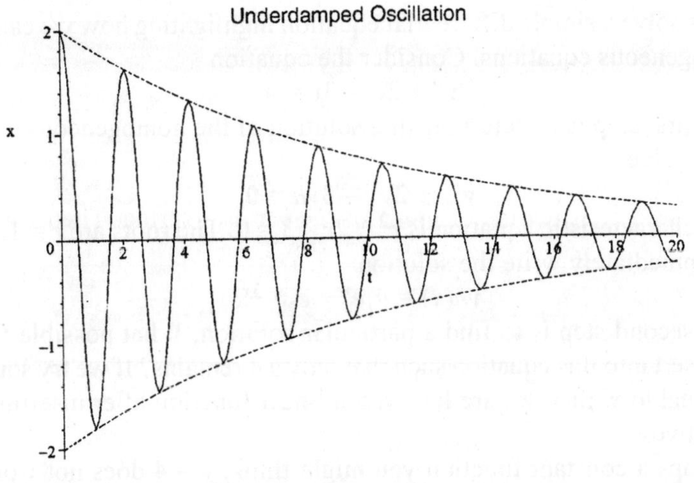

Fig. A Plot of Underdamped Oscillation given by $x(t) = 2e^{0.1t} \cos 3t$. The Dashed Lines are given by $x(t) = \pm 2e^{0.1t}$, Indicating the Bounds on the Amplitude of the Motion

It provides what is called a source term. Such terms can also arise in the mass-spring and pendulum systems.

One can drive such systems by periodically pushing the mass, or having the entire system moved, or impacted by an outside force. Such systems are called forced, or driven.

Typical systems in physics can be modeled by nonhomogenous second order equations. Thus, we want to find solutions of equations of the form

$$Ly(x) = a(x)y''(x) + b(x)y'(x) + c(x)y(x) = f(x).$$

Earlier we saw that the solution of equations are found in two steps.

1. First you solve the homogeneous equation for a general solution of $Ly_h = 0$, $y_h(x)$.
2. Then, you obtain a particular solution of the nonhomogeneous equation, $y_p(x)$.

To date, we only know how to solve constant coefficient, homogeneous equations. So, by adding a nonhomogeneous to such equations we need to figure out what to do with the extra term. In other words, how does one find the particular solution?

You could guess a solution, but that is not usually possible without a little bit of experience. So we need some other methods. There are two main methods.

In the first case, the Method of Undetermined Coefficients, one makes an intelligent guess based on the form of $f(x)$. In the second method, one can systematically developed the particular solution.

Method of Undetermined Coefficients

Let's solve a simple differential equation highlighting how we can handle nonhomogeneous equations. Consider the equation

$$y'' + 2y' - 3y = 4.$$

The first step is to determine the solution of the homogeneous equation. Thus, we solve

$$y''_h + 2y'_h - 3yh = 0.$$

The characteristic equation is $r^2 + 2r - 3 = 0$. The roots are $r = 1, -3$. So, we can immediately write the solution

$$yh(x) = c_1e^x + c_2e^{-3x}.$$

The second step is to find a particular solution. What possible function can we insert into this equation such that only a 4 remains? If we try something proportional to x, then we are left with a linear function after inserting x and its derivatives.

Perhaps a constant function you might think. $y = 4$ does not work. But, we could try an arbitrary constant, $y = A$.

Let's see. Inserting $y = A$ into equation, we obtain

$$-3A = 4.$$

Ah ha! We see that we can choose $A = -\frac{4}{3}$ and this works. So, we have a particular solution, $y_p(x) = -\frac{4}{3}$. This step is done.

Combining our two solutions, we have the general solution to the original nonhomogeneous equation. Namely,

$$y(x) = y_h(x) + y_p(x) = c_1e^x + c_2e^{-3x} - \frac{4}{3}.$$

Insert this solution into the equation and verify that it is indeed a solution. If we had been given initial conditions, we could now use them to determine our arbitrary constants.

What if we had a different source term? Consider the equation

$$y'' + 2y' - 3y = 4x.$$

The only thing that would change is our particular solution. So, we need a guess.

We know a constant function does not work by the last example. So, let's try $y_p = Ax$. Inserting this function into equation, we obtain

$$2A - 3Ax = 4x.$$

Picking $A = -4/3$ would get rid of the x terms, but will not cancel everything. We still have a constant left. So, we need something more general.

Let's try a linear function, $y_p(x) = ax + B$. Then we get after substitution into

$$2A - 3(Ax + B) = 4x.$$

Equating the coefficients of the different powers of x on both sides, we find a system of equations for the undetermined coefficients.

$$2A - 3B = 0$$
$$-3A = 4.$$

These are easily solved to obtain

$$A = -\frac{4}{3}$$

$$B = \frac{2}{3}A = -\frac{8}{9}.$$

$f(x)$	*Guess*
$a_n x^n + a_{n-1}x^{n-1} + \cdots + a_1 x + a_0$	$A_n x^n + A_{n-1}x^{n-1} + \cdots + A_1 x + A_0$
ae^{bx}	Ae^{bx}
$a \cos \omega x + b \sin \omega x$	$A\cos \omega x + B \sin \omega x$

So, our particular solution is

$$y_p(x) = -\frac{4}{3}x - \frac{8}{9}.$$

This gives the general solution to the nonhomogeneous problem as

$$y(x) = y_h(x) + y_p(x) = c_1 e^x + c_2 e^{-3x} - \frac{4}{3}x - \frac{8}{9}.$$

There are general forms that you can guess based upon the form of the driving term, $f(x)$. More general applications are covered in a standard text on differential equations.

However, the procedure is simple. Given $f(x)$ in a particular form, you make an appropriate guess up to some unknown parameters, or coefficients. Inserting the guess leads to a system of equations for the unknown coefficients. Solve the system and you have your solution. This solution is then added to the general solution of the homogeneous differential equation.

As a final example, let's consider the equation

$$y'' + 2y' - 3y = 2e^{-3x}.$$

According to the above, we would guess a solution of the form $y_p = Ae^{-3x}$. Inserting our guess, we find

$$0 = 2e^{-3x}.$$

Oops! The coefficient, A, disappeared! We cannot solve for it. What went wrong?

The answer lies in the general solution of the homogeneous problem. Note that e^x and e^{-3x} are solutions to the homogeneous problem. So, a multiple of e^{-3x} will not get us anywhere. It turns out that there is one further modification of the method. If our driving term contains terms that are solutions of the homogeneous problem, then we need to make a guess consisting of the smallest possible power of x times the function which is no longer a solution of the homogeneous problem. Namely, we guess $y_p(x) = Axe^{-3x}$. We compute the derivative of our guess, $y'_p = A(1 - 3x)e^{-3x}$ and $y''_p = A(9x - 6)e^{-3x}$. Inserting these into the equation, we obtain

$$[(9x - 6) + 2(1 - 3x) - 3x]Ae^{-3x} = 2e^{-3x},$$

or

$$-4A = 2.$$

So, $A = -1/2$ and $y_p(x) = -\frac{1}{2}xe^{-3x}$.

Method of Variation of Parameters

A more systematic way to find particular solutions is through the use of the Method of Variation of Parameters. The derivation is a little messy and the solution is sometimes messy, but the application of the method is straight forward if you can do the required integrals. We will first derive the needed equations and then do some examples.

We begin with the nonhomogeneous equation. Let's assume it is of the standard form

$$a(x)y''(x) + b(x)y'(x) + c(x)y(x) = f(x).$$

We know that the solution of the homogeneous equation can be written in terms of two linearly independent solutions, which we will call $y_1(x)$ and $y_2(x)$.

$$yh(x) = c_1y_1(x) + c_2y_2(x).$$

If one replaces the constants with functions, then you now longer have a solution to the homogeneous equation. Is it possible that you could stumble across the right functions with which to replace the constants and somehow end up with $f(x)$ when inserted into the left side of the differential equation? It turns out that you can.

So, let's assume that the constants are replaced with two unknown functions, which we will call $c_1(x)$ and $c_2(x)$. This change of the parameters is where the name of the method derives. Thus, we are assuming that a particular solution takes the form

$$y_p(x) = c_1(x)y_1(x) + c_2(x)y_2(x).$$

If this is to be a solution, then insertion into the differential equation should make it true. To do this we will first need to compute some derivatives.

The first derivative is given by

$$y'_p(x) = c_1(x)y'_1(x) + c_2(x)y'_2(x) + c'_1(x)y_1(x) + c'_2(x)y_2(x).$$

Next we will need the second derivative. But, this will give use eight terms. So, we will first make an assumption. Let's assume that the last two terms add to zero.

$$c'_1(x)y_1(x) + c'_2(x)y_2(x) = 0.$$

It turns out that we will get the same results in the end if we did not assume this. The important thing is that it works!

So, we now have the first derivative as

$$y'_p(x) = c_1(x)y'_1(x) + c_2(x)y'_2(x).$$

The second derivative is then only four terms.

$$y'_p(x) = c_1(x)y''_1(x) + c_2(x)y''_2(x) + c'_1(x)y'_1(x) + c'_2(x)y'_2(x).$$

Now that we have the derivatives, we can insert our guess into the differential equation. Thus, we have

$$\begin{aligned} f(x) &= a(x)(c_1(x)y''_1(x) + c_2(x)y''_2(x) + c'_1(x)y'_1(x) + c'_2(x)y'_2(x)) \\ &\quad + b(x)(c_1(x)y'_1(x) + c_2(x)y'_2(x)) \\ &\quad + c(x)(c_1(x)y_1(x) + c_2(x)y_2(x)). \end{aligned}$$

Regrouping the terms, we obtain

$$\begin{aligned} f(x) &= c_1(x)\,(a(x)y''_1(x) + b(x)y'_1(x) + c(x)y_1(x)) \\ &\quad c_2(x)\,(a(x)y''_2(x) + b(x)y'_2(x) + c(x)y_2(x)) \\ &\quad + a(x)\,(c'_1(x)y'_1(x) + c'_2(x)y'_2(x)). \end{aligned}$$

Note that the first two rows vanish since y_1 and y_2 are solutions of the homogeneous problem. This leaves the equation

$$c'_1(x)y'_2(x) + c'_2(x)y'_2(x) = \frac{f(x)}{a(x)}.$$

In summary, we have assumed a particular solution of the form

$$y_p(x) = c_1(x)y_1(x) + c_2(x)y_2(x).$$

This is only possible if the unknown functions $c_1(x)$ and $c_2(x)$ satisfy the system of equations

$$\begin{aligned} c'_1(x)y_1(x) + c'_2(x)y_2(x) &= 0 \\ c'_1(x)y'_1(x) + c'_2(x)y'_2(x) &= \frac{f(x)}{a(x)}. \end{aligned}$$

It is standard to solve this system for the derivatives of the unknown functions and then present the integrated forms. However, one could just start from here.

Example: Consider the problem. $y'' - y = e^{2x}$. We want the general solution of this nonhomogeneous problem.

The general solution to the homogeneous problem $y''_h - yh = 0$ is

$$y_h(x) = c_1e^x + c_2e^{-x}.$$

In order to use the Method of Variation of Parameters, we seek a solution of the form

$$y_p(x) = c_1(x)e^x + c_2(x)e^{-x}.$$

We find the unknown functions by solving the system, which in this case becomes

$$c_1'(x)e^x + c_2'(x)e^{-x} = 0$$

$$c_1'(x)e^x + c_2'(x)e^{-x} = e^{2x}.$$

Adding these equations we find that

$$2c'_1 e^x = e^{2x} \rightarrow c_1' = \frac{1}{2}e^x.$$

Solving for $c_1(x)$ we find

$$c_1(x) = \frac{1}{2}\int e^x dx = \frac{1}{2}e^x.$$

Subtracting the equations in the system yields

$$2c'_2 e^{-x} = -e^{2x} \rightarrow c_2' = \frac{1}{2}e^{3x}.$$

Thus,

$$c_2(x) = -\frac{1}{2}\int e^{3x} dx = -\frac{1}{6}e^{3x}.$$

The particular solution is found by inserting these results into y_p.

$$y_p(x) = c_1(x)y_1(x) + c_2(x)y_2(x)$$

$$= \left(\frac{1}{2}e^x\right)e^x + \left(-\frac{1}{6}e^{3x}\right)e^{-x}$$

$$= \frac{1}{3}e^{2x}.$$

Thus, we have the general solution of the nonhomogeneous problem as

$$y(x) = c_1e^x + c_2e^{-x} + \frac{1}{3}e^{2x}.$$

Example: Now consider the problem. $y'' + 4y = \sin x$. The solution to the homogeneous problem is

$$y_h(x) = c_1 \cos 2x + c_2 \sin 2x.$$

We now seek a particular solution of the form

$$yh(x) = c_1(x) \cos 2x + c_2(x) \sin 2x.$$

We let $y_1(x) = \cos 2x$ and $y_2(x) = \sin 2x$, $a(x) = 1$, $f(x) = \sin x$ in system:

$$c_1'(x) \cos 2x + c_2'(x) \sin 2x = 0$$

$$-2c_1'(x) \sin 2x + 2c_2'(x) \cos 2x = \sin x.$$

Now, use your favourite method for solving a system of two equations and two unknowns. In this case, we can multiply the first equation by 2 sin $2x$ and the second equation by cos $2x$. Adding the resulting equations will eliminate the c_1 terms. Thus, we have

$$c_1'(x) = \frac{1}{2}\sin x \cos 2x = \frac{1}{2}(2\cos^2 x - 1)\sin x .$$

Inserting this into the first equation of the system, we have

$$c_2'(x) = -c_2'(x)\frac{\sin 2x}{\cos 2x} = -\frac{1}{2}\sin x \cos 2x = -\sin^2 x \cos x$$

These can easily be solved.

$$c_2(x) = \frac{1}{2}\int (2\cos^2 x - 1)\sin x \, dx = \frac{1}{2}\left(\cos x - \frac{2}{3}\cos^3 x\right)$$

$$c_1(x) = \int \sin^x \cos x \, dx = -\frac{1}{3}\sin^3 x .$$

The final step in getting the particular solution is to insert these functions into $y_p(x)$. This gives

$$y_p(x) = c_1(x)y_1(x) + c_2(x)y_2(x)$$

$$= \left(-\frac{1}{3}\sin^3 x\right)\cos 2x + \left(\frac{1}{2}\cos x - \frac{1}{3}\cos^3 x\right)\sin x$$

$$= \frac{1}{3}\sin x .$$

So, the general solution is

$$y(x) = c_1 \cos 2x + c_2 \sin 2x + \frac{1}{3}\sin x .$$

NUMERICAL SOLUTIONS OF ODES

So far we have seen some of the standard methods for solving first and second order differential equations. However, we have had to restrict ourselves to very special cases in order to get nice analytical solutions to our initial value problems. While these are not the only equations for which we can get exact results, there are many cases in which exact solutions are not possible. In such cases we have to rely on approximation techniques, including the numerical solution of the equation at hand.

The use of numerical methods to obtain approximate solutions of differential equations and systems of differential equations has been known for some time. However, with the advent of powerful computers and desktop computers, we can now solve many of these problems with relative ease. The simple ideas used to solve first order differential equations can be extended to the solutions of more complicated systems of partial differential equations, such as the large scale problems of modelling ocean dynamics, weather systems and even cosmological problems stemming from general relativity. In this section we will look at the simplest method for solving first order equations, Euler's Method. While it is not the most efficient method, it does provide us with a picture of how one proceeds and can be improved by introducing better techniques, which are typically covered in a numerical analysis text.

Let's consider the class of first order initial value problems of the form

$$\frac{dy}{dx} = f(x, y),\ y(x_0) = y_0.$$

We are interested in finding the solution $y(x)$ of this equation which passes through the initial point (x_0, y_0) in the xy-plane for values of x in the interval $[a, b]$, where $a = x_0$. We will seek approximations of the solution at N points, labeled x_n for n = 1,..., N. For equally spaced points we have $\Delta x = x_1 - x_0 = x_2 - x_1$, etc. Then, $x_n = x_0 + n\Delta x$. In figure we show three such points on the x-axis. We will develop a simple numerical method, called Euler's Method. The interval of interest into N subintervals with $N + 1$ points x_n. We already know a point on the solution $(x_0, y(x_0)) = (x_0, y_0)$. How do we find the solution for other x values?

We first note that the differential equation gives us the slope of the tangent line at $(x, y(x))$ of our solution $y(x)$. The slope is $f(x, y(x))$. The tangent line drawn at (x_0, y_0).

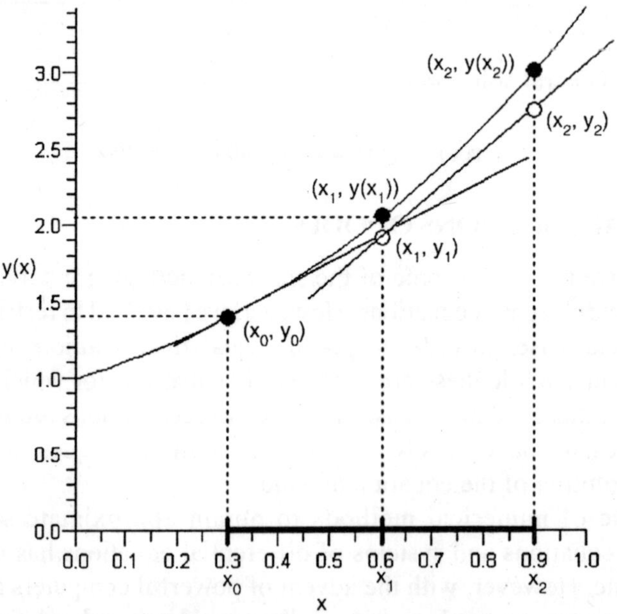

Fig. The basics of Euler's Method

An interval of the x axis is broken into N subintervals. The approximations to the solutions are found using the slope of the tangent to the solution, given by $f(x, y)$. Knowing previous approximations at (x_{n-1}, y_{n-1}), one can determine the next approximation, y_n.

Look now at $x = x_1$. A vertical line intersects both the solution curve and the tangent line. While we do not know the solution, we can determine the tangent line and find the intersection point. This intersection point is in theory close to the point on the solution curve. So, we will designate y_1 as the approximation of our solution $y(x_1)$. We just need to determine y_1.

The idea is simple. We approximate the derivative in our differential equation by its difference quotient:

$$\frac{dy}{dx} = \frac{y_1 - y_0}{x_1 - x_0} = \frac{y_1 - y_0}{\Delta x}.$$

But, we have by the differential equation that the slope of the tangent to the curve at (x_0, y_0) is

$$y'(x_0) = f(x_0, y_0).$$

Thus,

$$\frac{y_1 - y_0}{\Delta x} \approx f(x_0, y_0).$$

So, we can solve this equation for y_1 to obtain

$$y_1 = y_0 + \Delta x f(x_0, y_0).$$

This give y_1 in terms of quantities that we know.

We now proceed to approximate $y(x_2)$. We see that this can be done by using the slope of the solution curve at (x_1, y_1). The corresponding tangent line is shown passing though (x_1, y_1) and we can then get the value of y_2. Following the previous argument, we find that

$$y_2 = y_1 + \Delta x f(x_1, y_1).$$

Continuing this procedure for all x_n, we arrive at the following numerical scheme for determining a numerical solution to Euler's equation:

$$y_0 = y(x_0),$$
$$y_n = y_{n-1} + \Delta x f(x_{n-1}, y_{n-1}),\ n = 1, \ldots, N.$$

Example: We will consider a standard example for which we know the exact solution. This way we can compare our results. The problem is given that

$$\frac{dy}{dx} = x + y,\ y(0) = 1,$$

find an approximation for $y(1)$.

First, we will do this by hand. We will break up the interval [0, 1], since we want our solution at $x = 1$ and the initial value is at $x = 0$.

Let $\Delta x = 0.50$. Then, $x_0 = 0$, $x_1 = 0.5$ and $x_2 = 1.0$. Note that $N = \dfrac{b-a}{\Delta x} = 2$.

Table: Application of Euler's Method for $y' = x + y$, $y(0) = 1$ and $\Delta x = 0.5$.

n	x_n	$y_n = y_{n-1} + \Delta x f(x_{n-1}, y_{n-1}) = 0.5x_{n-1} + 1.5y_{n-1}$
0	0	1
1	0.5	$0.5(0) + 1.5(1.0) = 1.5$
2	1.0	$0.5(0.5) + 1.5(1.5) = 2.5$

Table: Application of Euler's Method for $y' = x + y$, $y(0) = 1$ and $\Delta x = 0.2$.

n	x_n	$y_n = 0.2x_{n-1} + 1.2y_{n-1}$
0	0	1
1	0.2	$0.2(0) + 1.2(1.0) = 1.2$
2	0.4	$0.2(0.2) + 1.2(1.2) = 1.48$
3	0.6	$0.2(0.4) + 1.2(1.48) = 1.856$
4	0.8	$0.2(0.6) + 1.2(1.856) = 2.3472$
5	1.0	$0.2(0.8) + 1.2(2.3472) = 2.97664$

We can carry out Euler's Method systematically. There is a column for each x_n and y_n. The first row is the initial condition. We also made use of the function $f(x, y)$ in computing the y_n's.

This sometimes makes the computation easier. As a result, we find that the desired approximation is given as $y_2 = 2.5$. Is this a good result? Well, we could make the spatial increments smaller. Let's repeat the procedure for $\Delta x = 0.2$, or $N = 5$.

Now we see that our approximation is $y_1 = 2.97664$. So, it looks like our value is near 3, but we cannot say much more. Decreasing Δx more shows that we are beginning to converge to a solution.

Table: Results of Euler's Method for $y' = x + y$, $y(0) = 1$ and varying Δx.

Δx	$y_N \approx y(1)$
0.5	2.5
0.2	2.97664
0.1	3.187484920
0.01	3.409627659
0.001	3.433847864
0.0001	3.436291854

The last computation would have taken 1000 lines in our table, one could use a computer to do this. A simple code in Maple would look like the following.

```
> Restart:
> f := (x, y) → y + x;
> a := 0. b := 1. N := 100: h := (b – a)/N,
> x[0] := 0 : y[0] := 1:
  for i from 1 to N do
  y[i] := y[i – 1] + h*f (x[i – 1], y[i – 1]):
  x[i] := x[0] + h*(i):
  od:
  evalf (y[N]);
```

In this case we could simply use the exact solution. The exact solution is easily found as

$$y(x) = 2e^x - x - 1.$$

So, the value we are seeking is

$$y(1) = 2e - 2 = 3.4365636....$$

Thus, even the last numerical solution was off by about 0.00027.

Adding a few extra lines for plotting, we can visually see how well our approximations compare to the exact solution.

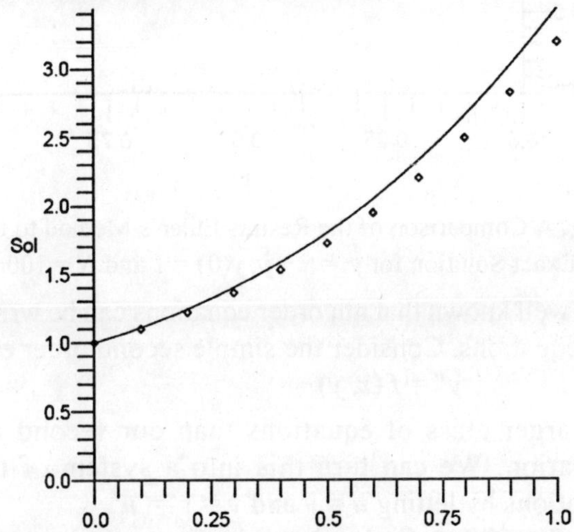

Fig. A Comparison of the Results Euler's Method to the Exact Solution for $y' = x + y$, $y(0) = 1$ and $N = 10$

We can see how quickly our numerical solution diverges from the exact solution. In figure we can see that visually the solutions agree, but we note from Table that for $\Delta x = 0.01$, the solution is still off in the second decimal place with a relative error of about 0.8%.

Why would we use a numerical method when we have the exact solution? Exact solutions can serve as test cases for our methods. We can make sure our code works before applying them to problems whose solution is not known.

There are many other methods for solving first order equations. One commonly used method is the fourth order Runge-Kutta method. This method has smaller errors at each step as compared to Euler's Method. It is well suited for programming and comes built-in in many packages like Maple and Matlab. Typically, it is set up to handle systems of first order equations.

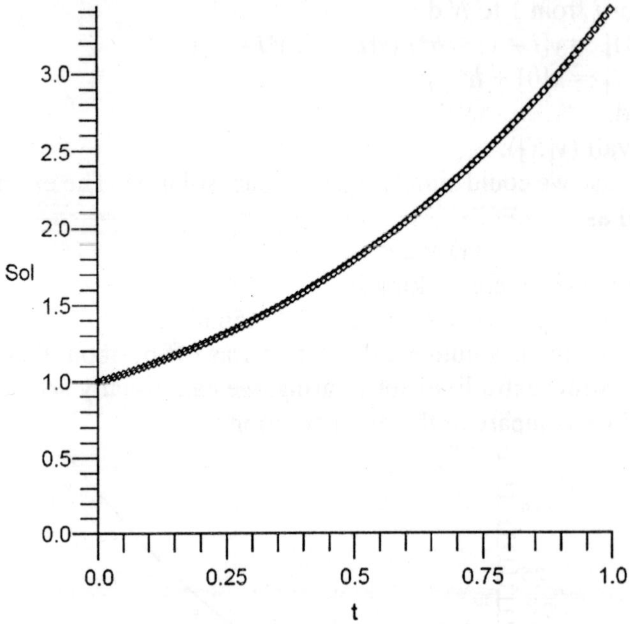

Fig. A Comparison of the Results Euler's Method to the Exact Solution for $y' = x + y$, $y(0) = 1$ and $N = 100$

In fact, it is well known that nth order equations can be written as a system of n first order equations. Consider the simple second order equation

$$y'' = f(x, y).$$

This is a larger class of equations than our second order constant coefficient equation. We can turn this into a system of two first order differential equations by letting $u = y$ and $v = y' = u'$.

Then, $v' = y'' = f(x, u)$. So, we have the first order system

$$u' = v,$$
$$v' = f(x, u).$$

We will not go further into the Runge-Kutta Method here. You can find more about it in a numerical analysis text. However, we will see that systems of differential equations do arise naturally in physics. Such systems are often coupled equations and lead to interesting behaviors.

COUPLED OSCILLATORS

In the last section we saw that the numerical solution of second order equations, or higher, can be cast into systems of first order equations. Such systems are typically coupled in the sense that the solution of at least one of the equations in the system depends on knowing one of the other solutions in the system.

In many physical systems this coupling takes place naturally. We will introduce a simple model in this section to illustrate the coupling of simple oscillators. However, we will reserve solving the coupled system until the next chapter after exploring the needed mathematics.

There are many problems in physics that result in systems of equations. This is because the most basic law of physics is given by Newton's Second Law, which states that if a body experiences a net force, it will accelerate.

Thus,

$$\sum \mathrm{F} = ma.$$

Since $a = \ddot{x}$ we have a system of second order differential equations in general for three dimensional problems, or one second order differential equation for one dimensional problems.

We have already seen the simple problem of a mass on a spring. Recall that the net force in this case is the restoring force of the spring given by Hooke's Law,

$$Fs = -kx,$$

where $k > 0$ is the spring constant and x is the elongation of the spring. When it is positive, the spring force is negative and when it is negative the spring force is positive. The equation for simple harmonic motion for the mass-spring system was found to be given by

$$m\ddot{x} + kx = 0.$$

This second order equation can be written as a system of two first order equations in terms of the unknown position and velocity. We first set

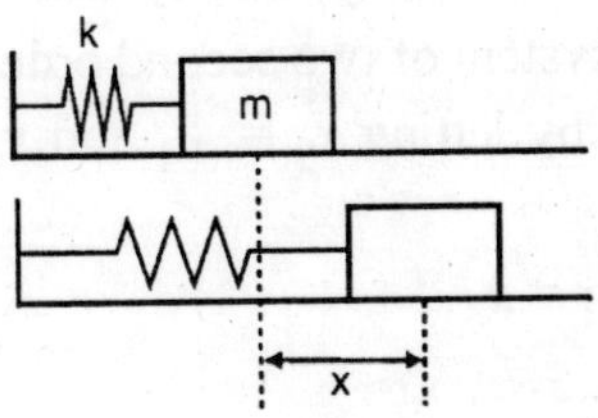

Fig. Spring-Mass System.

$y = \dot{x}$ and then rewrite the second order equation in terms of x and y. Thus, we have

$$\dot{x} = y$$
$$\dot{y} = -\frac{k}{m}x.$$

The coefficient matrix for this system is $\begin{pmatrix} 0 & 1 \\ -\omega^2 & 0 \end{pmatrix}$, where $\omega^2 = \dfrac{k}{m}$.

One can look at more complicated spring-mass systems. Consider two blocks attached with two springs. In this case we apply Newton's second law for each block. We will designate the elongations of each spring from equilibrium as x_1 and x_2.

For mass m_1, the forces acting on it are due to each spring. The first spring with spring constant k_1 provides a force on m_1 of $-k_1x_1$. The second spring is stretched, or compressed, based upon the relative locations of the two masses. So, it will exert a force on m_1 of $k_2 (x_2 - x_1)$.

Similarly, the only force acting directly on mass m_2 is provided by the restoring force from spring 2. So, that force is given by $-k_2 (x_2 - x_1)$. The reader should think about the signs in each case.

Putting this all together, we apply Newton's Second Law to both masses. We obtain the two equations

$$\begin{aligned} m_1\ddot{x}_1 &= -k_1x_1 + k_2(x_1 - x_2) \\ m_2\ddot{x}_2 &= -k_2(x_1 - x_2) \end{aligned}$$

Thus, we see that we have a coupled system of two second order differential equations.

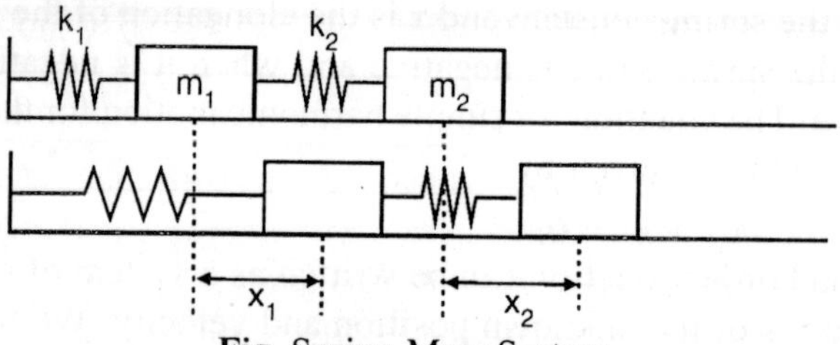

Fig. Spring-Mass System.

One can rewrite this system of two second order equations as a system of four first order equations by letting $x_3 = \dot{x}_1$ and $x_4 = \dot{x}_2$. This leads to the system

$$\begin{aligned} \dot{x}_1 &= x_3 \\ \dot{x}_2 &= x_4 \\ \dot{x}_3 &= -\frac{k_1}{m_1}x_1 + \frac{k_2}{m_1}(x_1 - x_2) \\ \dot{x}_4 &= \frac{k_2}{m_2}(x_1 - x_2). \end{aligned}$$

As we will see, this system can be written more compactly in matrix form:

$$\frac{d}{dt}\begin{pmatrix} x_1 \\ x_2 \\ x_3 \\ x_4 \end{pmatrix} = \begin{pmatrix} 0 & 0 & 1 & 0 \\ 0 & 0 & 0 & 1 \\ \frac{k_1 - k_2}{m_1} & -\frac{k_2}{m_1} & 0 & 0 \\ \frac{k_2}{m_2} & -\frac{k_2}{m_2} & 0 & 0 \end{pmatrix}\begin{pmatrix} x_1 \\ x_2 \\ x_3 \\ x_4 \end{pmatrix}$$

However, before we can solve this system of first order equations, we need to recall a few things from linear algebra.

THE NONLINEAR PENDULUM OPTIONAL

We can also make the system more realistic by adding damping. This could be due to energy loss in the way the string is attached to the support or due to the drag on the mass, etc. Assuming that the damping is proportional to the angular velocity, we have equations for the damped nonlinear and damped linear pendula:

$$L\ddot{\theta} + b\dot{\theta} + g\sin\theta = 0.$$

$$L\ddot{\theta} + b\dot{\theta} + g\theta = 0.$$

Finally, we can add forcing. Imagine that the support is attached to a device to make the system oscillate horizontally at some frequency. Then we could have equations such as

$$L\ddot{\theta} + b\dot{\theta} + g\sin\theta = F\cos\omega t.$$

Before returning to studying the equilibrium solutions of the nonlinear pendulum, we will look at how far we can get at obtaining analytical solutions. First, we investigate the simple linear pendulum.

The linear pendulum equation is a constant coefficient second

order $r = \pm\sqrt{\frac{q}{L}}i$. Thus, the general solution takes the form

$$\theta(t) = c_1 \cos\left(\sqrt{\frac{g}{L}}t\right) + c_2 \sin\left(\sqrt{\frac{g}{L}}t\right).$$

We note that this is usually simplified by introducing the angular frequency

$$\omega \equiv \sqrt{\frac{g}{L}}.$$

One consequence of this solution, which is used often in introductory physics, is an expression for the period of oscillation of a simple pendulum. The period is found to be

$$T = \frac{2\pi}{\omega} = 2\pi\sqrt{\frac{g}{L}}.$$

As we have seen, this value for the period of a simple pendulum was derived assuming a small angle approximation. How good is this approximation? What is meant by a small angle? We could recall from calculus that the Taylor series approximation of sin θ about θ = 0.

$$\sin\theta = \theta - \frac{\theta^3}{3!} + \frac{\theta^5}{5!} + \cdots.$$

One can obtain a bound on the error when truncating this series to one term after taking a numerical analysis course. But we can just simply plot the relative error, which is defined as

$$\text{Relative Error} = \frac{\sin\theta - \theta}{\sin\theta}.$$

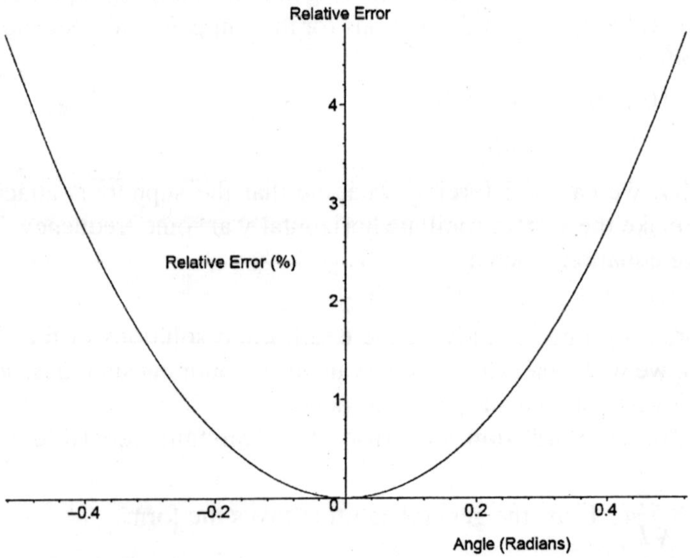

Fig. The Relative Error in Percent when Approximating sin θ by θ..

A plot of the relative error is given in figure. Thus for $\theta \approx 0.4$ radians (or, degrees) we have that the relative error is about 4%.

We would like to do better than this. So, we now turn to the nonlinear pendulum. We first rewrite equation is the simpler form

$$\ddot{\theta} + \omega^2\theta = 0\cdot$$

We next employ a technique that is useful for equations of the form

$$\ddot{\theta} + F(\theta) = 0$$

when it is easy to integrate the function F(θ): Namely, we note that

$$\frac{d}{dt}\left[\frac{1}{2}\dot{\theta}^2 + \int^{\theta(t)} F(\phi)d\phi\right] = (\ddot{\theta} + F(\theta))\dot{\theta}.$$

For our problem, we multiply equation by $\dot{\theta}$,

$$\ddot{\theta}\dot{\theta} + \omega^2\dot{\theta}\theta = 0$$

and note that the left side of this equation is a perfect derivative. Thus,

$$\frac{d}{dt}\left[\frac{1}{2}\dot{\theta}^2 + \omega^2 \cos\theta\right] = 0.$$

Therefore, the quantity in the brackets is a constant. So, we can write

$$\frac{1}{2}\dot{\theta}^2 - \omega^2 \cos\theta = c.$$

Solving for $\dot{\theta}$, we obtain

$$\frac{d\theta}{dt} = \sqrt{2(c + \omega^2 \cos\theta)}.$$

This equation is a separable first order equation and we can rearrange and integrate the terms to find that

$$t = \int dt = \int \frac{d\theta}{\sqrt{2(c + \omega^2 \cos\theta)}}.$$

When one gets a solution in this implicit form, one says that the problem has been solved by quadratures.

Namely, the solution is given in terms of some integral.

In fact, the above integral can be transformed into what is know as an elliptic integral of the first kind. We will rewrite our result and then use it to obtain an approximation to the period of oscillation of our nonlinear pendulum, leading to corrections to the linear result found earlier.

The swinging of a mass on a string, assuming no energy loss at the pivot point, is a conservative process. Namely, the total mechanical energy is conserved. Thus, the total of the kinetic and gravitational potential energies is a constant. The kinetic energy of the masses on the string is given as

$$T = \frac{1}{2}mv^2 = \frac{1}{2}mL^2\dot{\theta}^2.$$

The potential energy is the gravitational potential energy. If we set the potential energy to zero at the bottom of the swing, then the potential energy is U = mgh, where h is the height that the mass is from the bottom of the swing.

A little trigonometry gives that $h = L(1 - \cos\theta)$. So,

$$U = mgL(1 - \cos\theta).$$

So, the total mechanical energy is

$$E = \frac{1}{2}mL^2\dot{\theta}^2 + mgL(1-\cos\theta).$$

We note that a little rearranging shows that we can relate this to equation.

$$\frac{1}{2}\dot{\theta}^2 - \omega^2\cos\theta = \frac{1}{mL^2}E - \omega^2 = c.$$

We can use equation to get a value for the total energy. At the top of the swing the mass is not moving, if only for a moment. Thus, the kinetic energy is zero and the total energy is pure potential energy. Letting θ_0 denote the angle at the highest position, we have that

$$E = mgL(1 - \cos\theta_0) = mL^2\omega^2(1 - \cos\theta_0).$$

Therefore, we have found that

$$\frac{1}{2}\dot{\theta}^2 - \omega^2\cos\theta = \omega^2(1-\cos\theta_0).$$

Using the half angle formula,

$$\sin^2\frac{\theta}{2} = \frac{1}{2}(1-\cos\theta_0),$$

we can rewrite equation as

$$\frac{1}{2}\dot{\theta}^2 = 2\omega^2\left[\sin^2\frac{\theta_0}{2} - \sin^2\frac{\theta}{2}\right]$$

Solving for θ', we have

$$\frac{d\theta}{dt} = 2\omega\left[\sin^2\frac{\theta_0}{2} - \sin^2\frac{\theta}{2}\right]^{1/2}.$$

One can now apply separation of variables and obtain an integral similar to the solution we had obtained previously.

Noting that a motion from $\theta = 0$ to $\theta = \theta_0$ is a quarter of a cycle, we have that

$$T = \frac{2}{\omega}\int_0^{\theta_0} \frac{d\phi}{\sin^2\frac{\theta_0}{2} - \sin^2\frac{\theta}{2}}.$$

This result is not much different than our previous result, but we can now easily transform the integral into an elliptic integral. We define

$$z = \frac{\sin\frac{\theta}{2}}{\sin\frac{\theta_0}{2}}$$

and

$$k = \sin\frac{\theta_0}{2}.$$

Then equation becomes

$$T = \frac{4}{\omega}\int_0^1 \frac{dz}{\sqrt{(1-z^2)(1-k^2z^2)}}.$$

This is done by noting that $dz = \frac{1}{2k}\cos\frac{\theta}{2}\,d\theta = \frac{1}{2k}(1-k^2z^2)^{1/2}\,d\theta$ and that $\sin^2\frac{\theta_0}{2} - \sin^2\frac{\theta}{2} = k^2(1 - z^2)$. The integral in this result is an elliptic integral of the first kind. In particular, the elliptic integral of the first kind is defined as

$$F(\phi, k) \equiv \int_0^{\phi} \frac{d\theta}{\sqrt{1-k^2\sin^2\theta}} = \int_0^{\sin\phi} \frac{dz}{\sqrt{(1-z^2)(1-k^2z^2)}}.$$

In some contexts, this is known as the incomplete elliptic integral of the first kind and $K(k) = F\left(\frac{\pi}{2}, k\right)$ is called the complete integral of the first kind.

There are table of values for elliptic integrals and now one can use a computer algebra system to compute values of such integrals. For small angles, we have that k is small. So, we can develop a series expansion for the period, T, for small k. This is simply done by first expanding using the binomial expansion.

$$(1 - k^2z^2)^{-1/2} = 1 + \frac{1}{2}k^2z^2 + \frac{3}{8}k^2z^4 + O((kz)^6)$$

Inserting the expansion in the integrand and integrating term by term, one finds that

$$T = 2\pi\sqrt{\frac{L}{g}}\left[1 + \frac{1}{4}k^2 + \frac{9}{64}k^4 + \cdots\right].$$

This expression gives further corrections to the linear result, which only provides the first term. We show the relative errors incurred when keeping the k^2 and k^4 terms versus not keeping them.

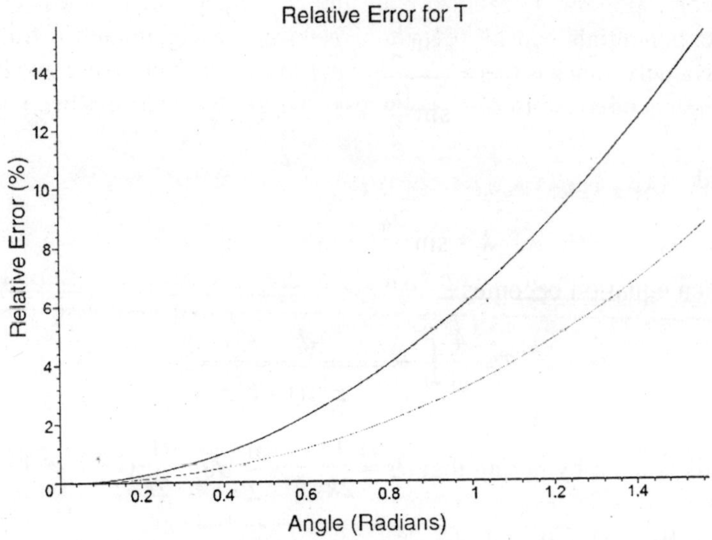

Fig. The Relative Error in Percent when Approximating the exact Period of a Nonlinear Pendulum with one, two, or three terms in equation

CAUCHY-EULER EQUATIONS - OPTIONAL

Another class of solvable second order differential equations that arc of interest are the Cauchy-Euler equations. These are given by

$$ax^2y''(x) + bxy'(x) + cy(x) = 0.$$

Note that in such equations the power of x in the coefficients matches the order of the derivative in that term.

These equations are solved in a manner similar to the constant coefficient equations.

One begins by making the guess $y(x) = x^r$. This leads to the characteristic equation

$$ar(r-1) + br + c = 0.$$

Again, one has a quadratic equation and the nature of the roots leads to three classes of solutions:

- Real, distinct roots. r_1, r_2. In this case the solutions corresponding to each root are linearly independent. Therefore, the general solution is simply $y(x) = c_1 x^{r1} + c_2 x^{r}2$.
- Real, equal roots. $r_1 = r_2 = r$. In this case the solutions corresponding to each root are linearly dependent. To find a second linearly independent solution, one uses the Method of Reduction of Order. This gives the second solution as $x^r \ln |x|$. Therefore, the general solution is found as $y(x) = (c_1 + c_2 \ln |x|)x^r$.
- Complex conjugate roots. $r_1, r_2 = \alpha \pm i\beta$. In this case the solutions

corresponding to each root are linearly independent. These complex exponentials can be rewritten in terms of trigonometric functions. Namely, one has that $x^\alpha \cos(\beta \ln |x|)$ and $x^\alpha \sin(\beta \ln |x|)$ are two linearly independent solutions. Therefore, the general solution becomes $y(x) = x^\alpha(c_1 \cos(\beta \ln |x|) + c_2 \sin(\beta \ln |x|))$.

FREE FALL AND TERMINAL VELOCITY

Some common differential equations that appear in physics. We will begin with the simplest types of equations and standard techniques for solving them We will end this part of the discussion by returning to the problem of free fall with air resistance. We will then turn to the study of oscillations, which are modelled by second order differential equations. Let us begin with a simple example from introductory physics. We recall that free fall is the vertical motion of an object under the force of gravity. It is experimentally determined that an object at some distance from the centre of the earth falls at a constant acceleration in the absence of other forces, such as air resistance. This constant acceleration is denoted by $-g$, where g is called the acceleration due to gravity. The negative sign is an indication that up is positive.

We will be interested in determining the position, $y(t)$, of the body as a function of time. From the definition of free fall, we have

$$\ddot{y}(t) = -g$$

Note that we will occasionally use a dot to indicate time differentiation.

This notation is standard in physics and we will begin to introduce you to this notation, though at times we might use the more familiar prime notation to indicate spatial differentiation, or general differentiation.

In equation we know g. It is a constant. Near the earth's surface it is about 9.81 m/s^2 or 32.2 ft/s^2. What we do not know is $y(t)$. This is our first differential equation. In fact it is natural to see differential equations appear in physics as Newton's Second Law, $F = ma$, plays an important role in classical physics.

So, how does one solve the differential equation ? We can do so by using what we know about calculus. It might be easier to see how if we put in a particular number instead of g. You might still be getting used to the fact that some letters are used to represent constants. We will come back to the more general form after we see how to solve the differential equation.

Consider

$$\ddot{y}(t) = 5.$$

Recalling that the second derivative is just the derivative of a derivative, we can rewrite the equation as

$$\frac{d}{dt}\left(\frac{dy}{dt}\right) = 5.$$

This tells s that the derivative of $dy = dt$ is 5. Can you think of a function whose derivative is 5? (Do not forget that the independent variable is t.) Yes, the derivative of $5t$ with respect to t is 5. Is this the only function whose derivative is 5? No! You can also differentiate 5t + 1, 5t + π, 5t – 6, etc. In general, the derivative of $5t + C$ is 5.

So, our equation can be reduced to

$$\frac{dy}{dt} = 5t + C.$$

Now we ask if you know a function whose derivative is $5t + C$. Well, you might be able to do this one in your head, but we just need to recall the Fundamental Theorem of Calculus, which relates integrals and derivatives. Thus, we have

$$y(t) = \frac{5}{2}t^2 + Ct + D,$$

where D is a second integration constant.

This is a solution to the original equation. That means it is a function that when placed into the differential equation makes both sides of the equal sign the same. You can always check your answer by showing that it satisfies the equation. In this case we have

$$\ddot{y}(t) = \frac{d^2}{dt^2}\left(\frac{5}{2}t^2 + Ct + D\right) = \frac{d}{dt}(5t + C) = 5 .$$

So, it is a solution.

We also see that there are two arbitrary constants, C and D. Picking any values for these gives a whole family of solutions. As we will see, our equation is a linear second order ordinary differential equation. We will see that the general solution of such an equation always has two arbitrary constants.

Let's return to the free fall problem. We solve it the same way. The only difference is that we can replace the constant 5 with the constant $-g$. So, we find that

$$\frac{dy}{dt} = -gt + C,$$

and

$$y(t) = \frac{1}{2}gt^2 + Ct + D .$$

Once you get down the process, it only takes a line or two to solve.

There seems to be a problem. Imagine dropping a ball that then undergoes free fall. We just determined that there are an infinite number of solutions to where the ball is at any time! Well, that is not possible. Experience tells us that if you drop a ball you expect it to behave the same way every time. Or does it? Actually, you could drop the ball from anywhere. You could also toss it up or throw it down. So, there are many ways you can release the ball before it is in free fall. That is where the constants come in. They have physical meanings.

If you set $t = 0$ in the equation, then you have that $y(0) = D$. Thus, D gives the initial position of the ball. Typically, we denote initial values with a subscript. So, we will write $y(0) = y_0$. Thus, $D = y_0$. That leaves us to determine C. It appears at first in equation. Recall $\frac{dy}{dt}$, the derivative of the position, is the vertical velocity, $v(t)$. It is positive when the ball moves upward.

Now, denoting the initial velocity $v(0) = v_0$, we see that equation becomes $\dot{y}(0) = C$. This implies that $C = v(0) = v_0$.

Putting this all together, we have the physical form of the solution for free fall as

$$y(t) = \frac{1}{2}gt^2 + v_0 t + y_0.$$

Doesn't this equation look familiar? Now we see that our infinite family of solutions consists of free fall resulting from initially dropping a ball at position y_0 with initial velocity v_0.

The conditions $y(0) = y_0$ and $\dot{y}(0) = v_0$ are called the initial conditions. A solution of a differential equation satisfying a set of initial conditions is often called a particular solution.

So, we have solved the free fall equation. Along the way we have begun to see some of the features that will appear in the solutions of other problems that are modelled with differential equation. We will extend our analysis to higher dimensions, in which we case will be faced with so-called partial differential equations, which involve the partial derivatives of functions of more that one variable.

But are we done with free fall ? We can relax some of the conditions that we have imposed. We can add air resistance.

Before we do that, we should also note that free fall at constant g only takes place near the surface of the Earth. What if a tile falls off the shuttle far from the surface? It will also fall to the earth. Actually, it may undergo projectile motion, which you may recall is a combination of horizontal motion and free fall.

To look at this problem we need to go to the origins of the acceleration due to gravity.

This comes out of Newton's Law of Gravitation. Consider a mass m at some distance $h(t)$ from the surface of the (spherical) Earth. Letting M and R be the Earth's mass and radius, respectively, Newton's Law of Gravitation states that

$$ma = F$$

$$m\frac{d^2h(t)}{dt^2} = G\frac{mM}{(R+h(t))^2}.$$

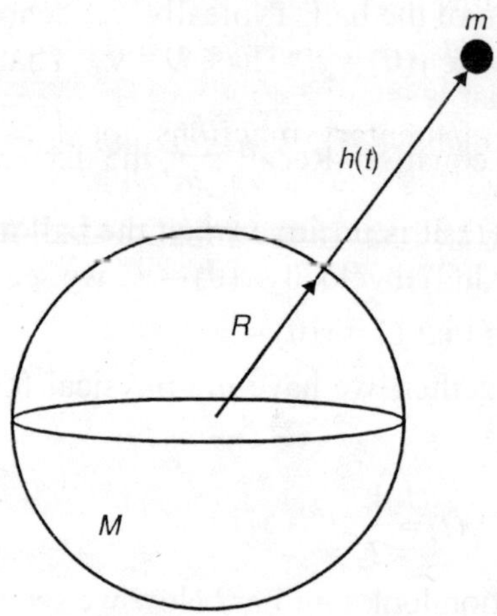

Fig. Free Fall Far from the Earth from a Height $h(t)$ from the Surface

Thus, we arrive at a differential equation

$$\frac{d^2h(t)}{dt^2} = \frac{GM}{(R+h(t))^2}.$$

This equation is not as easy to solve. We will leave it as a homework exercise for the reader.

First Order Differential Equations

Before moving on, we first define an n-th order ordinary differential equation. This is is an equation for an unknown function $y(x)$ that expresses a relationship between the unknown function and its first n derivatives. One could write this generally as

$$F(y^{(n)}, y^{(n-1)}, \ldots\ldots\, y', y, x) = 0.$$

An initial value problem consists of the differential equation plus the values of the first $n-1$ derivatives at a particular value of the independent variable, say x_0:

$$y^{(n-1)}(x_0) = y_{n-1},\ y^{(n-2)}(x_0) = y_{n-2}, \ldots,\ y(x_0) = y_0.$$

A linear nth order differential equation takes the form

$$a_n(x)y^{(n)}(x) + a_{n-1}(x)y^{(n-1)}(x) + \ldots + a_1(x)y'(x) + a_0(x)y(x) = f(x).$$

If $f(x) \equiv 0$, then the equation is said to be homogeneous, otherwise it is nonhomogeneous.

We will return to these definitions as we explore a variety of examples. However, we will start with the simplest of ordinary differential equations.

Typically, the first differential equations encountered are first order equations. A first order differential equation takes the form

$$F(y', y, x) = 0.$$

There are two general forms for which one can formally obtain a solution. The first is the separable case and the second is a linear first order equation. We indicate that we can formally obtain solutions, as one can indicate the needed integration that leads to a solution. However, these integrals are not always reducible to elementary functions nor does one necessarily obtain explicit solutions when the integrals are doable.

A first order equation is separable if it can be written the form

$$\frac{dy}{dx} = f(x)g(y).$$

Special cases result when either $f(x) = 1$ or $g(y) = 1$. In the first case the equation is said to be autonomous.

The general solution to equation is obtained in terms of two integrals:

$$\int \frac{dy}{g(y)} = \int f(x)\,dx + C,$$

where C is an integration constant. This yields a family of solutions to the differential equation corresponding to different values of C. If one can solve equation for $y(x)$, then one obtains an explicit solution. Otherwise, one has a family of implicit solutions. If an initial condition is given as well, then one might be able to find a member of the family that satisfies this condition, which is often called a particular solution.

Example: $y' = 2xy$, $y(0) = 2$.

Applying, one has

$$\int \frac{dy}{y} = \int 2x\,dx + C.$$

Integrating yields

$$\ln |y| = x^2 + C.$$

Exponentiating, one obtains the general solution,

$$y(x) = e^{x^2+C} = Ae^{x^2}, \text{ where } A = e^C.$$

(Note that since C is arbitrary, then so is e^C. Thus, there is no loss in generality using A instead of e^C.). Next, one seeks a particular solution satisfying the initial condition. For $y(0) = 2$, one finds that $A = 2$. So, the particular solution is $y(x) = 2e^{x^2}$.

Example: $yy' = -x$.

Following the same procedure as in the last example, one obtains.

$$\int y\,dy = -\int x\,dx + C \Rightarrow y^2 = -x^2 + A, \text{ where } A = 2C.$$

Thus, we obtain an implicit solution. Writing the solution as $x^2 + y^2 = A$, we see that this is a family of circles for $A > 0$ and the origin for $A = 0$. Thesecond type of first order equation encountered is the linear first order differential equation in the form

$$y_0(x) + p(x)y(x) = q(x).$$

In this case one seeks an integrating factor, $\mu(x)$, which is a function that one can multiply through the equation making the left side a perfect derivative. Multiplying the equation by μ, the resulting equation becomes

$$\frac{d}{dx}(\mu y) = \mu q.$$

The integrating factor that works is $\mu(x) = \exp(\int p(x)dx)$. The resulting equation is then easily integrated to obtain

$$y(x) = \frac{1}{\mu(x)}\left[\int^x \mu(\xi)q(\xi)d\xi + C\right].$$

Example: $xy' + y = x,\ x > 0\ y(1) = 0.$

One first notes that this is a linear first order differential equation. Solving for y', one can see that it is not separable. However, it is not in the standard form. So, we first rewrite the equation as

$$\frac{dy}{dx} + \frac{1}{x}y = 1.$$

Next, we determine the integrating factor

$$\mu(x) = \exp\left[\int^x \frac{d\xi}{\xi}\right] = e^{\ln x} = x.$$

Multiplying equation by the $\mu(x) = x$, we actually get back the original equation! In this case we have found that $xy' + y$ must have been the derivative of something to start. In fact,

$(xy)' = xy' + x$. Therefore, equation becomes

$(xy)' = x$.

Integrating one obtains

$$xy = \frac{1}{2}x^2 + C,$$

or

$$y(x) = \frac{1}{2}x + \frac{C}{x}.$$

Inserting this solution into the initial condition, $0 = \frac{1}{2} + C$.

Therefore, $C=-\frac{1}{2}$. Thus, the solution of the initial value problem is $y(x) = \frac{1}{2}\left(x-\frac{1}{x}\right)$.

There are other first order equations that one can solve for closed form solutions. However, many equations are not solvable, or one is simply interested in the behaviour of solutions. In such cases one turns to direction fields.

Terminal Velocity

Now let's return to free fall. What if there is air resistance? We first need to model the air resistance. As an object falls faster and faster, the drag force becomes greater. So, this resistive force is a function of the velocity. There are a couple of standard models that people use to test this. The idea is to write $F = ma$ in the form

$$m\ddot{y} = -mg + f(v),$$

where $f(v)$ gives the resistive force and mg is the weight. Recall that this applies to free fall near the Earth's surface. Also, for it to be resistive, $f(v)$ should oppose the motion. If the body is falling, then $f(v)$ should be positive. If it is rising, then $f(v)$ would have to be negative to indicate the opposition to the motion.

On common determination derives from the drag force on an object moving through a fluid. This force is given by

$$f(v) = \frac{1}{2}CA\rho v^2,$$

where C is the drag coefficient, A is the cross sectional area and ρ is the fluid density. For laminar flow the drag coefficient is constant.

Unless you are into aerodynamics, you do not need to get into the details of the constants. So, it is best to absorb all of the constants into one to simplify the computation. So, we will write $f(v) = bv^2$. Our equation can then be rewritten as

$$\dot{v} = kv^2 - g,$$

where $k = b/m$. Note that this is a first order equation for $v(t)$. It is separable too!

Formally, we can separate the variables and integrate the time out to obtain

$$t + K = \int^v \frac{dz}{kz^2 - g}.$$

(Note. We used an integration constant of K since C is the drag coefficient in this problem.) If we can do the integral, then we have a solution for v.

In fact, we can do this integral. You need to recall another common method of integration, which we have not reviewed yet. Do you remember Partial Fraction Decomposition? It involves factoring the denominator in our integral. Of course, this is ugly because our constants are represented by letters and are not specific numbers. Letting $\alpha^2 = g/k$, we can write the integrand as

$$\frac{1}{kz^2 - g} = \frac{1}{k}\frac{1}{z^2 - \alpha^2} = \frac{1}{2\alpha k}\left[\frac{1}{z-\alpha} - \frac{1}{z+\alpha}\right].$$

Now, the integrand can be easily integrated giving

$$t + K = \frac{1}{2\alpha k}\ln\left|\frac{v-\alpha}{v+\alpha}\right|.$$

Solving for v, we have

$$v(t) = \frac{1 - Ae^{2\alpha kt}}{1 + Ae^{2\alpha kt}}\alpha,$$

where $A \equiv e^K$. A can be determined using the initial velocity.

There are other forms for the solution in terms of a tanh function, which the reader can determine as an exercise. One important conclusion is that for large times, the ratio in the solution approaches –1.

Thus, $v \to -\alpha = -\sqrt{\frac{g}{k}}$. This means that the falling object will reach a terminal velocity.

As a simple computation, we can determine the terminal velocity. We will take an 80 kg skydiver with a cross sectional area of about 0.093 m^2. (The skydiver is falling head first.) Assume that the air density is a constant 1.2 kg/m^3 and the drag coefficient is $C = 2.0$. We first note that

$$v_{\text{terminal}} = -\sqrt{\frac{g}{k}} = -\sqrt{\frac{2mg}{CA\rho}}.$$

So,

$$v_{\text{terminal}} = -\sqrt{\frac{2(70)(9.8)}{(2.0)(0.093)(1.2)}} = 78m/s.$$

This is about 175 mph, which is slightly higher than the actual terminal velocity of a sky diver. One would need a more accurate determination of C.

Chapter 4

Linear Algebra

Calculus has its roots in physics and has become a very useful tool for modelling the physical world. Another very important area of mathematics is linear algebra.

VECTOR SPACES

Much of the discussion and terminology that we will use comes from the theory of vector spaces. Until now you may only have dealt with finite dimensional vector spaces.

Even then, you might only be comfortable with vectors in two and three dimensions. We will review a little of what we know about finite dimensional vector spaces.

The notion of a vector space is a generalization of the three dimensional vector spaces that you have seen in introductory physics and calculus. In three dimensions, we have objects called vectors, which you first visualized as arrows of a specific length and pointing in a given direction. To each vector, we can associate a point in a three dimensional Cartesian system. We just attach the tail of the vector v to the origin and the head lands at some point, (x, y, z). We then used the unit vectors i, j and k along the coordinate axes to write the vector in the form

$$v = xi + yj + zk.$$

Having defined vectors, we then learned how to add vectors and multiply vectors by numbers, or scalars.

Under these operations, we expected to get back new vectors. Then we learned that there were two types of multiplication of vectors. We could multiply two vectors to get either a scalar or a vector. This lead to the operations of dot and cross products, respectively.

The dot product was useful for determining the length of a vector, the angle between two vectors, or if the vectors were orthogonal.

In physics you first learned about vector products when you defined work, $W = F \cdot$ r. Cross products were useful in describing things like torque, $\tau = r \times F$, or the force on a moving charge in a magnetic field, $F = qv \times B$. We will return to these more complicated vector operations later when reviewing

Maxwell's equations of electrodyn-amics. The basic concept of a vector can be generalized to spaces of more than three dimensions.

You may first have seen this in your linear algebra class. The properties outlined roughly above need to be preserved.

So, we have to start with a space of vectors and the operations between them. We also need a set of scalars, which generally come from some field. However, in our applications the field will either be the set of real numbers or the set of complex numbers.

A vector space V over a field F is a set that is closed under addition and scalar multiplication and satisfies the following conditions: For any $u, v, w \in V$ and $a, b \in F$

- $u + v = v + u$.
- $(u + v) + w = u + (v + w)$.
- There exists a 0 such that $0 + v = v$.
- There exists a $-v$ such that $v + (-v) = 0$:
- $a(bv) = (ab)v$.
- $(a + b)v = av + bv$.
- $a(u + v) = au + bv$.
- $1(v) = v$.

In three dimensions the unit vectors i, j and k play an important role. Any vector in the three dimensional space can be written as a linear combination of these vectors,

$$v = xi + yj + zk.$$

In fact, given any three non-coplanar vectors, fa_1, a_2, a_3g, all vectors can be written as a linear combination of those vectors,

$$v = c_1a_1 + c_2a_1 + c_3a_1.$$

Such vectors are said to span the space and are called a basis for the space. We can generalize these ideas. In an n-dimensional vector space any vector in the space can be represented as the sum over n linearly independent vectors (the equivalent of non-coplanar vectors). Such a linearly independent set of vectors $\{v_j\}_{j=1}^{n}$ satisfies the condition

$$\sum_{j=1}^{n} c_j v_j = 0 \quad \Leftrightarrow \quad c_j = 0$$

Note that we will often use summation notation instead of writing out all of the terms in the sum.

The standard basis in an n-dimensional vector space is a generalization of the standard basis in three dimensions (i, j and k). We define

$$e_k = \Big(0,\cdots,0,\underbrace{1}_{k\text{ th space}},0,\cdots,0\Big), k = 1,\cdots,n.$$

Then, we can expand any $v \in V$ as

$$v = \sum_{k=1}^{n} u_k e_k,$$

where the v_k's are called the components of the vector in this basis.

Sometimes we will write v as an n-tuple $(v_1, v_2,...,v_n)$. This is similar to the ambiguous use of (x, y, z) to denote both vectors in as well as to represent points in the three dimensional space.

The only other thing we will need at this point is to generalize the dot product. Recall that there are two forms for the dot product in three dimensions. First, one has that

$$u \cdot v = uv \cos \theta,$$

where u and v denote the length of the vectors. The other form is the component form:

$$\text{u} \cdot v = u_1 v_1 + u_2 v_2 + u_3 v_3 = \sum_{k=1}^{3} u_k u_k.$$

Of course, this form is easier to generalize. So, we define the scalar product between two n-dimensional vectors as

$$< u,v > = \sum_{k=1}^{3} u_k u_k.$$

Actually, there are a number of notations that are used in other texts. One can write the scalar product as (u, v) or even in the Dirac bra-ket notation $< u|v >$.

We note that the (real) scalar product satisfies some simple properties. For vectors v, w and real scalar α we have

- $< v, v > \geq 0$ and $< v, v > = 0$ if and only if $v = 0$.
- $< v, w > = < w, v >$.
- $< \alpha\, v, w > = \alpha < v, w >$.

While it does not always make sense to talk about angles between general vectors in higher dimensional vector spaces, there is one concept that is useful. It is that of orthogonality, which in three dimensions is another way of saying the vectors are perpendicular to each other. So, we also say that vectors u and v are orthogonal if and only if $< u, v >= 0$. If $\{a_k\}_{k=1}^{n}$, is a set of basis vectors such that

$$< a_j, a_k >= 0, k \neq j,$$

then it is called an orthogonal basis.

If in addition each basis vector is a unit vector, then one has an orthonormal basis. This generalization of the unit basis can be expressed more compactly. We will denote such a basis of unit vectors by e_j for $j = 1... n$. Then,

$$< e_j, e_k > = \delta_{jk},$$

where we have introduced the Kronecker delta

$$\delta_{jk} \equiv \begin{cases} 0, & j \neq k \\ 1, & j = k \end{cases}$$

The process of making basis vectors have unit length is called normalization. This is simply done by dividing by the length of the vector. Recall that the length of a vector, v, is obtained as $v = \sqrt{v \cdot v}$. So, if we want to find a unit vector in the direction of v, then we simply normalize it as

$$v = \frac{v}{u}.$$

Notice that we used a hat to indicate that we have a unit vector. Furthermore, if $\{a_j\}_{j=1}^{n}$, is a set of orthogonal basis vectors, then

$$e_j = \frac{a_i}{\sqrt{< a_j, a_j >}}, j = 1...n.$$

Let $\{a_k\}_{k=1}^{n}$, be a set of orthogonal basis vectors for vector space V. We know that any vector v can be represented in terms of this basis, $v = \sum_{k=1}^{n} u_k a_k$ If we know the basis and vector, can we find the components? The answer is yes. We can use the scalar product of v with each basis element a_j. Using the properties of the scalar product, we have for $j = 1,..., n$

$$< a_j, v > = < a_j, \sum_{k=1}^{n} u_k a_k >$$

$$= \sum_{k=1}^{n} u_k < a_j, a_k >.$$

Since we know the basis elements, we can easily compute the numbers

$$A_{jk} \equiv < a_j, a_k >$$

and

$$b_j \equiv < a_j, v >.$$

Therefore, the system for the v_k's is a linear algebraic system, which takes the form

$$b_j = \sum_{k=1}^{n} A_{jk} u_k.$$

We can write this set of equations in a more compact form. The set of numbers A_{jk}, $j, k = 1,... n$ are the elements of an $n \times n$ matrix A with A_{jk} being an element in the jth row and kth column. Also, v_j and b_j can be written as column vectors, v and b, respectively. Thus, system can be written in matrix form as

$$Av = b.$$

However, if the basis is orthogonal, then the matrix $A_{jk} \equiv < a_j, a_k >$ is diagonal and the system is easily solvable. Recall that two vectors are orthogonal if and only if

$$< a_i, a_j > = 0, i \neq j.$$

Thus, in this case we have that

$$< a_j, v > = v_j < a_j, a_j >, j = 1,..., n.$$

or

$$v_j = \frac{<a_j, v>}{<a_j, a_j>}.$$

In fact, if the basis is orthonormal, i.e., the basis consists of an orthogonal set of unit vectors, then A is the identity matrix and the solution takes on a simpler form:

$$v_j = < a_j, v >.$$

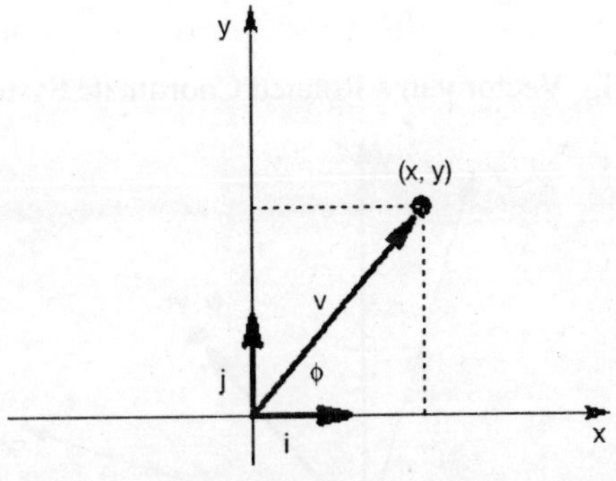

Fig. Vector v in a standard coordinate system.

LINEAR TRANSFORMATIONS

A main theme in linear algebra is to study linear transformations between vector spaces. These come in many forms and there are an abundance of applications in physics. For example, the transformation between the spacetime coordinates of observers moving in inertial frames in the theory of special relativity constitute such a transformation.

A simple example often encountered in physics courses is the rotation by a fixed angle. This is the description of points in space using two different coordinate bases, one just a rotation of the other by some angle. We begin with a vector v as described by a set of axes in the standard orientation. To find the coordinates (x, y), one needs only draw a perpendicular to the axes and read the coordinate off the axis.

In order to derive the needed transformation we will make use of polar coordinates.

The vector makes an angle of A with respect to the positive x-axis. The components (x, y) of the vector can be determined from this angle and the magnitude of v as

$$x = v \cos \phi$$
$$y = v \sin \phi.$$

We now consider another set of axes at an angle of θ to the old. We will designate these axes as x' and y'. Note that the basis vectors are different in this system. Projections to the axes are shown.

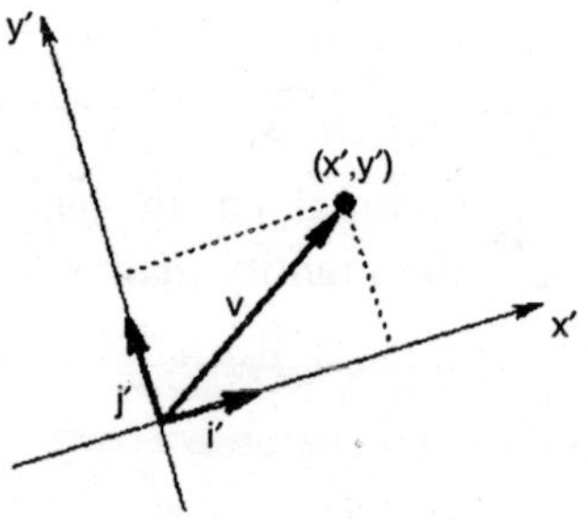

Fig. Vector v in a Rotated Coordinate System

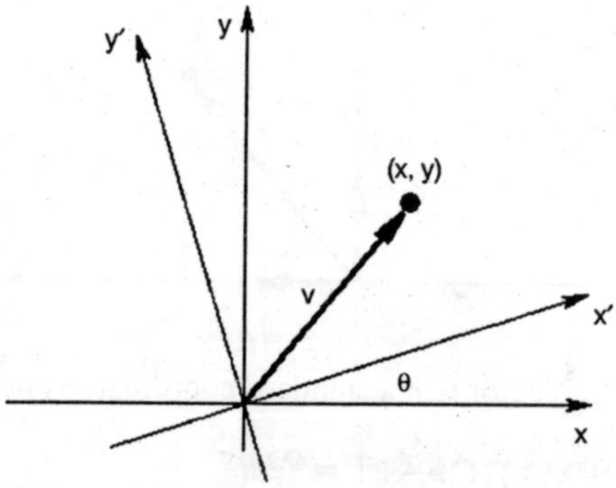

Fig. Comparison of the Coordinate Systems

The primed coordinates are not the same as the unprimed ones. The polar form for the primed system is given by

$$x' = v \cos(\phi + \theta)$$
$$y' = v \sin(\phi + \theta).$$

We can use this form to find a relationship between the two systems. Namely, we use the addition formula for trigonometric functions to obtain

$$x' = v \cos \phi \cos \theta - v \sin \phi \sin \theta$$

$$y' = v \sin \phi \cos \theta + v \cos \phi \sin \theta:$$

Noting that these expressions involve products of v with cos A and sin A, we can use the polar form for x and y to find the desired form:

$$x' = x\cos \theta - y \sin \theta$$

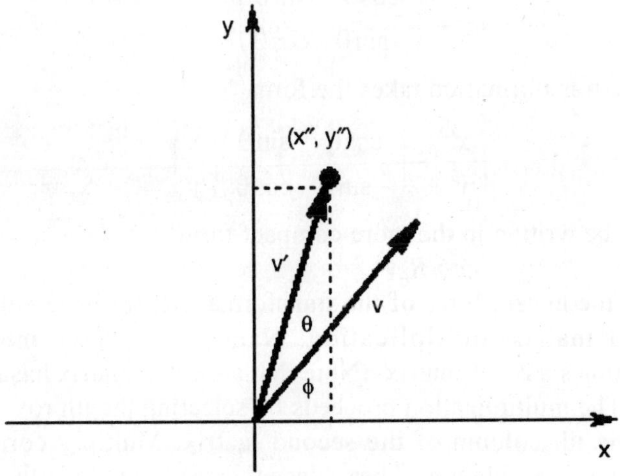

Fig. Rotation of Vector *v*

$$y' = x\sin \theta + y \cos \theta.$$

This is an example of a transformation between two coordinate systems. It is called a rotation by θ. We can designate it generally by

$$(x', y') = \hat{R}_\theta (x, y).$$

It is referred to as a passive transformation, because it does not affect the vector.

An active rotation is one in which one rotates the vector. One can derive a similar transformation for how the coordinate of the vector change under such a transformation. Denoting the new vector as v^0 with new coordinates (x^{00}, y^{00}), we have

$$x'' = x\cos \theta + y \sin \theta$$

$$y'' = -x\sin \theta + y \cos \theta.$$

We note that the active and passive rotations are related. Namely,

$$(x'', y'') = \hat{R}_{-\theta} (x, y) = R_\theta (x, y)$$

MATRICES

Linear transformations such as the rotation in the last section can be represented by matrices. Such matrix representations often become the core of a linear algebra class to the extent that one loses sight of their meaning.

We begin with the rotation transformation as applied to a vector in equation. We write vectors like v as a column matrix

$$v = \begin{pmatrix} x \\ y \end{pmatrix}$$

We can also write the trigonometric functions in a 2 × 2 matrix form as

$$R_\theta = \begin{pmatrix} \cos\theta & \sin\theta \\ -\sin\theta & \cos\theta \end{pmatrix}.$$

Then, the transformation takes the form

$$\begin{pmatrix} x' \\ y' \end{pmatrix} = \begin{pmatrix} \cos\theta & \sin\theta \\ -\sin\theta & \cos\theta \end{pmatrix} \begin{pmatrix} x \\ y \end{pmatrix}.$$

This can be written in the more compact form

$$v' = R_\theta v$$

In using the matrix form of the transformation, we have employed the definition of matrix multiplication. Namely, we have multiplied a 2 × 2 matrix times a 2 × 1 matrix. (Note that an $n \times m$ matrix has n rows and m columns.) The multiplication proceeds by selecting the ith row of the first matrix and the jth column of the second matrix. Multiply corresponding elements of each and add them. Then, place the result into the ijth entry of the product matrix. This operation can only be performed if the number of columns of the first matrix is the same as the number of columns of the second matrix.

As an example, we multiply a 3 × 2 matrix times a 2 × 2 matrix to obtain a 3 × 2 matrix.

$$\begin{pmatrix} 1 & 2 \\ 5 & -1 \\ 3 & 2 \end{pmatrix} \begin{pmatrix} 3 & 2 \\ 1 & 4 \end{pmatrix} = \begin{pmatrix} 1(3)+2(1) & 1(2)+2(4) \\ 5(3)+(-1)(1) & 5(2)+(-1)(4) \\ 3(3)+2(1) & 3(2)+2(4) \end{pmatrix}$$

$$= \begin{pmatrix} 5 & 10 \\ 14 & 6 \\ 11 & 14 \end{pmatrix}$$

In the above example, we have the row cos θ, sin θ and column x, y. Combining these we obtain $x \cos\theta + y \sin\theta$. This is x'. We perform the same operation for the second row:

$$\begin{pmatrix} x' \\ y' \end{pmatrix} = \begin{pmatrix} \cos\theta & \sin\theta \\ -\sin\theta & \cos\theta \end{pmatrix} \begin{pmatrix} x \\ y \end{pmatrix} = \begin{pmatrix} x\cos\theta + y\sin\theta \\ -x\sin\theta + y\cos\theta \end{pmatrix}.$$

In the last section we also introduced active rotations. These were rotations of vectors keeping the coordinate system fixed. Thus, we start with a vector $\boldsymbol{v}$ and rotate it by θ to get a new vector $\boldsymbol{u}$. That transformation can be written as

$$u = \hat{R}_\theta v,$$

where

$$\hat{R}_\theta = \begin{pmatrix} \cos\theta & -\sin\theta \\ \sin\theta & \cos\theta \end{pmatrix}.$$

Now consider a rotation by $-\theta$. Due to the symmetry properties of the sines and cosines, we have

$$\hat{R}_{-\theta} = \begin{pmatrix} \cos\theta & \sin\theta \\ -\sin\theta & \cos\theta \end{pmatrix}.$$

We see that if the 12 and 21 elements of this matrix are interchanged we recover $\hat{R}_\theta$. This is an example of what is called the transpose of $\hat{R}_\theta$. Given a matrix, A, its transpose A^T is the matrix obtained by interchanging the rows and columns of A. Formally, let A_{ij} be the elements of A. Then

$$A_{ij}^T = A_{ji}.$$

It is also the case that these matrices are inverses of each other. We can understand this in terms of the nature of rotations. We first rotate the vector by θ as $u = \hat{R}_\theta v$ and then rotate u by $-\theta$ obtaining $w = \hat{R}_\theta u$ Thus, the "composition" of these two transformations leads to

$$w = \hat{R}_{-\theta} u = \hat{R}_{-\theta}(\hat{R}_\theta v).$$

We can view this as a net transformation from v to w given by

$$w = \left(\hat{R}_{-\theta}\hat{R}_\theta\right)v,$$

where the transformation matrix for the composition is given by $\hat{R}_{-\theta}\hat{R}_\theta$. Actually, if you think about it, we should end up with the original vector. We can compute the resulting matrix by carrying out the multiplication. We obtain

$$\hat{R}_{-\theta}\hat{R}_\theta = \begin{pmatrix} \cos\theta & \sin\theta \\ -\sin\theta & \cos\theta \end{pmatrix}\begin{pmatrix} \cos\theta & -\sin\theta \\ \sin\theta & \cos\theta \end{pmatrix} = \begin{pmatrix} 1 & 0 \\ 0 & 1 \end{pmatrix}.$$

This is the 2×2 identity matrix. We note that the product of these two matrices yields the identity. This is like the multiplication of numbers. If $ab = 1$, then a and b are multiplicative inverses of each other. So, we see here that $\hat{R}_\theta$ and $\hat{R}_{-\theta}$ are inverses of each other as well. In fact, we have determined that

$$\hat{R}_{-\theta} = \hat{R}_\theta^{-1} = \hat{R}_\theta^T,$$

where the T designates the transpose. We note that matrices satisfying the relation $A^T = A^{-1}$ are called orthogonal matrices.

We can generalize what we have seen with this simple example. We begin with a vector v in an n-dimensional vector space. We can consider a transformation L that takes v into a new vector u as

$$u = L(v).$$

We will restrict ourselves to linear transformations. A linear transformation satisfies the following condition:

$$L(\alpha a) + \beta(b) = \alpha L(a) + \beta L(b)$$

for any vectors a and b and scalars α and β.

Such linear transformations can be represented by matrices. Take any vector v. It can be represented in terms of a basis. Let's use the standard basis fe$_i$g, $i = 1,\dots n$. Then we have

$$v = \sum_{i=1}^{n} v_i e_i.$$

Now consider the effect of the transformation L on v, using the linearity property:

$$L(v) = L\left(\sum_{i=1}^{n} v_i e_i\right) = \sum_{i=1}^{n} v_i L(e_i).$$

Thus, we see that determining how L acts on v requires that we know how L acts on the basis vectors. Namely, we need $L(e_i)$: Since e_i is a vector, this produces another vector in the space. But the resulting vector can be expanded in the basis. Let's assume that the resulting vector takes the form

$$L(e_i) = \sum_{j=1}^{n} L_{ji} e_j,$$

where L_{ji} is the jth component of $L(e_i)$ for each $i = 1,\dots, n$. The matrix of L_{ji}'s is called the matrix representation of the operator L.

Typically, in a linear algebra class you start with matrices and do not see this connection to linear operators. However, there will be times that you will need this connection to understand why matrices are involved. Furthermore, the matrix representation depends on the basis used. We used the standard basis above. However, you could have started with a different basis, such as dictated by another coordinate system. We will not go further into this point at this time and just stick with the standard basis.

Now that we know how L acts on basis vectors, what does this have to say about how L acts on any other vector in the space? Then we find

$$\begin{aligned} L(v) &= \sum_{i=1}^{n} v_i L(e_i) \\ &= \sum_{i=1}^{n} v_i \left(\sum_{j=1}^{n} L_{ji} e_j\right) \\ &= \sum_{j=1}^{n}\left(\sum_{i=1}^{n} v_i L_{ji}\right) e_j. \end{aligned}$$

Since $L(v) = u$, we see that the *j*th component of u can be written as

$$u_j = \sum_{i=1}^{n} L_{ji} v_i, j = 1...n.$$

This equation can be written in matrix form as

$$u = Lv,$$

where L now takes the role of a matrix. It is similar to the multiplication of the rotation matrix times a vector as seen in the last section. We will just work with matrix representations from here on.

Next, we can compose transformations like we had done with the two rotation matrices. Let $u = A(v)$ and $w = B(u)$ for two transformations A and B. (Thus, $v \to u \to$ w:) Then a composition of these transformations is given by

$$w = B(u) = B(Av).$$

This can be viewed as a transformation from v to w as

$$w = BA(v),$$

where the matrix representation of BA is given by the product of the matrix representations of A and B.

To see this, we look at the ijth element of the matrix representation of BA. We first note that the transformation from v to w is given by

$$w_i = \sum_{j=1}^{n} (BA)_{ij} v_j.$$

However, if we use the successive transformations, we have

$$w_i = \sum_{k=1}^{n} B_{ik} u_k.$$

$$= \sum_{k=1}^{n} B_{ik} \left(\sum_{j=1}^{n} A_{kj} v_j \right) = \sum_{j=1}^{n} \left(\sum_{k=1}^{n} B_{ik} A_{kj} \right) u_j.$$

We have two expressions for w_i as sums over v_j. So, the coefficients must be equal. This leads to our result:

$$(BA)_{ij} = \sum_{k=1}^{n} B_{ik} A_{kj} .$$

Thus, we have found the component form of matrix multiplication, which resulted from the composition of two linear transformations. This agrees with our earlier example of matrix multiplication: The ij -th component of the product is obtained by multiplying elements in the *i*th row of B and the *j*th column of A and summing.

There are many other properties of matrices and types of matrices that one will encounter. We will list a few.

First of all, there is the $n \times n$ identity matrix, I. The identity is defined as that matrix satisfying

$$IA = AI = A$$

for any $n \times n$ matrix A. The $n \times n$ identity matrix takes the form

$$I = \begin{pmatrix} 1 & 0 & \cdots & 0 \\ 0 & 1 & \cdots & 0 \\ \vdots & \vdots & \ddots & \vdots \\ 0 & 0 & \vdots & 1 \end{pmatrix}$$

A component form is given by the Kronecker delta. Namely, we have that

$$I_{ij} = \delta_{ij} \equiv \begin{cases} 0, & i \neq j \\ 1, & i = j \end{cases}$$

The inverse of matrix A is that matrix A^{-1} such that

$$AA^{-1} = A^{-1}A = I.$$

While there is a systematic method for determining the inverse in terms of cofactors, we will not cover it here. It suffices to note that the inverse of a 2×2 matrix is easily obtained. Let

$$A = \begin{pmatrix} a & b \\ c & d \end{pmatrix}.$$

Now consider the matrix

$$B = \begin{pmatrix} d & -b \\ -c & a \end{pmatrix}$$

Multiplying these matrices, we find that

$$AB = \begin{pmatrix} a & b \\ c & d \end{pmatrix} \begin{pmatrix} d & -b \\ -c & a \end{pmatrix} = \begin{pmatrix} ad-bc & 0 \\ 0 & ad-bc \end{pmatrix}.$$

This is not quite the identity, but it is a multiple of the identity. We just need to divide by $ad - bc$. So, we have found the inverse matrix:

$$A^{-1} = \frac{1}{ad-bc} \begin{pmatrix} d & -b \\ -c & a \end{pmatrix}.$$

We leave it to the reader to show that $A^{-1}A = I$.

The factor $ad - bc$ is the difference in the products of the diagonal and off-diagonal elements of matrix A. This factor is called the determinant of A. It is denoted as $\det(A)$ or $|A|$. Thus, we define

$$\det(A) = \begin{vmatrix} a & b \\ c & d \end{vmatrix} = ad - bc.$$

For higher dimensional matrices one can write the definition of the determinant. We will for now just indicate the process for 3 × 3 matrices. We write matrix A as

$$A = \begin{pmatrix} a_{11} & a_{12} & a_{13} \\ a_{21} & a_{22} & a_{23} \\ a_{31} & a_{32} & a_{33} \end{pmatrix}.$$

The determinant of A can be computed in terms of simpler 2 × 2 determinants. We define

$$\det A = \begin{vmatrix} a_{11} & a_{12} & a_{13} \\ a_{21} & a_{22} & a_{23} \\ a_{31} & a_{32} & a_{33} \end{vmatrix}$$

$$= a_{11}\begin{vmatrix} a_{22} & a_{23} \\ a_{32} & a_{33} \end{vmatrix} - a_{12}\begin{vmatrix} a_{21} & a_{23} \\ a_{31} & a_{33} \end{vmatrix} + a_{13}\begin{vmatrix} a_{21} & a_{22} \\ a_{31} & a_{32} \end{vmatrix}.$$

There are many other properties of determinants. For example, if $\det A = 0$, A is called a singular matrix. Otherwise, it is called nonsingular. If two rows, or columns, of a matrix are multiples of each other, then $\det A = 0$.

A standard application of determinants is the solution of a system of linear algebraic equations using Cramer's Rule. As an example, we consider a simple system of two equations and two unknowns. Let's consider this system of two equations and two unknowns, x and y, in the form

$$ax + by = e,$$
$$cx + dy = f.$$

The standard way to solve this is to eliminate one of the variables. (Just imagine dealing with a bigger system!). So, we can eliminate the x's. Multiply the first equation by c and the second equation by a and subtract. We then get

$$(bc - ad)y = (ec - fa).$$

If $bc - ad \neq 0$, then we can solve to y, getting

$$y = \frac{ec - fa}{bc - ad}$$

Similarly, we find

$$x = \frac{ed - bf}{ad - bc}.$$

We note the the denominators can be replaced with the determinant of the matrix of coefficients, Γ!

$$\begin{pmatrix} a & b \\ c & d \end{pmatrix}.$$

In fact, we can also replace each numerator with a determinant. Thus, our solutions may be written as

$$x = \frac{\begin{vmatrix} e & b \\ f & d \end{vmatrix}}{\begin{vmatrix} a & b \\ c & d \end{vmatrix}}$$

$$y = \frac{\begin{vmatrix} a & e \\ c & f \end{vmatrix}}{\begin{vmatrix} a & b \\ c & d \end{vmatrix}}.$$

This is Cramer's Rule for writing out solutions of systems of equations. Note that each variable is determined by placing a determinant with e and f placed in the column of the coefficient matrix corresponding to the order of the variable in the equation. The denominator is the determinant of the coefficient matrix. This construction is easily extended to larger systems of equations.

Another operation that we have seen earlier is the transpose of a matrix. The transpose of a matrix is a new matrix in which the rows and columns are interchanged. If write an $n \times m$ matrix A in standard form as

$$A = \begin{pmatrix} a_{11} & a_{12} & \cdots & a_{1m} \\ a_{21} & a_{22} & \cdots & a2m \\ \vdots & \vdots & \ddots & \vdots \\ a_{n1} & a_{n2} & \cdots & a_{nm} \end{pmatrix}$$

then the transpose is defined as

$$A^T = \begin{pmatrix} a_{11} & a_{21} & \cdots & a_{1n} \\ a_{12} & a_{22} & \cdots & a2n \\ \vdots & \vdots & \ddots & \vdots \\ a_{1m} & a_{2m} & \cdots & a_{mn} \end{pmatrix}.$$

In index form, we have

$$(A^T)_{ij} = A_{ji},\ i, j = 1,\ldots, n.$$

As we had seen in the last section, a matrix satisfying

$$A^T = A^{-1}, \text{ or } AA^T = A^TA = I,$$

is called an orthogonal matrix. One also can show that

$$(AB)^T = B^TA^T.$$

Finally, the trace of a square matrix is the sum of its diagonal elements:

$$Tr(A) = a_{11} + a_{22} + ... + a_{nn} = \sum_{i=1}^{n} a_{ii}.$$

We can show that for two square matrices

$$Tr(AB) = Tr(BA).$$

EIGENVALUE PROBLEMS

An Introduction to Coupled Systems

Recall that one of the reasons we have seemingly digressed into topics in linear algebra and matrices is to solve a coupled system of differential equations. The simplest example is a system of linear differential equations of the form

$$\frac{dx}{dt} = ax + by$$

$$\frac{dy}{dt} = cx + dy.$$

We note that this system is coupled. We cannot solve either equation without knowing either $x(t)$ or $y(t)$: A much easier problem would be to solve an uncoupled system like

$$\frac{dx}{dt} = \lambda_1 x$$

$$\frac{dy}{dt} = \lambda_2 y.$$

The solutions are quickly found to be

$$x(t) = c_1 e^{\lambda 1} t,$$

$$y(t) = c_2 e^{\lambda 2} t.$$

Here c_1 and c_2 are two arbitrary constants.

We can determine particular solutions of the system by specifying $x(t_0) = x_0$ and $y(t_0) = y_0$ at some time t_0. Thus,

$$x(t) = x_0 e^{\lambda 1} t,$$

$$y(t) = y_0 e^{\lambda 2} t.$$

Wouldn't it be nice if we could transform the more general system into one that is not coupled? Let's write our systems in more general form. We write the coupled system as

$$\frac{d}{dt} x = Ax$$

and the uncoupled system as

$$\frac{d}{dt} y = \Lambda y,$$

where

$$\Lambda = \begin{pmatrix} \lambda_1 & 0 \\ 0 & \lambda_2 \end{pmatrix}.$$

We note that × is a diagonal matrix.

Now, we seek a transformation between x and y that will transform the coupled system into the uncoupled system. Thus, we define the transformation

$$x = Sy.$$

Inserting this transformation into the coupled system we have

$$\frac{d}{dt}x = Ax \Rightarrow$$

$$\frac{d}{dt}Sy = ASy \Rightarrow$$

$$S\frac{d}{dt}y = ASy.$$

Multiply both sides by S^{-1}. [We can do this if we are dealing with an invertible transformation, i.e., a transformation in which we can get y from x as $y = S^{-1}x$.] We obtain

$$\frac{d}{dt}y = S^{-1}ASy.$$

Noting that

$$\frac{d}{dt}y = \Lambda y,$$

we have

$$\Lambda = S^{-1}AS.$$

The expression $S^{-1}AS$ is called a similarity transformation of matrix A. So, in order to uncouple the system, we seek a similarity transformation that results in a diagonal matrix. This process is called the diagonalization of matrix A. We do not know S, nor do we know Λ. We can rewrite this equation as

$$AS = S\Lambda.$$

We can solve this equation if S is real symmetric, i.e, $S^T = S$. [In the case of complex matrices, we need the matrix to be Hermitian, $S^T = S$ where the bar denotes complex conjugation. We first show that $S\Lambda = \Lambda S$. We look at the ijth component of $S\Lambda$ and rearrange the terms in the matrix product.

$$(S\,\Lambda)_{ij} = \sum_{k=1}^{n} S_{ik}\Lambda_{kj}$$

$$= \sum_{k=1}^{n} S_{ik}\lambda_j I_{kj}$$

$$= \sum_{k=1}^{n} \lambda_j I_{jk} S_{ki}^T$$

$$= \sum_{k=1}^{n} \Lambda_{jk} S_{ki} = (\Lambda S)_{ij}$$

This result leads us to the fact that S satisfies the equation

$$AS = \Lambda S:$$

Therefore, one has that the columns of S (denoted v) satisfy an equation of the form

$$Av = \lambda v.$$

This is an equation for vectors v and numbers λ given matrix A. It is called an eigenvalue problem. The vectors are called eigenvectors and the numbers, λ, are called eigenvalues. In principle, we can solve the eigenvalue problem and this will lead us to solutions of the uncoupled system of differential equations.

Example of an Eigenvalue Problem

We will determine the eigenvalues and eigenvectors for

$$A = \begin{pmatrix} 1 & -2 \\ -3 & 2 \end{pmatrix}$$

In order to find the eigenvalues and eigenvectors of this equation, we need to solve

$$Av = \lambda v.$$

Let $v = \begin{pmatrix} v_1 \\ v_2 \end{pmatrix}$. Then the eigenvalue problem can be written out. We have that

$$Av = \lambda v.$$

$$\begin{pmatrix} 1 & -2 \\ -3 & 2 \end{pmatrix}\begin{pmatrix} v_1 \\ v_2 \end{pmatrix} = \lambda \begin{pmatrix} v_1 \\ v_2 \end{pmatrix}$$

$$\begin{pmatrix} v_1 - 2v_2 \\ -3v_1 + 2v_2 \end{pmatrix} = \begin{pmatrix} \lambda v_1 \\ \lambda v_2 \end{pmatrix}.$$

So, we see that our system becomes

$$v_1 - 2v_2 = \lambda v^1,$$
$$-3v_1 + 2v_2 = \lambda v_2.$$

This can be rewritten as

$$(1 - \lambda)v_1 - 2v_2 = 0,$$
$$-3v_1 + (2 - \lambda)v_2 = 0.$$

This is a homogeneous system. We can try to solve it using elimination, as we had done earlier when deriving Cramer's Rule. We find that multiplying the first equation by $2 - \lambda$, the second by 2 and adding, we get

$$[(1 - \lambda)(2 - \lambda) - 6]v_1 = 0:$$

If the factor in the brackets is not zero, we obtain $v_1 = 0$. Inserting this into the system gives $v_2 = 0$ as well. Thus, we find v is the zero vector. However, this does not get us anywhere. We could have guessed this solution. This simple solution is the solution of all eigenvalue problems and is called the trivial solution. When solving eigenvalue problems, we only look for nontrivial solutions!

So, we have to stipulate that the factor in the brackets is zero. This means that v_1 is still unknown. This situation will always occur for eigenvalue problems. The general eigenvalue problem can be written as

$$Av - \lambda v = 0,$$

or by inserting the identity matrix,

$$Av - \lambda Iv = 0.$$

Finally, we see that we always get a homogeneous system,

$$(A - \lambda I)v = 0.$$

The factor that has to be zero can be seen now as the determinant of this system. Thus, we require

$$\det(A - \lambda I) = 0.$$

We write out this condition for the example at hand. We have that

$$\begin{vmatrix} 1-\lambda & -2 \\ -3 & 2-\lambda \end{vmatrix} = 0.$$

This will always be the starting point in solving eigenvalue problems. Note that the matrix is A with λ's subtracted from the diagonal elements.

Computing the determinant, we have

$$(1 - \lambda)(2 - \lambda) - 6 = 0,$$

or

$$\lambda^2 - 3\lambda - 4 = 0.$$

We therefore have obtained a condition on the eigenvalues! It is a quadratic and we can factor it:

$$(\lambda - 4)(\lambda + 1) = 0.$$

So, our eigenvalues are $\lambda = 4, -1$.

The second step is to find the eigenvectors. We have to do this for each eigenvalue. We first insert $\lambda = 4$ into our system:

$$-3v_1 - 2v_2 = 0,$$
$$-3v_1 - 2v_2 = 0.$$

Note that these equations are the same. So, we have one equation in two unknowns. We will not get a unique solution. This is typical of eigenvalue problems. We can pick anything we want for v_2 and then determine v_1. For example, $v_2 - 1$ gives $v_1 = -2/3$. A nicer solution would be $v_2 = 3$ and $v_1 = -2$: These vectors are different, but they point in the same direction in the v_1 v_2 plane.

For $\lambda = -1$, the system becomes

$$2v_1 - 2v_2 = 0,$$
$$-3v_1 + 3v_2 = 0.$$

While these equations do not at first look the same, we can divide out the constants and see that once again we get the same equation,

$$v_1 = v_2.$$

Picking $v_2 = 1$, we get $v_1 = 1$.

In summary, the solution to our eigenvalue problem is

$$\lambda = 4,\ v = \begin{pmatrix} -1 \\ 3 \end{pmatrix}$$

$$\lambda = -1,\ v = \begin{pmatrix} 1 \\ 1 \end{pmatrix}$$

Eigenvalue Problems

In the last subsection we were introduced to eigenvalue problems as a way to obtain a solution to a coupled system of linear differential equations. Eigenvalue problems appear in many contexts in physical applications. In the next subsection we will look at another problem that is a bit more geometric and will give us more insight into the process of diagonalization. We seek nontrivial solutions to the eigenvalue problem

$$Av = \lambda v.$$

We note that $v = 0$ is an obvious solution. Furthermore, it does not lead to anything useful. So, it is a trivial solution. Typically, we are given the matrix A and have to determine the eigenvalues, λ, and the associated eigenvectors, v, satisfying the above eigenvalue problem.

For now we begin to solve the eigenvalue problem for $v = \begin{pmatrix} v_1 \\ v_2 \end{pmatrix}$. Inserting this into equation, we obtain the homogeneous algebraic system

$$(a - \lambda)v_1 + bv_2 = 0,$$
$$cv_1 + (d - \lambda)v_2 = 0.$$

The solution of such a system would be unique if the determinant of the system is not zero. However, this would give the trivial solution $v_1 = 0$, $v_2 = 0$. To get a nontrivial solution, we need to force the determinant to be zero. This yields the eigenvalue equation

$$0 = \begin{vmatrix} a - \lambda & b \\ c & d - \lambda \end{vmatrix} = (a - \lambda)(d - \lambda) - bc.$$

This is a quadratic equation for the eigenvalues that would lead to nontrivial solutions. If we expand the right side of the equation, we find that

$$\lambda^2 - (a + d)\lambda + ad - bc = 0.$$

This is the same equation as the characteristic equation for the general constant coefficient differential equation. Thus, the eigenvalues correspond to the solutions of the characteristic polynomial for the system.

Once we find the eigenvalues, then there are possibly an infinite number solutions to the algebraic system. We will see this in the examples.

The method for solving eigenvalue problems, as you have seen, consists of just a few simple steps. We list these steps as follows:

- Write the coefficient matrix,
- Find the eigenvalues from the equation $\det(A - 1I) = 0$, and,
- Solve the linear system $(A - \lambda I)v = 0$ for each λ:

Rotations of Conics

You may have seen the general form for the equation of a conic in Cartesian coordinates in your calculus class. It is given by

$$Ax^2 + 2Bxy + Cy^2 + Ex + Fy = D.$$

This equation can describe a variety of conics (ellipses, hyperbolae and parabolae) depending on the constants. The E and F terms result from a translation of the origin and the B term is the result of a rotation of the coordinate system. We leave it to the reader to show that coordinate translations can be made to eliminate the linear terms. So, we will set $E = F = 0$ in our discussion and only consider quadratic equations of the form

$$Ax^2 + 2Bxy + Cy^2 = D.$$

If $B = 0$, then the resulting equation could be an equation for the standard ellipse or hyperbola with centre at the origin. In the case of an ellipse, the semimajor and semiminor axes lie along the coordinate axes. However, you could rotate the ellipse and that would introduce a B term, as we will see.

This conic equation can be written in matrix form. We note that

$$\begin{pmatrix} x & y \end{pmatrix}\begin{pmatrix} A & B \\ B & C \end{pmatrix}\begin{pmatrix} x \\ y \end{pmatrix} = Ax^2 + 2Bxy + Cy^2:$$

In short hand matrix form, we thus have for our equation

$$x^T Qx = D,$$

where Q is the matrix of coefficients A, B, and C.

We want to determine the transformation that puts this conic into a coordinate system in which there is no B term. Our goal is to obtain an equation of the form

$$A'x'^2 + C'y'^2 = D'$$

in the new coordinates $y^T = (x', y')$. The matrix form of this equation is given as

$$y^T\begin{pmatrix} A' & 0 \\ 0 & C' \end{pmatrix}y = D'.$$

We will denote the diagonal matrix by Λ.

So, we let

$$x = Ry,$$

where R is a rotation matrix. Inserting this transformation into our equation we find that

$$x^T Qx = (Ry)^T QRy$$
$$= y^T (R^T QR)y.$$

Comparing this result to to desired form, we have

$$\Lambda = R^T QR.$$

Recalling that the rotation matrix is an orthogonal matrix, $R^T = R^{-1}$, we have

$$\Lambda = R^{-1} QR.$$

Thus, the problem reduces to that of trying to diagonalize the matrix Q. The eigenvalues of Q will lead to the constants in the rotated equation and the eigenvectors, as we will see, will give the directions of the principal axes (the semimajor and semiminor axes). We will first show this in an example.

Example: Determine the principle axes of the ellipse given by

$$13x^2 - 10xy + 13y^2 - 72 = 0.$$

A plot of this conic in figure shows that it is an ellipse. However, we might not know this without plotting it. If the equation were in standard form, we could identify its general shape. So, we will use the method outlined above to find a coordinate system in which the ellipse appears in standard form.

The coefficient matrix for this equation is given by

$$Q = \begin{pmatrix} 13 & -5 \\ -5 & 13 \end{pmatrix}.$$

We seek a solution to the eigenvalue problem: $Qv = \lambda v$. Recall, the first step is to get the eigenvalue equation from $\det(Q - \lambda I) = 0$: For this

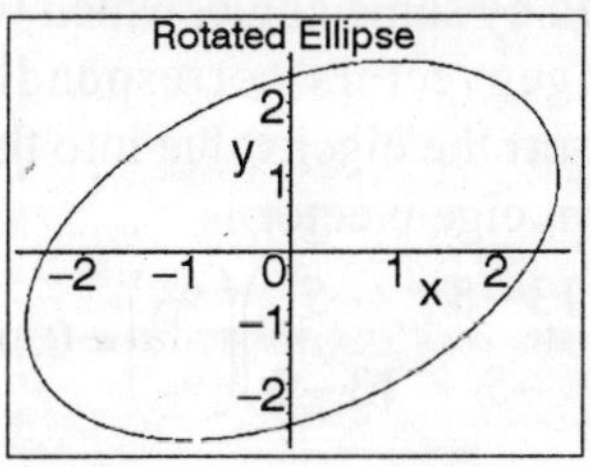

Fig. Plot of the Ellipse Given by $13x^2 - 10xy + 13y^2 - 72 = 0$

problem we have

$$\begin{vmatrix} 13-\lambda & -5 \\ -5 & 13-\lambda \end{vmatrix} = 0.$$

So, we have to solve

$$(13 - \lambda)^2 - 25 = 0.$$

This is easily solved by taking square roots to get

$$\lambda - 13 = \pm 5,$$

or

$$\lambda = 13 \pm 5 = 18, 8.$$

Thus, the equation in the new system is

$$8x'^2 + 18y'^2 = 72.$$

Dividing out the 72 puts this into the standard form

$$\frac{x'^2}{9} + \frac{y'^2}{4} = 1.$$

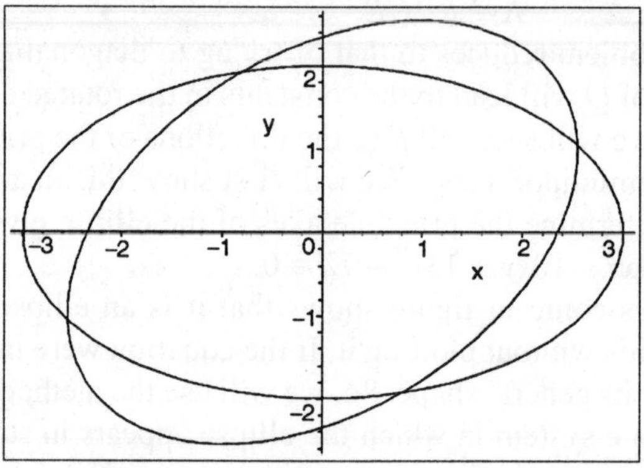

Fig. Plot of the ellipse given by $13x^2 - 10xy + 13y^2 - 72 = 0$ and the ellipse $\frac{x'^2}{9} + \frac{y'^2}{4} = 1$ showing that the first ellipse is a rotated version of the second ellipse

Now we can identify the ellipse in the new system. We note that the given ellipse is the new one rotated by some angle, which we still need to determine.

Next, we seek the eigenvectors corresponding to each eigenvalue. Eigenvalue 1. $\lambda = 8$. We insert the eigenvalue into the equation $(Q - \lambda I)v = 0$. The system for the unknown eigenvector is

$$\begin{pmatrix} 13-8 & -5 \\ -5 & 13-8 \end{pmatrix}\begin{pmatrix} v_1 \\ v_2 \end{pmatrix} = 0.$$

The first equation is

$$5v_1 - 5v_2 = 0,$$

or $v_1 = v_2$. Thus, we can choose our eigenvector to be

$$\begin{pmatrix} v_1 \\ v_2 \end{pmatrix} = \begin{pmatrix} 1 \\ 1 \end{pmatrix}.$$

Eigenvalue 2: $\lambda = 18$. In the same way, we insert the eigenvalue into the equation $(Q - \lambda I)v = 0$ and obtain

$$\begin{pmatrix} 13-18 & -5 \\ -5 & 13-18 \end{pmatrix}\begin{pmatrix} v_1 \\ v_2 \end{pmatrix} = 0.$$

The first equation is

$$-5v_1 - 5v_2 = 0,$$

or $v_1 = -v_2$. Thus, we can choose our eigenvector to be

$$\begin{pmatrix} v_1 \\ v_2 \end{pmatrix} = \begin{pmatrix} -1 \\ 1 \end{pmatrix}.$$

We superimpose the eigenvectors on our original ellipse. We see that the eigenvectors point in directions along the semimajor and semiminor axes and indicate the angle of rotation. Eigenvector one is at a 45° angle. Thus, our ellipse is a rotated version of one in standard position. Or, we could define new axes that are at 45° to the standard axes and then the ellipse would take the standard form in the new coordinate system. A general rotation of any conic can be performed. Consider the general equation:

$$Ax^2 + 2Bxy + Cy^2 + Ex + Fy = D.$$

We would like to find a rotation that puts it in the form

$$\lambda_1 x'^2 + \lambda_2 y'^2 + E'x' + F'y' = D.$$

We use the rotation matrix

$$\hat{R}_\theta = \begin{pmatrix} \cos\theta & -\sin\theta \\ \sin\theta & \cos\theta \end{pmatrix}$$

and define $x' = \hat{R}_\theta^T\ x$, or $x = R_\theta x'$.

The general equation can be written in matrix form:

$$x^T Qx + fx \quad = D,$$

where Q is the usual matrix of coefficients A, B, and C and $f = (E, F)$. Transforming this equation gives

$$x'^T R^{-1}{}_\theta\, QR_\theta x' + fR_\theta x' = D$$

The resulting equation is of the form

$$A'x'^2 + 2B'x'y' + C'\,y'^2 + E'\,x' + F'\,y' = D,$$

where

$$B' = 2(C - A)\sin\theta\cos\theta \qquad + 2B(2\cos\theta^2 - 1).$$

If we want the nonrotated form, then we seek an angle θ such that $B' = 0$. Noting that $2\sin\theta\cos\theta = \sin 2\theta$ and $2\cos\theta^2 - 1 = \cos 2\theta$, this gives $\tan(2\theta) = A - C$

$$\tan(2\theta) = \frac{A-C}{B}.$$

So, in our previous example, with $A = C = 13$ and $B = -5$, we have $\tan(2\theta) = \infty$. Thus, $2\theta = \pi/2$, or $\theta = \pi/4$:

Finally, we had noted that knowing the coefficients in the general quadratic is enough to determine the type of conic represented without doing any plotting.

This is based on the fact that the determinant of the coefficient matrix is invariant under rotation.

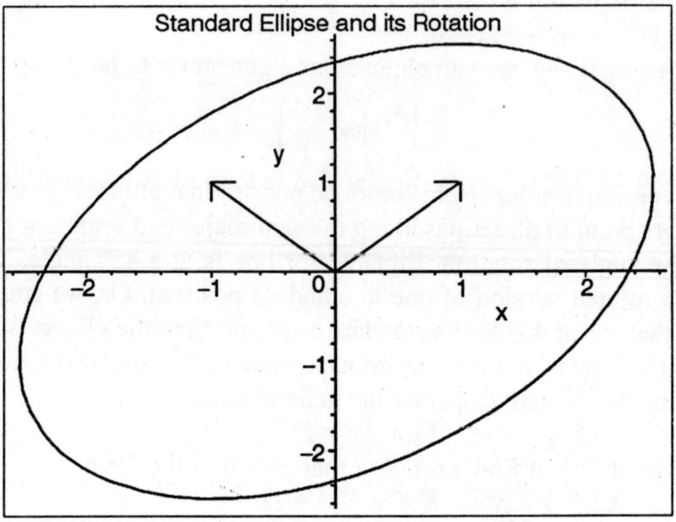

Fig. Plot of the Ellipse given by $13x^2 - 10xy + 13y^2 - 72 = 0$ and the Eigenvectors. Note that they are Along the Semimajor and Semiminor Axes and Indicate the Angle of Rotation

We see this from the equation for diagonalization

$$\begin{aligned}\det(\Lambda) &= \det(R^{-1}{}_\theta\, Q R_\theta)\\ &= \det(R^{-1}{}_\theta)\det(Q)\det(R_\theta)\\ &= \det(R^{-1}{}_\theta\, R_\theta)\det(Q)\\ &= \det(Q).\end{aligned}$$

Therefore, we have

$$\lambda_1\lambda_2 = AC - B^2:$$

Looking at equation, we have three cases:

- Ellipse $\lambda_1\,\lambda_2 > 0$ or $B^2 - AC < 0$.
- Hyperbola $\lambda_1\lambda_2 < 0$ or $B^2 - AC > 0$.
- Parabola $\lambda_1\lambda_2 = 0$ or $B^2 - AC = 0$. and one eigenvalue is nonzero. Otherwise the equation degenerates to a linear equation.

Example:

$$xy = 6.$$

As a final example, we consider this simple equation. We can see that this is a rotated hyperbola by plotting $y = 6/x$.

The coefficient matrix for this equation is given by

$$A = \begin{pmatrix} 0 & -0.5 \\ 0.5 & 0 \end{pmatrix}.$$

The eigenvalue equation is

$$\begin{vmatrix} -\lambda & -0.5 \\ -0.5 & -\lambda \end{vmatrix} = 0.$$

Thus,

$$\lambda^2 - 0.25 = 0,$$

or

$$\lambda = \pm 0.5.$$

Once again, $\tan(2\theta) = \infty$, so the new system is at 45° to the old. The equation in new coordinates is $0.5x^2 + (-0.5)y^2 = 6$, or $x^2 - y^2 = 12$.

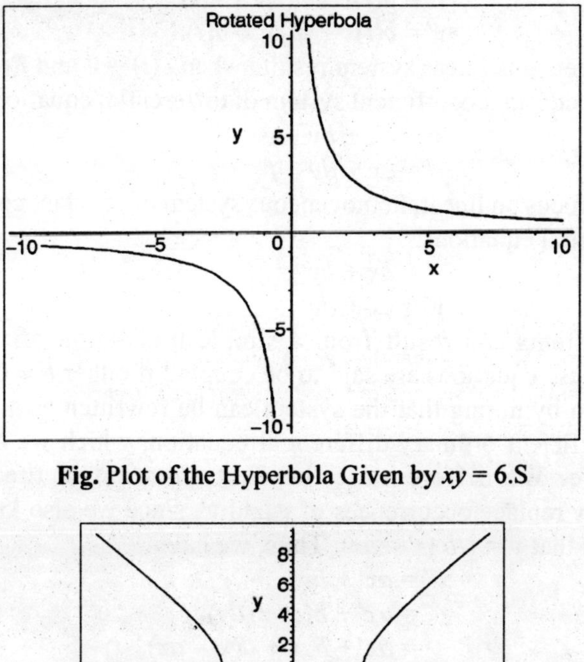

Fig. Plot of the Hyperbola Given by $xy = 6$.S

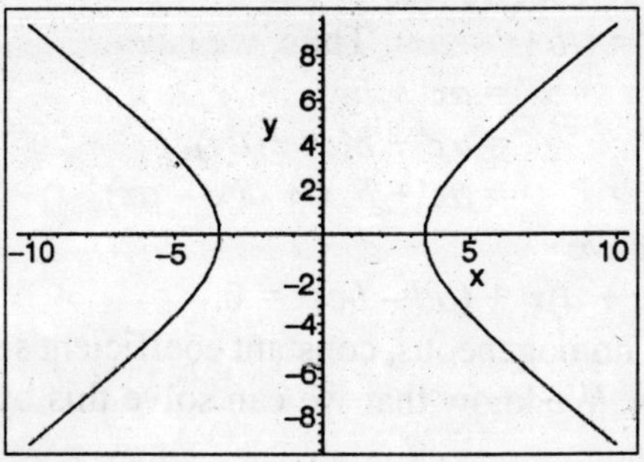

Fig. Plot of the Rotated Hyperbola Given by $x^2 - y^2 = 12$

A RETURN TO COUPLED SYSTEMS

We now return to examples of solving a coupled system of equations. We will review some theory of linear systems with constant coefficients. While the general techniques have already been covered, we present a bit more detail for the interested reader.

A general form for first order systems in the plane is given by a system of two equations for unknowns $x(t)$ and $y(t)$

$$x'(t) = P(x, y, t)$$
$$y'(t) = Q(x, y, t).$$

An autonomous system is one in which there is no explicit time dependence:

$$x'(t) = P(x, y)$$
$$y'(t) = Q(x, y).$$

Otherwise the system is called nonautonomous. A linear system takes the form

$$x' = a(t)x + b(t)y + e(t)$$
$$y' = c(t)x + d(t)y + f(t).$$

A homogeneous linear system results when $e(t) = 0$ and $f(t) = 0$.

A linear, constant coefficient system of differential equations is given by

$$x' = ax + by + e$$
$$y' = cx + dy + f.$$

We will focus on linear, homogeneous system of constant coefficient first order differential equations:

$$x' = ax + by$$
$$y' = cx + dy$$

Such systems can result from a simple translation of the unknown functions. These equations are said to be coupled if either $b \neq 0$ or $c \neq 0$.

We begin by noting that the system can be rewritten as a second order constant coefficient ordinary differential equation, which we already know how to solve. We differentiate the first equation in the system and systematically replace occurrences of y and y', since we also know from the first equation that $y = 1/b\ (x' - ax)$: Thus, we have

$$x'' = ax' + by'$$
$$= ax' + b(cx + dy)$$
$$= ax' + bcx + d(x' - ax).$$

Therefore, we have

$$x'' - (a + d)x' + (ad - bc)x = 0.$$

This is a linear, homogeneous, constant coefficient second order ordinary differential equation. We know that we can solve this by first looking at the roots of the equation

$$r^2 - (a + d)r + ad - bc = 0$$

and writing down the appropriate general solution for $x(t)$: Then we find $y(x) = 1/b\ (x' - ax)$: We now demonstrate this for a specific example.

Example:

$$x' = -x + 6y$$
$$y' = x - 2y.$$

Carrying out the above steps, we have that $x'' + 3x' - 4x = 0$. This has a characteristic equation of $r^2 + 3r - 4 = 0$. The roots of this equation are $r = 1, -4$.

Therefore, $x(t) = c_1e^t + c_2e^{-4t}$. But, we still need $y(t)$: From the first equation of the system we have

$$y(t) = \frac{1}{6}(x' + x) = \frac{1}{6}\left(2c_1e^t - 3c_2e^{-4t}\right).$$

Thus, the solution to our system is

$$x(t) = c_1e^t + c_2e^{-4t},$$

$$y(t) = \frac{1}{3}c_1e^t - \frac{1}{2}c_2e^{-4t}.$$

Sometimes one needs initial conditions. For these systems we would specify conditions like $x(0) = x_0$ and $y(0) = y_0$. These would allow the determination of the arbitrary constants as before.

We will next recast our system in matrix form and present a different analysis, which can easily be extended to systems of first order differential equations of more than two unknowns.

We start with the usual system in equation. Let the unknowns be represented by the vector

$$\mathbf{x}(t) = \begin{pmatrix} x(t) \\ y(t) \end{pmatrix}.$$

Then we have that

$$\mathbf{x}' = \begin{pmatrix} x' \\ y' \end{pmatrix} = \begin{pmatrix} ax + by \\ cx + dy \end{pmatrix} = \begin{pmatrix} a & b \\ c & d \end{pmatrix}\begin{pmatrix} x \\ y \end{pmatrix} \equiv A\mathbf{x}.$$

Here we have introduced the coefficient matrix A. This is a first order vector equation, $\mathbf{x}' = A\mathbf{x}$. Formerly, we can write the solution as $\mathbf{x} = \mathbf{x}_0e^{At}$. We will make some sense out of the exponential of a matrix.

We would like to investigate the solution of our system. Our investigations will lead to new techniques for solving linear systems using matrix methods.

We begin by recalling the solution to the specific problem. We obtained the solution to this system as

$$x(t) = c_1e^t + c_2e^{-4t},$$

$$y(t) = \frac{1}{3}c_1e^t - \frac{1}{2}c_2e^{-4t}.$$

This can be rewritten using matrix operations. Namely, we first write the solution in vector form.

$$\mathbf{x} = \begin{pmatrix} x(t) \\ y(t) \end{pmatrix} = \begin{pmatrix} c_1e^t + c_2e^{-4t} \\ \frac{1}{3}c_1e^t - \frac{1}{2}c_2e^{4t} \end{pmatrix}$$

$$= \begin{pmatrix} c_1 e^t \\ \frac{1}{3} c_1 e^t \end{pmatrix} + \begin{pmatrix} c_2 e^{-4t} \\ -\frac{1}{2} c_2 e^{-p4t} \end{pmatrix}$$

$$= c_1 \begin{pmatrix} 1 \\ \frac{1}{3} \end{pmatrix} e^t + c_2 \begin{pmatrix} 1 \\ -\frac{1}{2} \end{pmatrix} e^{-4t}.$$

We see that our solution is in the form of a linear combination of vectors of the form

$$x = ve^{\lambda t}$$

with v a constant vector and λ a constant number. This is similar to how we began to find solutions to second order constant coefficient equations. So, for the general problem we insert this guess. Thus,

$$x' = Ax \Rightarrow$$
$$\lambda ve^{\lambda t} = Ave^{\lambda t}.$$

For this to be true for all t, we then have that

$$Av = \lambda v.$$

This is an eigenvalue problem. A is a 2×2 matrix for our problem, but could easily be generalized to a system of n first order differential equations. We will confine our remarks for now to planar systems. However, we need to recall how to solve eigenvalue problems and then see how solutions of eigenvalue problems can be used to obtain solutions to our systems of differential equations.

Often we are only interested in equilibrium solutions. For equilibrium solutions the system does not change in time. Therefore, we consider $x' = 0$ and $y' = 0$: Of course, this can only happen for constant solutions. Let x_0 and y_0 be equilibrium solutions. Then, we have

$$0 = ax_0 + by_0,$$
$$0 = cx_0 + dy_0.$$

This is a linear system of homogeneous algebraic equations. One only has a unique solution when the determinant of the system is not zero, $ad - bc \neq 0$: In this case, we only have the origin as a solution, i.e., $(x_0, y_0) = (0, 0)$. However, if $ad - bc = 0$, then there are an infinite number of solutions. Studies of equilibrium solutions and their stability occur more often in systems that do not readily yield to analytic solutions. Such is the case for many nonlinear systems. Such systems are the basis of research in nonlinear dynamics and chaos.

SOLVING CONSTANT COEFFICIENT SYSTEMS IN 2D

Before proceeding to examples, we first indicate the types of solutions that could result from the solution of a homogeneous, constant coefficient system of first order differential equations.

We begin with the linear system of differential equations in matrix form.

$$\frac{dx}{dt} = \begin{pmatrix} a & b \\ c & d \end{pmatrix} x = Ax\,.$$

The type of behaviour depends upon the eigenvalues of matrix A. The procedure is to determine the eigenvalues and eigenvectors and use them to construct the general solution.

If you have an initial condition, $x(t_0) = x_0$, you can determine your two arbitrary constants in the general solution in order to obtain the particular solution.

Thus, if $x_1\,(t)$ and $x_2\,(t)$ are two linearly independent solutions, then the general solution is given as $x(t) = c_1\, x_1\,(t) + c_2\, x_2(t)$. Then, setting $t = 0$, you get two linear equations for c_1 and c_2: $c_1x_1(0) + c_2x_2\,(0) = x_0$.

The major work is in finding the linearly independent solutions. This depends upon the different types of eigenvalues that you obtain from solving the eigenvalue equation, $\det(x - \lambda I) = 0$: The nature of these roots indicate the form of the general solution.

Case I: Two real, distinct roots. Solve the eigenvalue problem $Av = \lambda v$ for each eigenvalue obtaining two eigenvectors v_1, v_2: Then write the general solution as a linear combination $x(t) = c_1e^{\lambda 1t}v_1 + c_2e^{\lambda 2t}v_2$

Case II: One Repeated Root. Solve the eigenvalue problem $Av = \lambda v$ for one eigenvalue λ, obtaining the first eigenvector v_1: One then needs a second linearly independent solution. This is obtained by solving the nonhomogeneous problem $Av_2 - \lambda v_2 = v_1$ for v_2.

The general solution is then given by $x(t) = c_1e^{\lambda t}v_1 + c_2e^{\lambda t}(v_2 + tv_1)$.

Case III: Two complex conjugate roots. Solve the eigenvalue problem $Ax = \lambda x$ for one eigenvalue, $\lambda = \alpha + i\beta$, obtaining one eigenvector v. Note that this eigenvector may have complex entries. Thus, one can write the vector $y(t) = e^{\lambda t}v = e^{\alpha t}(\cos \beta t + i \sin \beta t)v$.

Now, construct two linearly independent solutions to the problem using the real and imaginary parts of $y(t)$: $y_1\,(t) = \text{Re}(y(t))$ and $y_2(t) = \text{Im}(y(t))$. Then the general solution can be written as $x(t) = c_1y_1(t) + c_2y_2(t)$.

The construction of the general solution in Case *I* is straight forward. However, the other two cases need a little explanation.

We first look at case III.

Note that since the original system of equations does not have any i's, then we would expect real solutions. So, we look at the real and imaginary parts of the complex solution. We have that the complex solution satisfies the equation

$$\frac{d}{dt}\,[\mathrm{Re}(y(t)) + i\mathrm{Im}(y(t))] = A[\mathrm{Re}(y(t)) + i\ \mathrm{Im}(y(t))].$$

Differentiating the sum and splitting the real and imaginary parts of the equation, gives

$$\frac{d}{dt}\mathrm{Re}(y(t)) + i\frac{d}{dt}\mathrm{Im}(y(t)) = A[\mathrm{Re}(y(t))] + iA[\mathrm{Im}(y(t))]:$$

Setting the real and imaginary parts equal, we have

$$\frac{d}{dt}\mathrm{Re}(y(t)) = A[\mathrm{Re}(y(t))],$$

and

$$\frac{d}{dt}\,\mathrm{Im}(y(t)) = A[\mathrm{Im}(y(t))].$$

Therefore, the real and imaginary parts each are linearly independent solutions of the system and the general solution can be written as a linear combination of these expressions.

We now turn to Case II. Writing the system of first order equations as a second order equation for $x(t)$ with the sole solution of the characteristic equation, $\lambda = \frac{1}{2}(a + d)$, we have that the general solution takes the form

$$x(t) = (c_1 + c_2 t)e^{\lambda t}.$$

This suggests that the second linearly independent solution involves a term of the form $vte^{\lambda t}$. It turns out that the guess that works is

$$x = te^{\lambda t}v_1 + e^{\lambda t}v_2.$$

Inserting this guess into the system $x^0 = \mathrm{Ax}$ yields

$$(te^{\lambda t}v_1 + e^{\lambda t}v_2)' = A[te^{\lambda t}v_1 + e^{\lambda t}v_2]$$

$$e^{\lambda t}v_1 + \lambda^t e^{\lambda t}v_1 + \lambda e^{\lambda t}v_2 = \lambda te^{\lambda t}v_1 + e^{\lambda t}Av_2.$$

Using the eigenvalue problem and noting this is true for all t, we find that

$$v_1 + \lambda v_2 = +Av_2.$$

Therefore, $(A - \lambda I)v_2 = v_1$. We know everything except for v_2. So, we just solve for it and obtain the second linearly independent solution.

EXAMPLES OF THE MATRIX METHOD

Here we will give some examples of constant coefficient systems of differential equations for the three cases mentioned in the previous section. These are also examples of solving matrix eigenvalue problems.

Example: $A = \begin{pmatrix} 4 & 2 \\ 3 & 3 \end{pmatrix}.$

Eigenvalues: We first determine the eigenvalues.

$$0 = \begin{vmatrix} 4-\lambda & 3 \\ 3 & 3-\lambda \end{vmatrix}$$

Therefore,

$$0 = (4-\lambda)(3-\lambda) - 6$$
$$0 = \lambda^2 - 7\lambda + 6$$
$$0 = (\lambda - 1)(\lambda - 6)$$

The eigenvalues are then $\lambda = 1, 6$: This is an example of Case I.

Eigenvectors: Next we determine the eigenvectors associated with each of these eigenvalues. We have to solve the system $Av = \lambda v$ in each case.

$\lambda = 1$.

$$\begin{pmatrix} 4 & 2 \\ 3 & 3 \end{pmatrix}\begin{pmatrix} v_1 \\ v_2 \end{pmatrix} = \begin{pmatrix} v_1 \\ v_2 \end{pmatrix}$$

$$\begin{pmatrix} 3 & 2 \\ 3 & 2 \end{pmatrix}\begin{pmatrix} v_1 \\ v_2 \end{pmatrix} = \begin{pmatrix} 0 \\ 0 \end{pmatrix}$$

This gives $3v_1 + 2v_2 = 0$. One possible solution yields an eigenvector of

$$\begin{pmatrix} v_1 \\ v_2 \end{pmatrix} = \begin{pmatrix} 2 \\ -3 \end{pmatrix}.$$

$\lambda = 6$.

$$\begin{pmatrix} 4 & 2 \\ 3 & 3 \end{pmatrix}\begin{pmatrix} v_1 \\ v_2 \end{pmatrix} = 6\begin{pmatrix} v_1 \\ v_2 \end{pmatrix}$$

$$\begin{pmatrix} -2 & 2 \\ 3 & -3 \end{pmatrix}\begin{pmatrix} v_1 \\ v_2 \end{pmatrix} = \begin{pmatrix} 0 \\ 0 \end{pmatrix}$$

For this case we need to solve $-2v_1 + 2v_2 = 0$. This yields

$$\begin{pmatrix} v_1 \\ v_2 \end{pmatrix} = \begin{pmatrix} 1 \\ 1 \end{pmatrix}.$$

General Solution: We can now construct the general solution.

$$x(t) = c_1 e^{\lambda 1 t} v_1 + c_2 e^{\lambda 2 t} v_2$$
$$= c_1 e^{t}\begin{pmatrix} 2 \\ -3 \end{pmatrix} + c_2 e^{6t}\begin{pmatrix} 1 \\ 1 \end{pmatrix}$$
$$= \begin{pmatrix} 2c_1 e^{t} + c_2 e^{6t} \\ -3c_1 e^{t} + c_2 e^{6t} \end{pmatrix}.$$

Example: $A = \begin{pmatrix} 3 & -5 \\ 1 & -1 \end{pmatrix}$.

Eigenvalues: Again, one solves the eigenvalue equation.

$$0 = \begin{vmatrix} 3-\lambda & -5 \\ 1 & -1-\lambda \end{vmatrix}$$

Therefore,

$$0 = (3 - \lambda)(-1 - \lambda) + 5$$
$$0 = \lambda^2 - 2\lambda + 2$$
$$= \frac{-(-2) \pm \sqrt{4 - 4(1)(2)}}{2} = 1 \pm i.$$

The eigenvalues are then $\lambda = 1 + i$, $1 - i$. This is an example of Case III.

Eigenvectors: In order to find the general solution, we need only find the eigenvector associated with $1 + i$.

$$\begin{pmatrix} 3 & -5 \\ 1 & -1 \end{pmatrix}\begin{pmatrix} v_1 \\ v_2 \end{pmatrix} = (1 + i)\begin{pmatrix} v_1 \\ v_2 \end{pmatrix}$$

$$\begin{pmatrix} 2-i & -5 \\ 1 & -2-i \end{pmatrix}\begin{pmatrix} v_1 \\ v_2 \end{pmatrix} = \begin{pmatrix} 0 \\ 0 \end{pmatrix}.$$

We need to solve $(2 - i)v_1 - 5v_2 = 0$. Thus,

$$\begin{pmatrix} v_1 \\ v_2 \end{pmatrix} = \begin{pmatrix} 2+i \\ 1 \end{pmatrix}.$$

Complex Solution: In order to get the two real linearly independent solutions, we need to compute the real and imaginary parts of $ve^{\lambda t}$.

$$e^{\lambda t}\begin{pmatrix} 2+i \\ 1 \end{pmatrix} = e^{(1+i)t}\begin{pmatrix} 2+i \\ 1 \end{pmatrix}$$

$$= e^t(\cos t + i \sin t)\begin{pmatrix} 2+i \\ 1 \end{pmatrix}$$

$$= e^t\begin{pmatrix} (2+i)(\cos t + i\sin t) \\ \cos t + i\sin t \end{pmatrix}$$

$$= e^t\begin{pmatrix} (2\cos t - i\sin t) + i(\cos t + 2\sin t) \\ \cos t + i\sin t \end{pmatrix}$$

$$= e^t\begin{pmatrix} 2\cos t - \sin t \\ \cos t \end{pmatrix} + ie^t\begin{pmatrix} \cos t + 2\sin t \\ \sin t \end{pmatrix}$$

General Solution: Now we can construct the general solution.

$$x(t) = c_1 e^t \begin{pmatrix} 2\cos t - \sin t \\ \cos t \end{pmatrix} + c_2 e^t \begin{pmatrix} \cos t + 2\sin t \\ \sin t \end{pmatrix}$$

$$= e^t \begin{pmatrix} c_1(2\cos t - \sin t) + c_2(\cos t + 2\sin t) \\ c_1 \cos t + c_2 \sin t \end{pmatrix}.$$

Note: This can be rewritten as

$$x(t) = e^t \cos t \begin{pmatrix} 2c_1 + c_2 \\ c_1 \end{pmatrix} + e^t \sin t \begin{pmatrix} 2c_2 - c_1 \\ c_2 \end{pmatrix}.$$

Example: $A = \begin{pmatrix} 7 & -1 \\ 9 & 1 \end{pmatrix}$.

Eigenvalues:

$$0 = \begin{vmatrix} 7-\lambda & -1 \\ 9 & 1-\lambda \end{vmatrix}$$

Therefore,

$$0 = (7-\lambda)(1-\lambda) + 9$$
$$0 = \lambda^2 - 8\lambda + 16$$
$$0 = (\lambda - 4)^2.$$

There is only one real eigenvalue, $\lambda = 4$. This is an example of Case II.

Eigenvectors: In this case we first solve for v_1 and then get the second linearly independent vector.

$$\begin{pmatrix} 7 & -1 \\ 9 & 1 \end{pmatrix} \begin{pmatrix} v_1 \\ v_2 \end{pmatrix} = 4 \begin{pmatrix} v_1 \\ v_2 \end{pmatrix}$$

$$\begin{pmatrix} 3 & -1 \\ 9 & -3 \end{pmatrix} \begin{pmatrix} v_1 \\ v_2 \end{pmatrix} = \begin{pmatrix} 0 \\ 0 \end{pmatrix}.$$

Therefore, we have

$$3v_1 - v_2 = 0, \quad \Rightarrow \quad \begin{pmatrix} v_1 \\ v_2 \end{pmatrix} = \begin{pmatrix} 1 \\ 3 \end{pmatrix}.$$

Second Linearly Independent Solution: Now we need to solve $Av_2 - \lambda v_2 = v_1$.

$$\begin{pmatrix} 7 & -1 \\ 9 & 1 \end{pmatrix} \begin{pmatrix} u_1 \\ u_2 \end{pmatrix} - 4 \begin{pmatrix} u_1 \\ u_2 \end{pmatrix} = \begin{pmatrix} 1 \\ 3 \end{pmatrix}$$

$$\begin{pmatrix} 3 & -1 \\ 9 & -3 \end{pmatrix} \begin{pmatrix} u_1 \\ u_2 \end{pmatrix} = \begin{pmatrix} 1 \\ 3 \end{pmatrix}.$$

We therefore need to solve the system of equations

$$3u_1 - u_2 = 1$$

$$9u_1 - 3u_2 = 3.$$

The solution is $\begin{pmatrix} u_1 \\ u_2 \end{pmatrix} = \begin{pmatrix} 1 \\ 2 \end{pmatrix}$.

General Solution: We construct the general solution as

$$y(t) = c_1 e^{\lambda t} v_1 + c_2 e^{\lambda t}(v_2 + tv_1).$$

$$= c_1 e^{4t}\begin{pmatrix} 1 \\ 3 \end{pmatrix} + c_2 e^{4t}\left[\begin{pmatrix} 1 \\ 2 \end{pmatrix} + t\begin{pmatrix} 1 \\ 3 \end{pmatrix}\right]$$

$$= e^{4t}\begin{pmatrix} c_1 + c_2(1+t) \\ 3c_1 + c_2(2+3t) \end{pmatrix}.$$

INNER PRODUCT SPACES OPTIONAL

The linear algebra background that is needed in undergraduate physics. We have only discussed finite dimensional vector spaces, linear transformations and their matrix representations, and solving eigenvalue problems. There is more that we could discuss and more rigor.

As we progress through the course we will return to the basics and some of their generalizations. An important generalization for physics is to infinite dimensional vector spaces, in particular - function spaces. This conceptual framework is very important in areas such as quantum mechanics because this is the basis of solving the eigenvalue problems that come up there so often with the Schroodinger equation.

We will also see in the next chapter that the appropriate background spaces are function spaces in which we can solve the wave and heat equations. While we do not immediately need this understanding to carry out our computations, it can later help in the overall understanding of the methods of solution of linear partial differential equations. In particular, one definitely needs to grasp these ideas in order to fully understand and appreciate quantum mechanics.

We will consider the space of functions of a certain type. They could be the space of continuous functions on [0,1], or the space of differentiably continuous functions, or the set of functions integrable from a to b. However, you can see that there are many types of function spaces. We will further need to be able to add functions and multiply them by scalars. Thus, the set of functions and these operations will provide us with a vector space of functions.

We will also need a scalar product defined on this space of functions.

There are several types of scalar products, or inner products, that we can define. For a real vector space, we define. An inner product $<, >$ on a real vector space V is a mapping from $V \times V$ into R such that for $u, v, w \in V$ and $\alpha \in R$ one has

1. $< v, v > \geq 0$ and $< v, v > = 0$ if and only if $v = 0$.
2. $< v, w > = < w,v >$.
3. $< \alpha v, w > = \alpha < v,w >$.
4. $< u + v, w > = < u,w > + < v,w >$.

A real vector space equipped with the above inner product leads to a real inner product space. A more general definition with the second property replaced with $< v, w >= < w, v >$ is needed for complex inner product spaces.

For the time being, we are dealing just with real valued functions. We need an inner product appropriate for such spaces. One such definition is the following. Let $f(x)$ and $g(x)$ be functions defined on $[a, b]$. Then, we define the inner product, if the integral exists, as

$$< f, g >= \int_a^b f(x) g(x) dx .$$

So, we have functions spaces equipped with an inner product. Can we find a basis for the space?

For an n-dimensional space we need n basis vectors. For an infinite dimensional space, how many will we need? How do we know when we have enough? We will consider the answers to these questions as we proceed through the text.

Let's assume that we have a basis of functions $\{\phi_n(x)\}_{n=1}^{\infty}$. Given a function $f(x)$, how can we go about finding the components of f in this basis? In other words, let

$$f(x) = \sum_{n=1}^{\infty} c_n \phi_n(x) .$$

How do we find the c_n's? Does this remind you of the problem, we had earlier?

Formally, we take the inner product of f with each ϕ_j, to find

$$< \phi_j, f > = < \phi_j, \sum_{n=1}^{\infty} c_n \phi_n >$$

$$= \sum_{n=1}^{\infty} c_n < \phi_j, \phi_n >.$$

If our basis is an orthogonal basis, then we have

$$< \phi_j, \phi_n > = N_j \pm \delta_{ij},$$

where δ_{ij} is the Kronecker delta. Thus, we have

$$<A_j, f> = \sum_{n=1}^{\infty} c_n <\phi_j, \phi_n>$$

$$= \sum_{n=1}^{\infty} c_n N_j \delta_{ij}$$

$$= c_j N_j.$$

So, the expansion coefficient is

$$c_j = \frac{<\phi_j, f>}{N_j} = \frac{<\phi_j, f>}{<\phi_j, \phi_j>}.$$

$$Nj < A_j, A_j >:$$

Let's determine if the set of functions $\phi_n(x) = \sin n_x$ for $n = 1, 2,...$ is orthogonal on the interval $[= \pi, \phi]$. We need to show that $< \phi_n, \phi_m > = 0$ for $n \neq m$. Thus, we have for $n \neq m$

$$< \phi_n, \phi_m > = \int_{-\pi}^{\pi} \sin nx \sin mx \, dx$$

$$= \frac{1}{2} \int_{-\pi}^{\pi} [\cos(n-m)x - \cos(n+m)x] dx$$

$$= \frac{1}{2} \left[\frac{\sin(n-m)x}{n-m} - \frac{\sin(n+m)x}{n+m} \right]_{-\pi}^{\pi} = 0.$$

Here we have made use of a trigonometric identity for the product of two sines.

So, we have determined that the set $\phi_n(x) = \sin nx$ for $n = 1, 2,...$ is an orthogonal set of functions on the interval $[= \pi, \pi]$. Just as with vectors in three dimensions, we can normalize our basis functions to arrive at an orthonormal basis. This is simply done by dividing by the length of the vector. Recall that the length of a vector was obtained as $v = \sqrt{v \cdot v}$.

In the same way, we define the norm of our functions by

$$\|f\| = \sqrt{< f, f >}.$$

Note, there are many types of norms, but this will be sufficient for us. For the above basis of sine functions, we want to first compute the norm of each function. Then we would like to find a new basis from this one such that each basis eigenfunction has unit length and is therefore an orthonormal basis. We first compute

$$\|\phi_n\|^2 = \int_{-\pi}^{\pi} \sin^2 nx \, dx$$

$$= \frac{1}{2}\int_{-\pi}^{\pi}[1-\cos 2nx]\,dx$$

$$= \frac{1}{2}\left[x - \frac{\sin 2nx}{2n}\right]_{-\pi}^{\pi} = \pi.$$

We have found from this computation that

$$< \phi_j, \phi_n > = \pi\delta_{ij}$$

and that $||\phi_n|| = \sqrt{\pi}$. Defining $\psi_n(x) = \frac{1}{\sqrt{\pi}}\phi_n(x)$, we have normalized the ϕ_n's and have obtained an orthonormal basis of functions on $[-\pi, \phi]$.

Expansions of functions in trigonometric bases occur often and originally resulted from the study of partial differential equations named Fourier series.

Chapter 5

Complex Representations of Functions

COMPLEX REPRESENTATIONS OF WAVES

We have seen that we can seek the frequency content of a function $f(t)$ defined on an interval $[0, T]$ by looking for the Fourier coefficients in the Fourier series expansion

$$f(t) = \frac{a_0}{2} + \sum_{n=1}^{\infty} a_n \cos\frac{2\pi nt}{T} + b_n \sin\frac{2\pi nt}{T}$$

The coefficients take forms like

$$a_n = \frac{2}{T}\int_0^T f(t)\cos\frac{2\pi nt}{T}dt$$

However, trigonometric functions can be written in a complex exponential form. This is based on Euler's formula (or, Euler's identity).

$$e^{i\theta} = \cos\theta + i\sin\theta.$$

The complex conjugate is found by replacing i with i to obtain

$$e^{-\theta} = \cos\theta - i\sin\theta.$$

Adding these expressions, we have

$$2\cos\theta = e^{i\theta} + e^{-i}\theta.$$

Subtracting the exponentials leads to an expression for the sine function. Thus, we have the important result that sines and cosines can be written as complex exponentials.

$$\cos\theta = \frac{e^{i\theta} + e^{-i\theta}}{2},$$

$$\sin\theta = \frac{e^{i\theta} - e^{-i\theta}}{2i}.$$

So, we can write

$$\cos\frac{2\pi\, nt}{T} = \frac{1}{2}\left(e^{\frac{2\pi int}{T}} + e^{-\frac{2\pi int}{T}}\right).$$

We can use this information to rewrite our series as a sum over complex exponentials in the form

$$f(t) = \sum_{n=-\infty}^{\infty} c_n e^{\frac{2\pi int}{T}}$$

where the Fourier coefficients now take the form

$$c_n = \int_0^T f(t) e^{\frac{2\pi int}{T}}$$

In fact, in order to connect our analysis to ideal signals over an infinite interval and containing a continuum of frequencies, we will see the above sum become an integral and we will naturally find ourselves needing to work with functions of complex variables and perform complex integrals.

We can extend these ideas to develop a complex representation for waves. We obtained the solution

$$u(x,\ t) = \frac{1}{2}\left[\sum_{n=1}^{\infty} A_n \sin k_n(x+ct) + \sum_{n=1}^{\infty} A_n \sin k_n(x-ct)\right]$$

We can replace the sines with their complex forms as

$$u(x,\ t) = \frac{1}{4i}\left[\sum_{n=1}^{\infty} A_n \left(e^{ikn(x+ct)} - e^{-ik_n(x+ct)}\right)\right. \left. + \sum_{n=1}^{\infty} A_n \left(e^{ikn(x-ct)} - e^{-ik_n(x+ct)}\right)\right]$$

Now, defining $k_{-n} = -k_n$, we can rewrite this solution in the form

$$u(x,\ t) = \sum_{n=-\infty}^{\infty} \left[c_n e^{ikn(x+ct)} + d_n e^{ik_n(x-ct)}\right]$$

Such representations are also possible for waves propagating over the entire real line.

In such cases we are not restricted to discrete frequencies and wave numbers. The sum of the harmonics will then be a sum over a continuous range, which means that our sums become integrals. So, we are then lead to the complex representation

$$u(x,\ t) = \int_{-\infty}^{\infty} \left[c(k) e^{ik(x+ct)} + d(k) e^{ik_n(x-ct)}\right] dk.$$

The forms $e^{ik(x+ct)}$ and $e^{ik(x-ct)}$ are complex representations of what are called plane waves in one dimension. The integral represents a general wave form consisting of a sum over plane waves, typically representing wave packets. The Fourier coefficients in the representation can be complex valued functions and the evaluation of the integral may be done using methods from complex analysis. We would like to be able to compute such integrals.

With the above ideas in mind, we will now take a tour of complex analysis. We will first review some facts about complex numbers and then introduce complex functions. This will lead us to the calculus of functions of a complex variable, including differentiation and complex integration.

COMPLEX NUMBERS

Complex numbers were first introduced in order to solve some simple problems. The history of complex numbers only extends about two equations such as $x^2 + 1 = 0$. The solution is $x = \pm\sqrt{-1}$. Due to the usefulness of this concept, which was not realized at first, a special symbol was introduced - the imaginary unit, $i = \sqrt{-1}$.

A complex number is a number of the form $z = x + iy$, where x and y are real numbers. x is called the real part of z and y is the imaginary part of z. Examples of such numbers are $3 + 3i$, $-1i$, $4i$ and 5. Note that $5 = 5 + 0i$ and $4i = 0 + 4i$.

There is a geometric representation of complex numbers in a two dimensional plane, known as the complex plane C. This is given by the Argand diagram, here we can think of the complex number $z = x + iy$ as a point (x, y) in the complex plane or as a vector. The magnitude, or length, of this vector is called the complex modulus of $|z| = \sqrt{x^2 + y^2}$.

We can also use the geometric picture to develop a polar representation of complex numbers. We can see that in terms of r and θ we have that

$$x = r\cos\theta,$$

$$y = r\sin\theta.$$

Thus,

$$z = x + iy = r(\cos\theta + i\sin\theta) = re^{i\theta}.$$

Here we have used Euler's formula.

So, given r and θ we have $z = r^{ei}\theta$. However, given the Cartesian form, $z = x + iy$, we can also determine the polar form, since

$$r = \sqrt{x^2 + y^2}$$

$$\tan\theta = \frac{y}{x}.$$

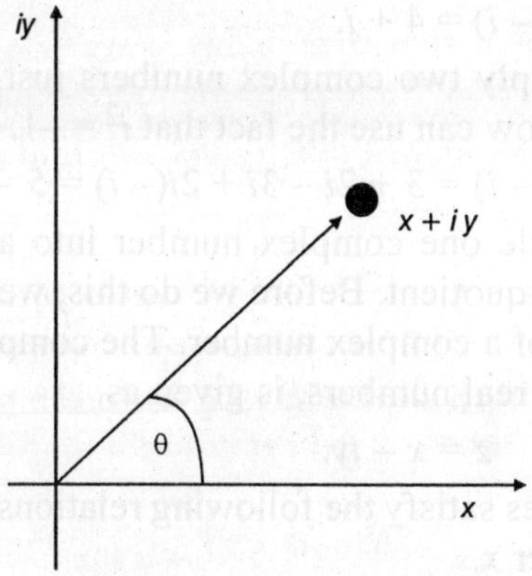

Fig. The Argand Diagram for Plotting Complex Numbers in the Complex z-plane

Note that $r = |z|$.

Example: Write $1 + i$ in polar form. If one locates $1 + i$ in the complex plane, then it might be possible to immediately determine the polar form from the angle and length of the "complex vector". If one did not see the polar form from the plot in the z-plane, then one can systematically determine the results. We want to write $1 + i$ in polar form. $1 + i = re^{i\theta}$ for some r and θ. Using the above relations, we have $r = \sqrt{x^2 + y^2} = \sqrt{2}$ and $\tan\theta = \frac{y}{x} = 1$. This gives $\theta = \frac{\pi}{4}$. So, we have found that

$$1 + i = \sqrt{2}e^{i\pi/4}$$

We also have the usual operations. We can add two complex numbers and obtain another complex number. This is simply done by adding the real parts and the imaginary parts. So,

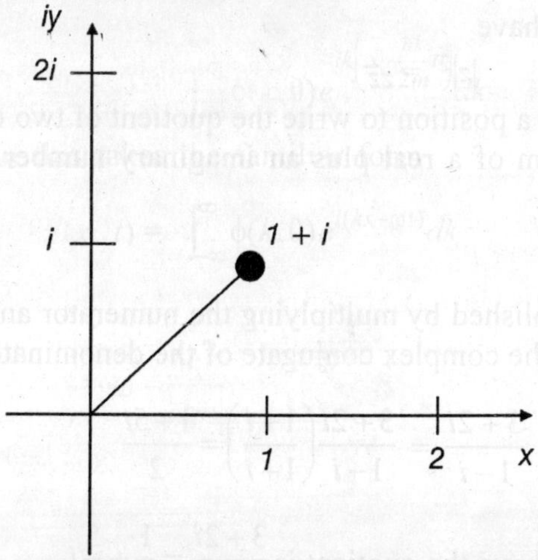

Fig. Locating $1 + i$ in the Complex z-plane

$$(3 + 2i) + (1 - i) = 4 + i.$$

We can also multiply two complex numbers just like we multiply any binomials, though we now can use the fact that $i^2 = -1$. For example, we have

$$(3 + 2i)(1 - i) = 3 + 2i - 3i + 2i(-i) = 5 - i.$$

We can even divide one complex number into another one and get a complex number as the quotient. Before we do this, we need to introduce the complex conjugate, $\bar{z}$, of a complex number. The complex conjugate of $z = x + iy$, where x and y are real numbers, is given as

$$\bar{z} = x - iy.$$

Complex conjugates satisfy the following relations for complex numbers z and w and real number x.

$$\overline{z + w} = \bar{z} + \bar{w}$$

$$\overline{zw} = \overline{zw}$$

$$\bar{\bar{z}} = z$$

$$\bar{x} = x.$$

One consequence is that the complex conjugate of $z = re^{i\theta}$ is

$$\bar{z} = \overline{re^{i\theta}} = \overline{\cos\theta + i\sin\theta} = \cos\theta - i\sin\theta = re^{-\theta}$$

Another consequence is that

$$z\bar{z} = re^{i\theta}re^{-i\theta} = r^2.$$

Thus, the product of a complex number with its complex conjugate is a real number. We can also write this result in the form

$$z\bar{z} = (x + iy)(x - iy) = x^2 + y^2 = |z|^2.$$

Therefore, we have

$$|z|^2 = z\bar{z}.$$

Now we are in a position to write the quotient of two complex numbers in the standard form of a real plus an imaginary number. As an example,

we $\dfrac{3+2i}{1-i}$.

This is accomplished by multiplying the numerator and denominator of this expression by the complex conjugate of the denominator:

$$\frac{3+2i}{1-i} = \frac{3+2i}{1-i}\left(\frac{1+i}{1+i}\right) = \frac{1+5i}{2}$$

Therefore, we have the quotient is $\dfrac{3+2i}{1-i} = \dfrac{1}{2} + \dfrac{5}{2}i.$

We can also look at powers of complex numbers. For example,

$$(1 + i)^2 = 2i,$$
$$(1 + i)^3 = (1 + i)(2i) = 2i - 2.$$

But, what is $(1 + i)^{1/2} = \sqrt{1+i}$?

In general, we want to find the *n*th root of a complex number. Let $t = z^{1=n}$. To find t in this case is the same as asking for the solution of

$$t^n - z = 0$$

given z. But, this is the root of an nth degree equation, for which we expect n roots. We can answer our question if we write z in polar form, $z = r^{ei\theta}$. Then,

$$\begin{aligned} z^{1/n} &= \left(re^{i\theta}\right)^{1/n} \\ &= r^{1/n}e^{i\theta/n} \\ &= r^{1/n}\left[\cos\frac{\theta}{n} + i\sin\frac{\theta}{n}\right] \end{aligned}$$

If we use this to obtain an answer to our problem, we get

$$(1 + i)^{1/2} = \left(\sqrt{2}e^{i\pi/4}\right)^{1/2} = 2^{1/4}e^{i\pi}e^{i\pi/8}$$

But this is only one solution. We expected two solutions→

The problem is that the polar representation for z is not unique. We note that

$$e^{2k\pi i} = 1,\ k = 0, \pm 1, \pm 2, \ldots.$$

So, we can rewrite z as $z = r^{ei\theta}\, e^{2k\pi i} = r^{ei(\theta+2k\pi)}$. Now, we have that

$$\begin{aligned} z^{1/n} &= r^{1/n}e^{i(\theta+2k\pi)/n} \\ &= r^{1/n}\left[\cos\left(\frac{\theta+2k\pi}{n}\right) + i\sin\left(\frac{\theta+2k\pi}{n}\right)\right] \end{aligned}$$

We note that we only get different values for $k = 0, 1, \ldots, n - 1$.

Now, we can finish our example.

$$\begin{aligned} (1 + i)^{1/2} &= \left(\sqrt{2}e^{i\pi/4} \underbrace{e^{2k\pi i}}_{\text{insert } 1 = e^{2k\pi i}}\right)^{1/2} \\ &= 2^{1/4}e^{i(\pi/8+k\pi)} \\ &= \left\{2^{1/4}e^{i\pi/8}, 2^{1/4}e^{9\pi i/8}\right\} \end{aligned}$$

Finally, what is $\sqrt[n]{1}$? Our first guess would be $\sqrt[n]{1} = 1$. But, we know that there should be n roots. These roots are called the *n*th roots of unity.

Using the above result in equation with $r = 1$ and $\theta = 0$, we have that

$$\sqrt[n]{1} = \left[\cos\frac{2\pi k}{n} + i\sin\frac{2\pi k}{n}\right], \quad k = 0, ..., n-1.$$

For example, we have

$$\sqrt[3]{1} = \left[\cos\frac{2\pi k}{n} + i\sin\frac{2\pi k}{n}\right], \quad k = 0, 1, 2.$$

These three roots can be written out as

$$\sqrt[3]{1} = 1, -\frac{1}{2} + \frac{\sqrt{3}}{2}i, -\frac{1}{2} - \frac{\sqrt{3}}{2}i.$$

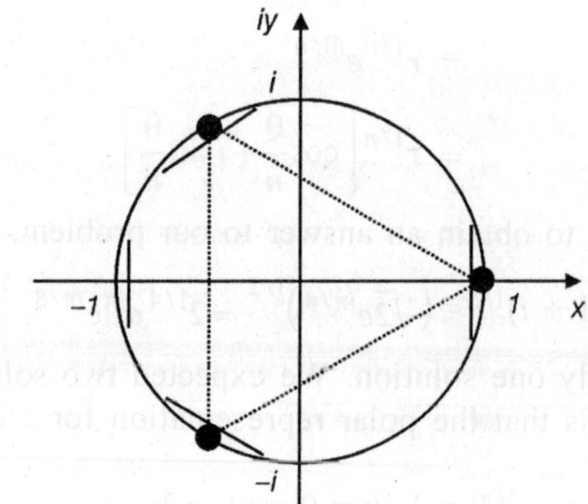

Fig. Locating the Cube roots of Unity in the Complex z-plane

Note, the reader can verify that these are indeed the cube roots of unity. We note that

$$\left(-\frac{1}{2} + \frac{\sqrt{3}}{2}i\right)^2 = -\frac{1}{2} + \frac{\sqrt{3}}{2}i$$

and

$$\left(-\frac{1}{2} + \frac{\sqrt{3}}{2}i\right)^3 = \left(-\frac{1}{2} + \frac{\sqrt{3}}{2}i\right)\left(-\frac{1}{2} + \frac{\sqrt{3}}{2}i\right) = 1.$$

We can locate these cube roots of unity in the complex plane. We see that these points lie on the unit circle and are at the vertices of an equilateral triangle. In fact, all nth roots of unity lie on the unit circle and are the vertices of a regular n-gon.

COMPLEX VALUED FUNCTIONS

We would like to next explore complex functions and the calculus of complex functions. We begin by defining a function that takes complex numbers into complex numbers, $f: C \to C$. It is difficult to visualize such functions.

One typically uses two copies of the complex plane to indicate how such functions behave. We will call the domain the z-plane and the image will lie in the w-plane.

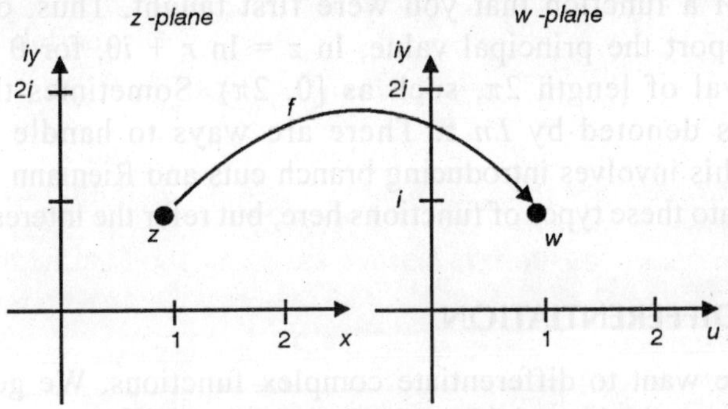

Fig. Defining a Complex Valued Function on C

We let $z = x + iy$ and $w = u + iv$. Then we can define our function as

$$w = f(z) = f(x + iy) = u(x, y) + iv(x, y).$$

We see that one can view this function as a function of z or a function of x and y. Often, we have an interest in writing out the real and imaginary parts of the function, which can be viewed as functions of two variables.

Example: $f(z) = z^2$.

For example, we can look at the simple function $f(z) = z^2$. It is a simple matter to determine the real and imaginary parts of this function. Namely, we have

$$z^2 = (x + iy)^2 = x^2 - y^2 + 2ixy.$$

Therefore, we have that

$$u(x, y) = x^2 - y^2, \; v(x, y) = 2xy.$$

Example: $f(z) = e^z$.

For this case, we make use of Euler's Formula.

$$\begin{aligned} f(z) &= e^z \\ &= e^{x+iy} \\ &= e^x e^{iy} \\ &= e^x(\cos y + i \sin y). \end{aligned}$$

Thus, $u(x, y) = e^x \cos y$ and $v(x, y) = e^x \sin y$.

Example: $f(z) = \ln z$.

In this case we make use of the polar form, $z = re^{i\theta}$. Our first thought would be to simply compute

$$\ln z = \ln r + i\theta.$$

However, the natural logarithm is multivalued, just like the nth root. Recalling that $e^{2\pi ik} = 1$ for k an integer, we have $z = re^{i(\theta+2\pi k)}$.

Therefore,

$$\ln z = \ln r + i(\theta + 2\pi k), \quad k = \text{integer}.$$

The natural logarithm is a multivalued function. In fact there are an infinite number of values for a given z. Of course, this contradicts the definition of a function that you were first taught. Thus, one typically will only report the principal value, $\ln z = \ln r + i\theta$, for θ restricted to some interval of length 2π, such as $[0, 2\pi)$. Sometimes the principal logarithm is denoted by $Ln\ z$. There are ways to handle multivalued functions. This involves introducing branch cuts and Riemann surfaces. We will not go into these types of functions here, but refer the interested reader to other texts.

COMPLEX DIFFERENTIATION

Next we want to differentiate complex functions. We generalize our definition from single variable calculus,

$$f'(z) = \lim_{\Delta z \to 0} \frac{f(z+\Delta z) - f(z)}{\Delta z},$$

provided this limit exists.

The computation of this limit is similar to what we faced in multivariable calculus. Letting $\Delta z \to 0$ means that we get closer to z. There are many paths that one can take that will approach z.

It is sufficient to look at two paths in particular. We first consider the path y = constant. Such a path. For this path,

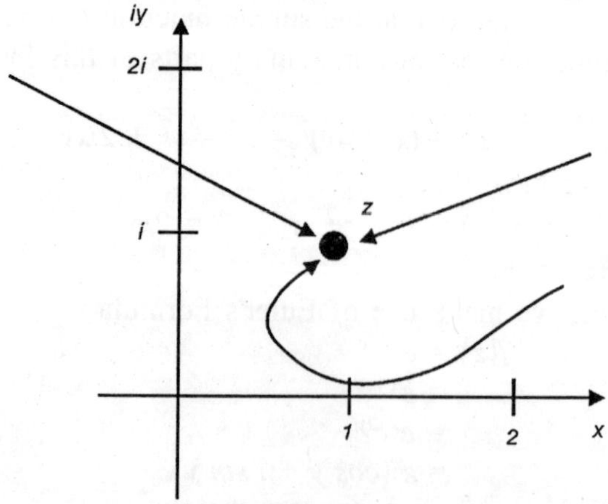

Fig. There are Many paths that Approach z as $\Delta z \to 0$

$\Delta z = \Delta x + i\Delta y = \Delta x$, since y does not change along the path. The derivative, if it exists, is then computed as

$$f'(z) = \lim_{\Delta z \to 0} \frac{f(z+\Delta z) - f(z)}{\Delta z}$$

$$= \lim_{\Delta x \to 0} \frac{u(x+\Delta x, y) + iv(x+\Delta x, y) - (u(x,y) + iv(x,y))}{\Delta x}$$

$$= \lim_{\Delta x \to 0} \frac{u(x+\Delta x, y) - u(x,y)}{\Delta x} + \lim_{\Delta x \to 0} i\frac{v(x+\Delta x, y) - v(x,y)}{\Delta x}.$$

The last two limits are easily identified as partial derivatives of real valued functions of two variables. Thus, we have shown that when $f'(z)$ exists,

$$f'(z) = \frac{\partial u}{\partial x} + i\frac{\partial v}{\partial x}.$$

A similar computation can be made if instead we take a path corresponding to x = constant. In this case $\Delta z = i\Delta y$ and

$$f'(z) = \lim_{\Delta z \to 0} \frac{f(z+\Delta z) - f(z)}{\Delta z}$$

$$= \lim_{\Delta y \to 0} \frac{u(x, y+\Delta y) + iv(x, y+\Delta y) - (u(x,y) + iv(x,y))}{i\Delta y}$$

$$= \lim_{\Delta y \to 0} \frac{u(x+\Delta x, y) - u(x,y)}{i\Delta y} + \lim_{\Delta y \to 0} \frac{v(x, y+\Delta y) - v(x,y)}{\Delta y}$$

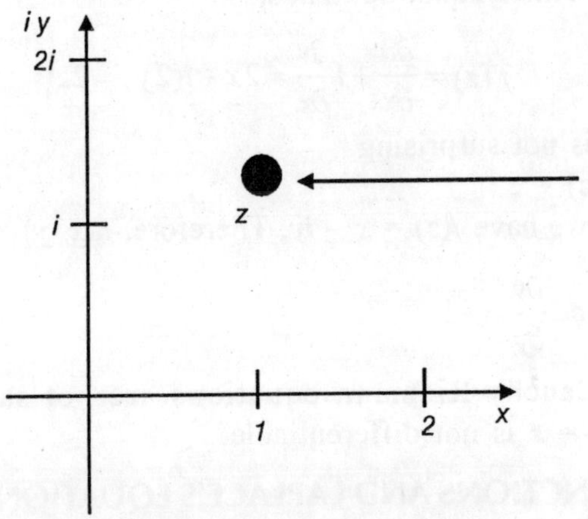

Fig. A path that Approaches z with y = Constant

Therefore,

$$f'(z) = \frac{\partial v}{\partial y} - i\frac{\partial u}{\partial y}.$$

We have found two different expressions for $f'(z)$ by following two different paths to z. If the derivative exists, then these two expressions must be the same. Equating the real and imaginary parts of these expressions, we have

$$\frac{\partial u}{\partial x} = \frac{\partial v}{\partial y}$$

$$\frac{\partial v}{\partial x} = -\frac{\partial u}{\partial y}.$$

These are known as the Cauchy-Riemann equations and we have the following theorem.

Theorem: $f(z)$ is holomorphic (differentiable) if and only if the Cauchy-Riemann equations are satisfied.

Example: $f(z) = z^2$.

In this case we have already seen that $z^2 = x^2 - y^2 + 2ixy$. Therefore, $u(x, y) = x^2 - y^2$ and $v(x, y) = 2xy$. We first check the Cauchy-Riemann equations.

$$\frac{\partial u}{\partial x} = 2x = \frac{\partial v}{\partial y}$$

$$\frac{\partial v}{\partial x} = 2y = -\frac{\partial u}{\partial y}.$$

Therefore, $f(z) = z^2$ is differentiable. We can further compute the derivative using either equation. Thus,

$$f'(z) = \frac{\partial u}{\partial x} + i\frac{\partial v}{\partial x} = 2x + i(2y) = 2z.$$

This result is not surprising.

Example: $f(z) = z$.

In this case we have $f(z) = x - iy$. Therefore, $u(x, y) = x$ and $v(x, y) = -y$. But, $\frac{\partial u}{\partial x} = 1$ and $\frac{\partial v}{\partial y} = -1$.

Thus, the Cauchy-Riemann equations are not satisfied and we conclude the $f(z) = z$ is not differentiable.

HARMONIC FUNCTIONS AND LAPLACE'S EQUATION

Another consequence of the Cauchy-Riemann equations is that both

$u(x, y)$ and $v(x, y)$ are harmonic functions. A real-valued function $u(x, y)$ is harmonic if it satisfies Laplace's equation in two dimensions, $\nabla^2 u = 0$, or

$$\frac{\partial^2 u}{\partial x^2} + \frac{\partial^2 u}{\partial y^2} = 0.$$

Theorem: f is differentiable if and only if u and v are harmonic functions.

This is easily proven using the Cauchy-Riemann equations.

$$\frac{\partial^2 u}{\partial x^2} = \frac{\partial}{\partial x}\frac{\partial u}{\partial x}$$

$$= \frac{\partial}{\partial x}\frac{\partial v}{\partial y}$$

$$= \frac{\partial}{\partial y}\frac{\partial v}{\partial x}$$

$$= -\frac{\partial}{\partial y}\frac{\partial u}{\partial y}$$

$$= -\frac{\partial^2 u}{\partial y^2}.$$

1. Is $u(x, y) = x^2 + y^2$ harmonic?

$$\frac{\partial^2 u}{\partial x^2} + \frac{\partial^2 u}{\partial y^2} = 2 + 2 \neq 0.$$

No, it is not.

2. Is $u(x, y) = x^2 - y^2$ harmonic?

$$\frac{\partial^2 u}{\partial x^2} + \frac{\partial^2 u}{\partial y^2} = 2 - 2 = 0.$$

Yes, it is.

Given a harmonic function $u(x, y)$, can one find a function, $v(x, y)$, such $f(z) = u(x, y) + iv(x, y)$ is differentiable? Such a v is called the harmonic conjugate.

Example: $u(x, y) = x^2 - y^2$ is harmonic, find the harmonic conjugate, $v(x, y)$, such that $u + iv$ is differentiable.

The Cauchy-Riemann equations tell us the following about the unknown function, $v(x, y)$:

$$\frac{\partial v}{\partial x} = -\frac{\partial u}{\partial y} = 2y,$$

$$\frac{\partial v}{\partial y} = \frac{\partial u}{\partial x} = 2x.$$

We can integrate the first of these equations to obtain

$$v(x, y) = \int 2y\,dx = 2xy + c(y).$$

Here $c(y)$ is an arbitrary function of y. One can check to see that this works by simply differentiating the result with respect to x. However, the second equation must also hold. So, we differentiate our result with respect to y to find that

$$\frac{\partial v}{\partial y} = 2x + c'(y).$$

Since we were supposed to get $2x$, we have that $c'(y) = 0$. Thus, $c(y) = k$ is a constant.

We have just shown that we get an infinite number of functions,

$$v(x, y) = 2xy + k,$$

such that

$$f(z) = x^2 - y^2 + i(2xy + k)$$

is differentiable. In fact, for $k = 0$ this is nothing other than $f(z) = z2$. So, we have $f(z) = z^2 + ik$.

COMPLEX INTEGRATION

In the last sections we were introduced to functions of a complex variable. We have also established when functions are differentiable as complex functions, or holomorphic. Now we will turn to integration in the complex plane.

We will learn how to compute complex path integrals, or contour integrals. We will see that contour integral methods are also useful in the computation of some real integrals.

Complex Path Integrals

We begin by investigating the computation of complex path integrals.

Given two points in the complex plane, connected by a path Γ, we would like to define the integral of $f(z)$ along Γ,

$$\int_\Gamma f(z)\,dz.$$

A natural procedure would be to work in real variables, by writing

$$\int_\Gamma f(z)\,dz = \int_\Gamma [u(x,y) + iv(x,y)]\,(dx + idy)$$

In order to carry out the integration, we then have to find a parametrization of the path and use methods from the third semester calculus class.

Before carrying this out with some examples, we will first provide some definitions.

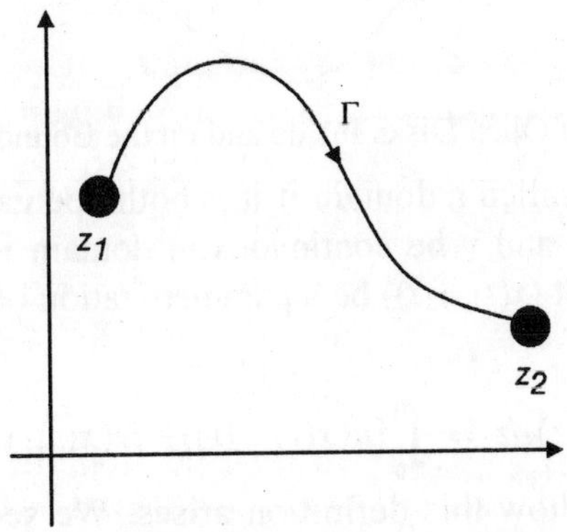

Fig. We Would Like to Integrate a Complex Function $f(z)$ over the path Γ in the Complex Plane

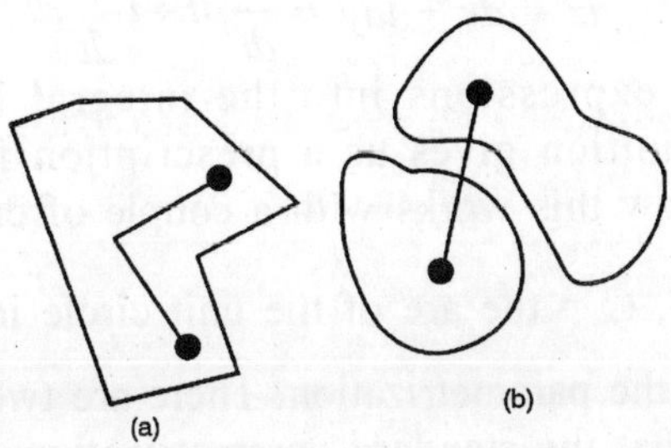

Fig. Examples of (a) a Connected set and (b) a Disconnected Set

Definition: A set D is connected if and only if for all z_1, and z_2 in D there exists a piecewise smooth curve connecting z_1 to z_2 and lying in D. Otherwise it is called disconnected.

Definition: A set D is open if and only if for all z_0 in D there exists an open disk $|z - z_0| < \rho$ in D. In figure we show a region with two disks. For all points on the interior of the region one can find at least one disk contained entirely in the region.

The closer one is to the boundary, the smaller the radii of such disks. However, for a point on the boundary, every such disk would contain points inside and outside the disk.

Thus, an open set in the complex plane would not contain any of its boundary points.

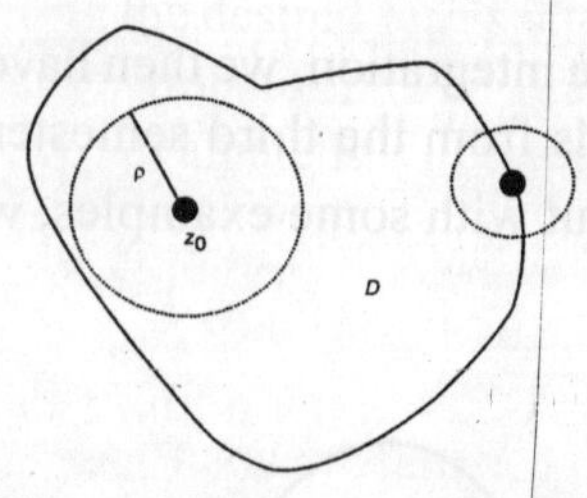

Fig. Locations of Open Disks Inside and on the Boundary of a Region

Definition: D is called a domain if it is both open and connected.

Definition: Let u and v be continuous in domain D, and Γ a piecewise smooth curve in D. Let $(x(t), y(t))$ be a parametrization of Γ for $t_0 \le t \le t_1$. Then

$$\int_\Gamma f(z)\,dz = \int_{t_0}^{t_1} [u(x(t), y(t)) + iv(x(t), y(t))]\left(\frac{dx}{dt} + i\frac{dy}{dt}\right)dt$$

It is easy to see how this definition arises. We see that we can write

$$f(z) = u(x, y) + iv(x, y)$$

and $z = x(t) + iy(t)$. Then,

$$dz = dx + idy = \frac{dx}{dt}dt + i\frac{dy}{dt}dt.$$

Inserting these expressions into the integral leads to the above definition. This definition gives us a prescription for computing path integrals. Let's see how this works with a couple of examples.

Example: $\int_C z^2 dz$, C = the arc of the unit circle in the first quadrant.

We first specify the parametrization. There are two ways we could do this. First, we note that the standard parametrization of the unit circle is

$$(x(\theta), y(\theta)) = (\cos\theta, \sin\theta), 0 \le \theta \le \frac{\pi}{2}.$$

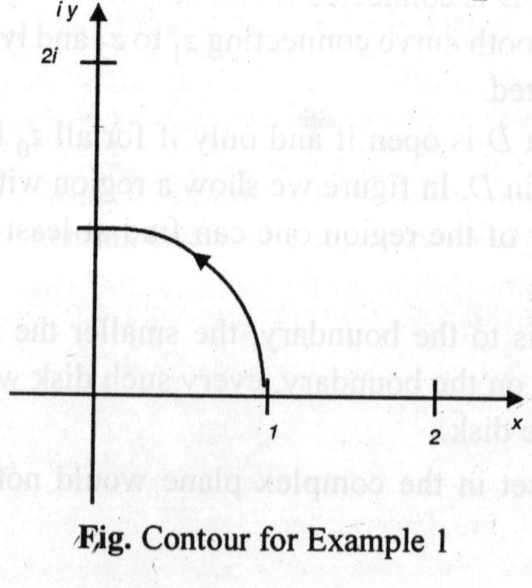

Fig. Contour for Example 1

This is simply the result of using the polar forms

$$x = r\cos\theta$$
$$y = r\sin\theta$$

for $r = 1$ and restricting θ to trace out a quarter of a circle. Then, we have

$$z = \cos\theta + i\sin\theta$$

and

$$dz = (-\sin\theta + i\cos\theta)d\theta.$$

Using this parametrization, the path integral becomes

$$\int_C z^2 dz = \int_0^{\frac{\pi}{2}} (\cos\theta + i\sin\theta)^2(-\sin\theta + i\cos\theta)\,d\theta.$$

We can multiply this out and integrate, having to perform some trigonometric integrations:

$$\int_0^{\frac{\pi}{2}} [\sin^3\theta - 3\cos^2\theta\sin\theta + i(\cos^3\theta - 3\cos\theta\sin^2\theta)\,d\theta.$$

While this is doable, there is a simpler procedure. We first note that $z = e^{i\theta}$ on C. So, $dz = ie^{i\theta}d\theta$. The integration then becomes

$$\int_C z^2 dz = \int_0^{\frac{\pi}{2}} (e^{i\theta})^2 ie^{i\theta}\,d\theta$$

$$= i\int_0^{\frac{\pi}{2}} e^{3i\theta}\,d\theta$$

$$= \left[\frac{ie^{3i\theta}}{3i}\right]_0^{\frac{\pi}{2}} = -\frac{1+i}{3}.$$

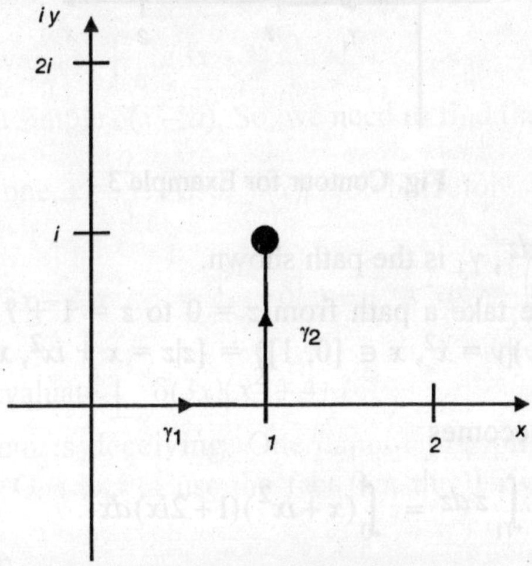

Fig. Contour for Example 2

Example: $\int_\Gamma - z\, dz$, Γ is the path shown.

In this problem we have a path that is a piecewise smooth curve. We can compute the path integral by computing the values along the two segments of the path and adding up the results. Let the two segments be called γ_1 and γ_2.

Over γ_1 we note that $y = 0$. Thus, $z = x$ for $x \in [0, 1]$. It is natural to take x as the parameter. So, dz = dx and we have

$$\int_{\gamma_1} z\,dz = \int_0^1 x\,dx = \frac{1}{2}.$$

For path γ_2 we have that $z = 1 + iy$ for $y \in [0, 1]$. Thus, $dz = i dy$. The integral becomes

$$\int_{\gamma_2} z\,dz = \int_0^1 (1+iy)i\,dy = i - \frac{1}{2}.$$

Combining these results, we have $\int_\Gamma z\, dz = \frac{1}{2} + (i - \frac{1}{2}) = i$.

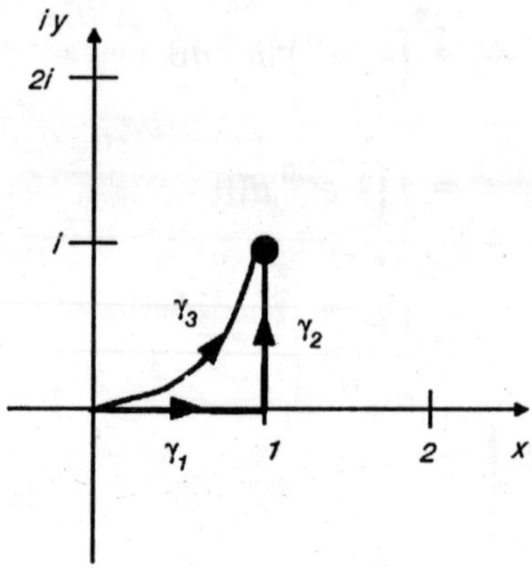

Fig. Contour for Example 3

Example: $\int_{\gamma_3} z\,dz$, γ_3 is the path shown.

In this case we take a path from $z = 0$ to $z = 1 + i$ along a different path. Let $\gamma_3 = f(x, y)|y = x^2, x \in [0, 1]\} = \{z|z = x + ix^2, x \in [0, 1]\}$. Then, $dz = (1 + 2ix)dx$.

The integral becomes

$$\int_{\gamma_1} z\,dz = \int_0^1 (x+ix^2)(1+2ix)\,dx$$

$$= \int_0^1 (x+2ix^2-2x^3)\,dx = i.$$

In the last case we found the same answer as in Example 2. But we should not take this as a general rule for all complex path integrals. In fact, it is not true that integrating over different paths always yields the same results. We will now look into this notion of path independence.

Definition: The integral $\int f(z)\, dz$ is path independent if

$$\int_{\Gamma_1} f(z)dz = \int_{\Gamma_2} f(z)dz$$

for all paths from z_1 to z^2.

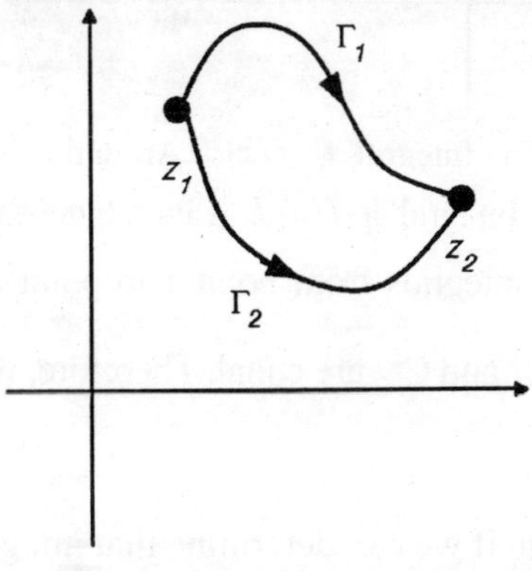

Fig. $\int_{\Gamma 1} f(z)\, dz = \int_{\Gamma 2} f(z)\, dz$ **for all paths from** z_1 **to** z_2 **when the Integral of** f **(z) is Path Independent**

If $\int f(z)\, dz$ is path independent, then the integral of f (z) over all closed loops is zero,

$$\int_{\text{closed loops}} f(z)dz = 0.$$

A common notation for integrating over closed loops is $\oint_C f(z)dz$. But first we have to define what we mean by a closed loop.

Definition: A simple closed contour is a path satisfying

- The end point is the same as the beginning point. (This makes the loop closed.)
- The are no self-intersections. (This makes the loop simple.)

A loop in the shape of a figure eight is closed, but it is not simple.

Now, consider an integral over a closed loop C. We pick two points on the loop breaking it into two contours, C_1 and C_2. Then we make use of the path independence by defining C_2 to be the path along C_2 but in the opposite direction. Then,

$$\oint_C f(z)dz = \int_{C_1} f(z)dz + \int_{C_2} f(z)dz$$

$$= \int_{C_1} f(z)dz - \int_{C_2^-} f(z)dz.$$

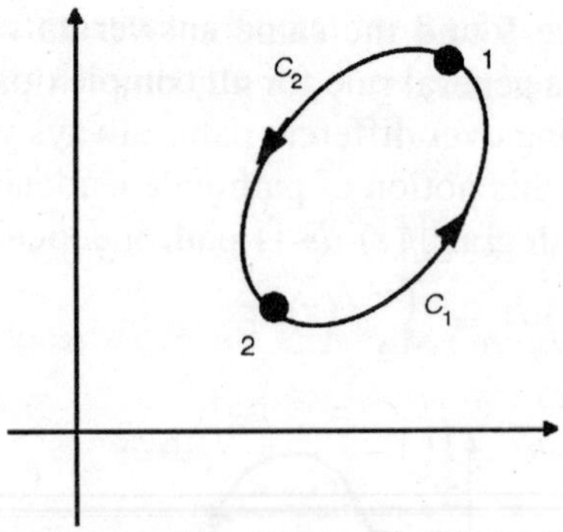

Fig. The Integral $\oint_C f(z)\,dz$ Around C is zero if the Integral $\int_\Gamma f(z)\,dz$ is Path Independent

Assuming that the integrals from point 1 to point 2 are path independent, then the integrals over *C1* and C^-_2 are equal. Therefore, we have $\oint_C f(z)\,dz = 0$.

Cauchy's Theorem

Next we investigate if we can determine that integrals over simple closed contours that vanish without doing all the work of parametrizing the contour. First, we need to establish the direction about which we traverse the contour.

Definition: A curve with parametrization $(x(t), y(t))$ has a normal $(n_x, n_y) = \left(-\frac{dx}{dt}, \frac{dy}{dt}\right)$.

Recall that the normal is a perpendicular to the curve. There are two such perpendiculars. The above normal points outward and the other normal points toward the interior of a closed curve. We will define a positively oriented contour as one that is traversed with the outward normal pointing to the right.

As one follows loops, the interior would then be on the left. As an example, the contours in figure are positively oriented.

We now consider $\oint_C (u+iv)\,dz$ over a simple closed contour. This can be

written in terms of two real integrals in the xy-plane.

$$\oint_C (u+iv)\,dz = \int_C (u+iv)(dx+idy)$$

$$= \int_C udx - vdy + i\int_C v\,dx + u\,dy.$$

These integrals in the plane can be evaluated using Green's Theorem in the Plane. Recall this theorem from your last semester of calculus.

Theorem: Green's Theorem in the Plane.

Let $P(x, y)$ and $Q(x, y)$ be continuously differentiable functions on and inside the simple closed curve C. Denoting the enclosed region S, we have

$$\int_C Pdx + Qdy = \iint_S \left(\frac{\partial Q}{\partial x} - \frac{\partial P}{dy} \right) dxdy.$$

Using Green's Theorem to rewrite the first integral, we have

$$\int_C u\,dx - v\,dy = \iint_S \left(\frac{-\partial v}{\partial x} - \frac{\partial u}{dy} \right) dxdy.$$

If u and v satisfy the Cauchy-Riemann equations,

$$\frac{\partial u}{\partial x} = \frac{\partial v}{dy}, \quad \frac{\partial v}{\partial x} = -\frac{\partial u}{dy},$$

then the integrand in the double integral vanishes. Therefore,

$$\int_C u\,dx - v\,dy = 0$$

In a similar fashion, one can show that

$$\int_C v\,dx - u\,dy = 0.$$

We have thus proven the following theorem.

Theorem If u and v satisfy the Cauchy-Riemann equations inside and on the simple closed contour C, then

$$\oint (u + iv)\,dz = 0.$$

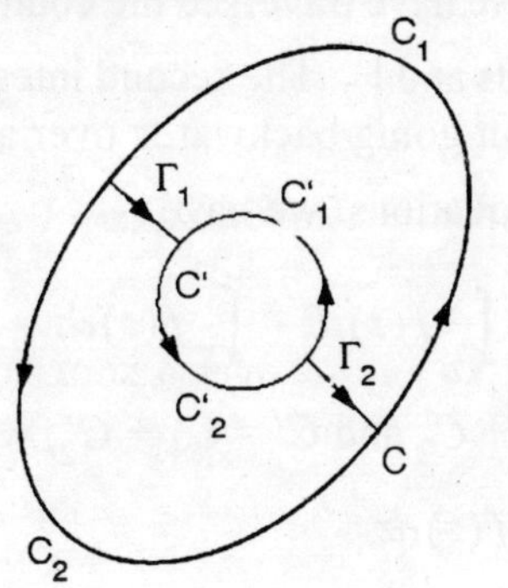

Fig. The Contours Needed to Prove that $\oint_C f(z)dz = \oint_{C'} f(z)dz$. when $f(z)$ is Holomorphic between the Contours C and C'

Corollary: $\oint_C f(z)dz = 0$ when f is differentiable in domain D with C ρ D. Either one of these is referred to as Cauchy's Theorem.

Example: Consider $\oint_{|z-1|=3} z^4\, dz$. Since $f(z) = z^4$ is differentiable inside the circle $|z - 1| = 3$, this integral vanishes.

We can use Cauchy's Theorem to show that we can deform one contour into another, perhaps simpler, contour.

Theorem: If $f(z)$ is holomorphic between two simple closed contours, C and C', then $\oint_C f(z)dz = \oint_{C'} f(z)dz$.

We consider the two curves, now connect the two contours with contours Γ_1 and Γ_2 as shown. This splits C into contours $C1$ and $C2$ and C' into contours C'_1 and C'_2.

$f(z)$ is differentiable inside the newly formed regions between the curves and the boundaries of these regions are now simple closed curves.

Therefore, Cauchy's Theorem tells us that integrals of $f(z)$ over these regions are zero. Noting that integrations over contours in the opposite direction introduce a negative sign, we have from Cauchy's Theorem that

$$\int_{C_1} f(z)dz + \int_{\Gamma_1} f(z)dz - \int_{C'_1} f(z)dz + \int_{\Gamma_2} f(z)dz = 0$$

and

$$\int_{C_2} f(z)dz - \int_{\Gamma_2} f(z)dz \quad \int_{C'_2} f(z)dz \quad \int_{\Gamma_1} f(z)dz = 0$$

In the first integral we have traversed the contours in the following order. C_1, Γ_1, C'_1 backwards and Γ_2. The second integral denotes the integration over the lower region, but going backwards over all contours except for C_2.

Adding these two equations, we have

$$\int_{C_1} f(z)dz + \int_{C_2} f(z)dz - \int_{C'_1} f(z)dz - \int_{C'_2} f(z)dz = 0$$

Noting that $C = C_1 + C_2$ and $C' = C'_1 + C'_2$, we have

$$\oint_C f(z)dz = \oint_{C'} f(z)dz,$$

as was to be proven.

Example: Compute $\oint_R \frac{dz}{z}$ for R the rectangle $[-2, 2]$ X $[-2i, 2i]$.

We can do this integral by computing four separate integrals over the sides of this square in the complex plane. One simply parametrizes each line segment, performs the integration and sums the four results. The last theorem tells us that we could instead integrate over a simpler contour by deforming the square into a circle as long as $f(z) = \frac{1}{z}$ is differentiable in the region bounded by the square and the circle.

The theorem tells us that

$$\oint_R \frac{dz}{z} = \oint_{|z|=1} \frac{dz}{z}$$

The latter integral can be computed using the parametrization $z = e^{i\theta}$ for $\theta \in [0, 2\pi]$. Thus,

$$\oint_{|z|=1} \frac{dz}{z} = \int_0^{2\pi} \frac{ie^{i\theta} d\theta}{e^{i\theta}}$$

$$= i \int_0^{2\pi} d\theta = 2\pi i \, .$$

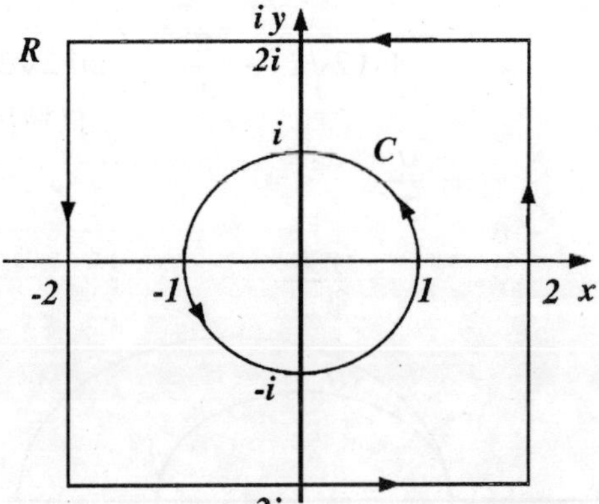

Fig. The Contours used to Compute $\oint_R \frac{dz}{z}$

Note that to compute the integral around R we can deform the contour to the circle C since $f(z)$ is differentiable in the region between the contours.

Therefore, we have found that $\oint_R \frac{dz}{z} = 2\pi i$R by deforming the original simple closed contour.

For fun, let's do this the long way to see how much effort was saved. We will label the contour. The lower segment, γ_1 of the square can be simply parametrized by noting that along this segment $z = x - 2i$ for $x \in [-2, 2]$. Then, we have

$$\int_{\gamma_1} \frac{dz}{z} = \int_{-2}^{2} \frac{dx}{x-2i}$$

$$= \ln|x-2i|_{-2}^{2}$$

$$= \left(\ln\left(2\sqrt{2}\right)+\frac{7\pi i}{4}\right)-\left(\ln\left(2\sqrt{2}\right)+\frac{5\pi i}{4}\right)$$

$$= \frac{\pi i}{2}.$$

We note that the arguments of the logarithms are determined from the angles made by the diagonals provided.

Similarly, the integral along the top segment is computed as

$$\int_{\gamma_3} \frac{dz}{z} = \int_{-2}^{2} \frac{dx}{x+2i}$$

$$= \ln|x-2i|_{-2}^{2}$$

$$= \left(\ln\left(2\sqrt{2}\right)+\frac{7\pi i}{4}\right)-\left(\ln\left(2\sqrt{2}\right)+\frac{5\pi i}{4}\right)$$

$$= \frac{\pi i}{2}.$$

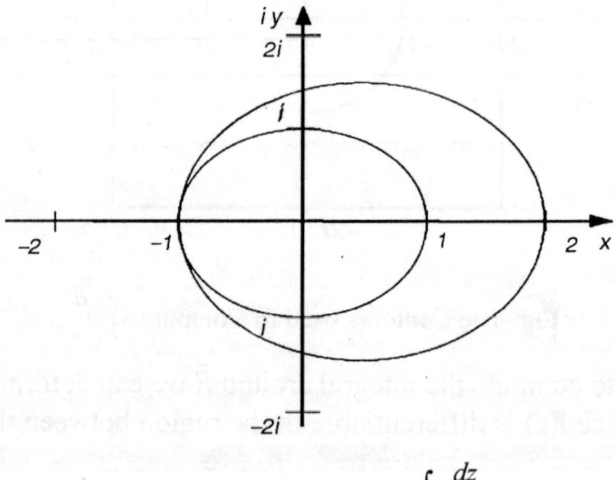

Fig. The Contours Used to Compute $\oint_R \frac{dz}{z}$ Tare Shown

The added diagonals are for the reader to easily see the arguments of the logarithms that appear in during the integration.

The integral over the right side is

$$\int_{\gamma_2} \frac{dz}{z} = \int_{-2}^{2} \frac{idy}{2+iy}$$

$$= \ln|2+iy|_{-2}^{2}$$

$$= \left(\ln(2\sqrt{2})+\frac{\pi i}{4}\right)-\left(\ln(2\sqrt{2})+\frac{\pi i}{4}\right)$$

$$= \frac{\pi i}{2}.$$

Finally, the integral over the left side is

$$\int_{\gamma_4}\frac{dz}{z} = \int_{-2}^{2}\frac{idy}{-2+iy}$$

$$= \ln|2+iy|_{-2}^{2}$$

$$= \left(\ln(2\sqrt{2})+\frac{5\pi i}{4}\right)-\left(\ln(2\sqrt{2})+\frac{3\pi i}{4}\right)$$

$$= \frac{\pi i}{2}.$$

Therefore, we have that

$$\oint_R \frac{dz}{z} = 4\left(\frac{\pi i}{2}\right) = 2\pi i.$$

Note that we had obtained the same result using Cauchy's Theorem. However, it took quite a bit of computation!

The converse of Cauchy's Theorem is not true, namely $\oint_C f(z)dz = 0$ does not imply that *f*(*z*) is differentiable. What we do have though, is

Morera's Theorem.

Theorem: Let *f* be continuous in a domain *D*. Suppose that for every simple closed contour *C* in *D*, $\oint_C f(z)dz = 0$. Then *f* is differentiable in *D*. The proof is a bit more detailed than we need to go into here. However, this result is used in the next section.

Analytic Functions and Cauchy's Integral Formula

In the previous section we saw that Cauchy's Theorem was useful for computing certain integrals without having to parametrize the contours, or deforming certain contours to simpler ones. The integrand needs to possess certain differentiability properties. In this section, we will generalize our integrand slightly so that we can integrate a larger family of complex functions. This will take the form of what is called Cauchy's Integral

Formula, which is different from Cauchy's Theorem. We first need to explore the concept of analytic functions.

Definition: $f(z)$ is analytic in D if for every open disk $|z - z_0| < \rho$ lying in D, $f(z)$ can be represented as a power series in z0. Namely,

$$f(z) = \sum_{n=0}^{\infty} c_n (z - z_0)^n$$

This series converges uniformly and absolutely inside the circle of convergence, $|z - z_0| < R$, with a radius of convergence R.

Since $f(z)$ can be written as a uniformly convergent power series, we can integrate it term by term over any simple closed contour in D containing z_0. In particular, we have to compute integrals like $\oint_C (z - z_0)^n \, dz$. As we will see in the homework exercises, these integrals evaluate to zero. Thus, we can show that.

Theorem: For $f(z)$ analytic in D and any C lying in D, $\oint_C f(z)dz = 0$.

Also, f is a uniformly convergent sum of continuous functions, so $f(z)$ is also continuous.

Thus, by Morera's Theorem, we have that $f(z)$ is differentiable if it is analytic. Often terms like analytic, differentiable and holomorphic are used interchangeably, though there is a subtle distinction due to their definitions.

Let's recall some manipulations from the study of series of real functions. The reader might need to recall how to sum geometric series. A review is given at the end of this section. Essentially, we will make repeated use of the result

$$\sum_{n=0}^{\infty} ar^n = \frac{a}{a - r}, \qquad |r| < 1.$$

Example: $f(z) = \dfrac{1}{1+z}$ for $z_0 = 0$.

This case is simple. $f(z)$ is the sum of a geometric series for $|z| < 1$. We have

$$f(z) = \frac{1}{1+z} = \sum_{n=0}^{\infty} (-z)^n$$

Thus, this series expansion converges inside the unit circle in the complex plane.

Example: $f(z) = \dfrac{1}{1+z}$ for $z_0 = \dfrac{1}{2}$.

We now look into an expansion about a different point. We could compute the expansion coefficients using Taylor' formula for the coefficients. However, we can also make use of the formula for geometric series after rearranging the function. We seek an expansion in powers of $z-\frac{1}{2}$. So, we rewrite the function in a form that has this term. Thus,

$$f(z)=\frac{1}{1+z}=\frac{1}{1+\left(z-\frac{1}{2}+\frac{1}{2}\right)}=\frac{1}{\frac{3}{2}+\left(z-\frac{1}{2}\right)}$$

This is not quite in the form we need. It would be nice if the denominator were of the form of one plus something.

We can get the denominator into such a form by factoring out the $\frac{3}{2}$. Then we would have

$$f(z)=\frac{2}{3}\frac{1}{1+\frac{3}{2}\left(z-\frac{1}{2}\right)}.$$

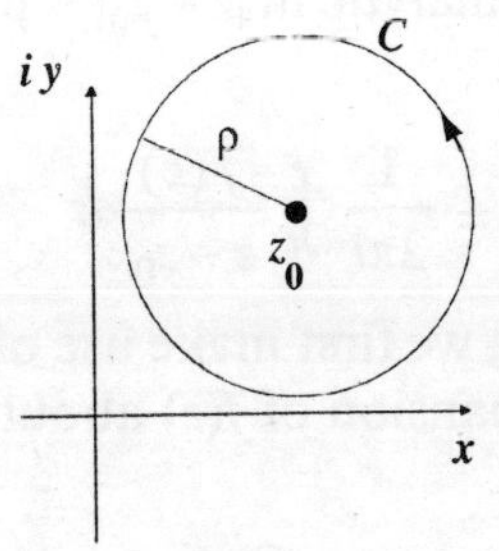

Fig. Regions of Convergence for Expansions of $f(z)=\frac{1}{1+z}$ $z=0$ and $z=\frac{1}{2}$

The second factor now has the form $\frac{1}{1-r}$, which would be the sum of a geometric series with first term $a=1$ and ratio $r=-\frac{2}{3}\left(z-\frac{1}{2}\right)$ provided that $|r|<1$. Therefore, we have found that

$$f(z)=\frac{2}{3}\sum_{n=0}^{\infty}\left[-\frac{2}{3}\left(z-\frac{1}{2}\right)\right]^{n}$$

for

$$\left|-\frac{2}{3}\left(z-\frac{1}{2}\right)\right|<1$$

This convergence interval can be rewritten as

$$\left|z-\frac{1}{2}\right|<\frac{3}{2}.$$

This is a circle centered at $z = \frac{1}{2}$ with radius $\frac{3}{2}$.

We show the regions of convergence for the power series expansions of $f(z) = \frac{1}{1+z}$ about $z = 0$ an d $z = \frac{1}{2}$. We note that the first expansion gives us that $f(z)$ is at least analytic inside the region $|z| < 1$.

The second expansion shows that $f(z)$ is analytic in a region even further outside to the region $\left|z-\frac{1}{2}\right|<\frac{3}{2}$. There are expansions outside of these regions, though some are series expansions involving negative powers of $z - z_0$.

We now present the Cauchy Integral Formula.

Theorem: Let $f(z)$ be analytic in $|z - z_0| < \rho$ and let C be the boundary (circle) of this disk. Then,

$$f(z_0) = \frac{1}{2\pi i}\oint_C \frac{f(z)}{z-z_0}dz$$

In order to prove this, we first make use of the analyticity of $f(z)$. We insert the power series expansion of $f(z)$ about z^0 into the integrand. Then we have

$$\begin{aligned}\frac{f(z)}{z-z_0} &= \frac{2}{z-z_0}\left[\sum_{n=0}^{\infty} c_n(z-z_0)^n\right] \\ &= \frac{2}{z-z_0}\left[c_0 + c_1(z-z_0) + c_2(z-z_0)^2 + \cdots\right] \\ &= \frac{c_0}{z-z_0} + \underbrace{c_0 + c_2(z-z_0) + \cdots}_{\text{analytic function}}\end{aligned}$$

As noted the integrand can be written as

$$\frac{f(z)}{z-z_0} = \frac{c_0}{z-z_0} + h(z),$$

where $h(z)$ is an analytic function, since it is representable as a Taylor series expansion about z_0. We have already shown that analytic functions are differentiable, so by Cauchy's Theorem $\oint_C h(z)\,dz = 0$. Noting also that $c_0 = f(z_0)$ is the first term of a Taylor series expansion about $z = z_0$, we have

$$\oint_C \frac{f(z)}{z-z_0}dz = \oint_C \left[\frac{c_0}{z-z_0} + h(z)\right]dz = f(z_0)\oint_C \frac{1}{z-z_0}dz$$

We need only compute the integral $\oint_C \frac{1}{z-z_0}dz$ to finish the proof of Cauchy's Integral Formula. This is done by parametrizing the circle, $|z - z_0| = \rho$. Let

$$z - z_0 = \rho^{ei\theta}.$$

(Note that this has the right complex modulus since $|e^{i\theta}| = 1$). Then $dz = i\rho e^{i\theta}d\theta$. Using this parametrization, we have

$$\oint_C \frac{f(z)}{z-z_0}dz = \int_0^{2\pi} \frac{i\rho e^{i\theta}\,d\theta}{\rho e^{i\theta}} = i\int_0^{2\pi} d\theta = 2\pi i$$

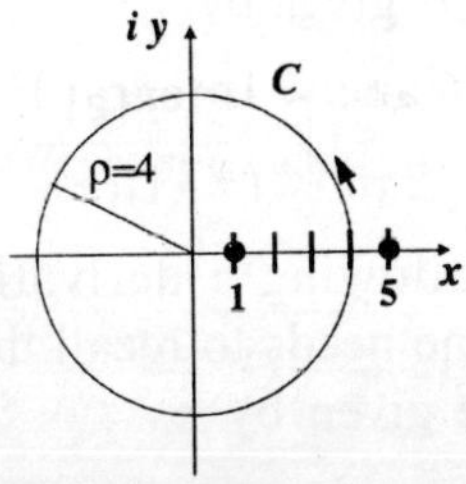

Fig. Circular Contour used in Proving the Cauchy Integral Formula

Therefore,

$$\oint_C \frac{f(z)}{z-z_0}dz = f(z_0)\int_C \frac{1}{z-z_0}dz = 2\pi f(z_0)$$

as was to be shown.

Example: Using the Cauchy Integral Formula

We now compute $\oint_{|z|=4} \frac{\cos z}{z^2 - 6z + 5}dz$.

In order to apply the Cauchy Integral Formula, we need to factor the denominator, $z^2 - 6z + 5 = (z - 1)(z - 5)$. We next locate the zeroes of the denominator. We see the contour and the points $z = 1$ and $z = 5$. The only point inside the region bounded by the contour is $z = 1$. Therefore, we can apply the Cauchy Integral Formula for $f(z) = \frac{\cos z}{z-5}$ to the integral

$$\int_{|z|=4} \frac{\cos z}{(z-1)(z-5)} dz = \int_{|z|=4} \frac{f(z)}{(z-1)} dz = 2\pi i f(1)$$

Therefore, we have

$$\int_{|z|=4} \frac{\cos z}{(z-1)(z-5)} dz = \frac{\pi i \cos(1)}{2}$$

We have shown that $f(z_0)$ has an integral representation for $f(z)$ analytic in $|z - z_0| < \rho$. In fact, all derivatives of an analytic function have an

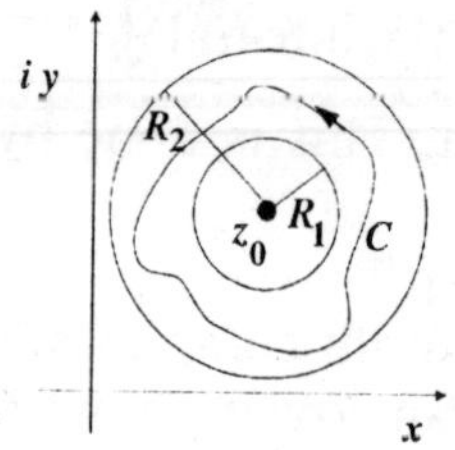

Fig. Circular Contour used in Computing $\int_{|z|=4} \frac{\cos z}{z^2 - 6z + 5} dz$

integral representation. This is given by

$$f^{(n)}(z_0) = \frac{n!}{2\pi i} \oint_C \frac{f(z)}{(z-1)(z-5)}$$

This can be proven following a derivation similar to that for the Cauchy Integral Formula. One needs to recall the coefficients of the Taylor series expansion for $f(z)$ are given by

$$c_n = \frac{f^{(n)}(z_0)}{n!}.$$

We also need the following lemma

Lemma:

$$\oint_C \frac{dz}{(z-z_0)^{n+1}} = \begin{cases} 0, & n \neq \\ 2\pi i, & n = 0 \end{cases}$$

This will be a homework problem. The integrals are similar to the $n = 0$ case above.

Geometric Series

In this section we have made use of geometric series. A geometric series is of the form

$$\sum_{n=0}^{\infty} ar^n = a + ar + ar^2 + ar^2 + \ldots + ar^n + \ldots.$$

Here a is the first term and r is called the ratio. It is called the ratio because the ratio of two consecutive terms in the sum is r. The sum of a geometric series, when it converges, can easily be determined. We consider the nth partial sum.

$$s_n = a + ar + ... + ar^{n-2} + ar^{n-1}.$$

Now, multiply this equation by r.

$$rs_n = ar + ar^2 + ... + ar^{n-1} + ar^n.$$

Subtracting these two equations, while noting the many cancellations, we have

$$(1 - r)s_n = a - ar^n.$$

Thus, the nth partial sums can be written in the compact form

$$s_n = \frac{a(1-r^n)}{1-r}.$$

Recalling that the sum, if it exists, is given by $S = \lim_{n\to\infty} s_n$. Letting n get large in the partial sum we need only evaluate $\lim_{n\to\infty} r^n$. From our special limits we know that this limit is zero for $|r| < 0$. Thus, we have the sum of the geometric series.

$$\sum_{n=0}^{\infty} ar^n = \frac{a}{1-r} \qquad |r| < 1.$$

The reader should verify that the geometric series diverges for all other values of r. Namely, consider what happens for the separate cases $|r| > 1$, $r = 1$ and $r = -1$.

Next, we present a few typical examples of geometric series.

Example: $\sum_{n=0}^{\infty} \frac{1}{2^n}$.

In this case we have that $a = 1$ and $r = 12$. Therefore, this infinite series converges and the sum is

$$S = \frac{1}{1-\frac{1}{2}} = 2.$$

Example: $\sum_{k=2}^{\infty} \frac{4}{3^k}$.

In this example we note that the first term occurs for $k = 2$. So, $a = \frac{4}{9}$. Also, $r = \frac{1}{3}$. So,

$$S = \frac{\frac{4}{9}}{1-\frac{1}{3}} = \frac{2}{3}$$

Example: $\sum_{n=1}^{\infty}\left(\frac{2}{2^n}-\frac{2}{5^n}\right)$.

Finally, in this case we do not have a geometric series, but we do have the difference of two geometric series. Of course, we need to be careful whenever rearranging infinite series. In this case it is allowed. Thus, we have

$$\sum_{n=1}^{\infty}\left(\frac{3}{2^n}-\frac{2}{5^n}\right)=\sum_{n=1}^{\infty}\frac{3}{2^n}-\sum_{n=1}^{\infty}\frac{2}{5^n}.$$

Now we can add both geometric series.

$$\sum_{n=1}^{\infty}\left(\frac{3}{2^n}-\frac{2}{5^n}\right)=\frac{\frac{2}{3}}{1-\frac{1}{2}}-\frac{\frac{2}{5}}{1-\frac{1}{5}}=3-\frac{1}{2}=\frac{5}{2}.$$

COMPLEX SERIES REPRESENTATIONS

Until this point we have only talked about series whose terms have nonnegative powers of $z - z_0$. It is possible to have series representations in which there are negative powers. In the last section we investigated expansions of $f(z)=\frac{1}{1+z}$ about $z = 0$ and $z=\frac{1}{2}$. The regions of convergence for each series. Let us reconsider each of these expansions, but for values of z outside the region of convergence previously found.

Example: $f(z)=\frac{1}{1+z}$ for $|z| > 1$.

As before, we make use of the geometric series. Since $|z| > 1$, we instead rewrite our function as

$$f(z)=\frac{1}{1+z}=\frac{1}{z}\frac{2}{1+\frac{1}{z}}.$$

We now have the function in a form of the sum of a geometric series with first term $a = 1$ and ratio $r=-\frac{1}{z}$. We note that $|z| > 1$ implies that $|r| < 1$. Thus, we have the geometric series

$$f(z)=\frac{1}{z}\sum_{n=0}^{\infty}\left(-\frac{1}{z}\right)^n=\sum_{n=0}^{\infty}(-1)^n z^{-n-1}$$

This can be re-indexed using $j = n + 1$ to obtain

$$f(z) = \sum_{j=1}^{\infty} (-1)^{j-1} z^{-j}.$$

Note that this series, which converges outside the unit circle, $|z| > 1$, has negative powers of z.

Example: $f(z) = \dfrac{1}{1+z}$ for $\left|z - \dfrac{1}{2}\right| > \dfrac{3}{2}$.

In this case, we write as before

$$f(z) = \frac{1}{1+z} = \frac{1}{1+\left(z - \frac{1}{2} + \frac{1}{2}\right)} = \frac{1}{\frac{3}{2} + \left(z - \frac{1}{2}\right)}.$$

Instead of factoring out the $\dfrac{3}{2}$ we factor out the $\left(z - \dfrac{1}{2}\right)$ term. Then, we obtain

$$f(z) = \frac{1}{1+z} = \frac{1}{\left(z - \frac{1}{2}\right)} = \frac{1}{\left(1 + \frac{3}{2} + \left(z - \frac{1}{2}\right)^{-1}\right)}$$

Again, we identify $a = 1$ and $r = -\dfrac{3}{2}\left(z - \dfrac{1}{2}\right)^{-1}$. This leads to the series

$$f(z) = \frac{1}{\left(z - \frac{1}{2}\right)} \sum_{n=0}^{\infty} \left(-\frac{3}{2}\left(z - \frac{1}{2}\right)^{-1}\right)^n.$$

This converges for $\left|z - \dfrac{1}{2}\right| > \dfrac{3}{2}$ and can also be re-indexed to verify that this series involves negative powers of $z - \dfrac{1}{2}$.

This leads to the following theorem.

Theore: Let $f(z)$ be analytic in an annulus, $R_1 < |z - z_0| < R_2$, with C a positively oriented simple closed curve around z0 and inside the annulus. Then,

$$f(z) = \sum_{j=0}^{\infty} a_j (z - z_0)^j + \sum_{j=1}^{\infty} b_j (z - z_0)^{-j},$$

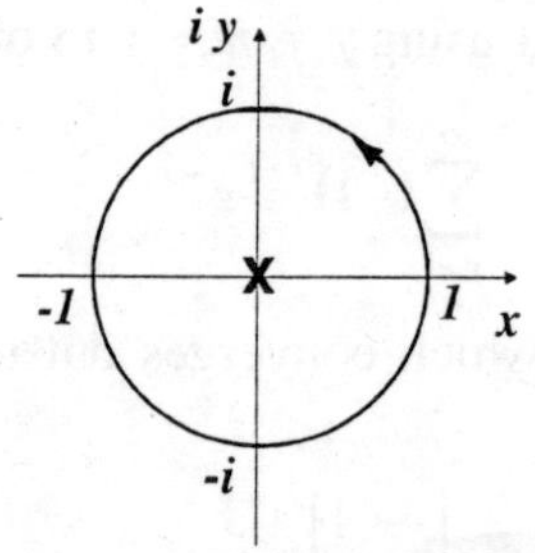

Fig. This figure shows an Annulus, $R_1 < |z - z_0| < R_2$, with C a Positively Oriented Simple closed Curve Around z_0 and Insidc thc Annulus

with

$$a_j = \frac{1}{2\pi i} \oint_C \frac{f(z)}{(z - z_0)^{j+1}} dz,$$

and

$$b_j = \frac{1}{2\pi i} \oint_C \frac{f(z)}{(z - z_0)^{-j+1}} dz .$$

The above series can be written in the more compact form

$$f(z) = \sum_{j=-\infty}^{\infty} c_j (z - z_0)^j .$$

Such a series expansion is called a Laurent expansion.

Example: Expand $f(z) = \dfrac{1}{(1-z)(2+z)}$ in the annulus $1 < |z| < 2$.

Using partial fractions, we can write this as

$$f(z) = \frac{1}{3}\left[\frac{1}{1-z} + \frac{1}{2+z}\right].$$

We can expand the first fraction, $\dfrac{1}{1-z}$ as an analytic function in the region $|z| > 1$ and the second fraction, $\dfrac{1}{2+z}$, as an analytic function in $|z| < 2$. This is done as follows. First, we write

$$\frac{1}{2+z} = \frac{1}{2\left[1 - \left(-\dfrac{z}{2}\right)\right]} = \frac{1}{2}\sum_{n=0}^{\infty}\left(-\frac{z}{2}\right)^n .$$

Then we write

$$\frac{1}{1-z}=-\frac{1}{z\left[1-\left(-\frac{1}{z}\right)\right]}=-\frac{1}{z}\sum_{n=0}^{\infty}\frac{1}{z^n}.$$

Therefore, in the common region, $1<|z|<2$, we have that

$$\frac{1}{(1-z)(2+z)}=\frac{1}{3}\left[\frac{1}{2}\sum_{n=0}^{\infty}\left(-\frac{z}{2}\right)^n-\sum_{n=0}^{\infty}\frac{1}{z^{n+1}}\right]$$

$$=\sum_{n=0}^{\infty}\frac{(-1)^n}{6(2^n)}z^n+\sum_{n=0}^{\infty}\frac{(-1)}{3}z^{-n}.$$

SINGULARITIES AND THE RESIDUE THEOREM

In the last section we found that we could integrate functions satisfying some analyticity properties along contours without using detailed parametrizations around the contours. We can deform contours if the function is analytic in the region between the original and new contour. In this section we will extend our tools for performing contour integrals.

The integrand in the Cauchy Integral Formula was of the form $g(z)=\frac{f(x)}{z-z_0}$ $f(z)$ where $f(z)$ is well behaved at z_0. The point $z=z_0$ is called a singularity of $g(z)$, as $g(z)$ is not defined there. As we saw from the proof of the Cauchy Integral Formula, $g(z)$ has a Laurent series expansion about $z=z_0$,

$$g(z)=\frac{f(z_0)}{z-z_0}+f'(z_0)+\frac{1}{2}f''(z_0)(z-z_0)^2+\cdots$$

We now define singularities and then classify isolated singularities.

Definition: A singularity of $f(z)$ is a point at which $f(z)$ fails to be analytic.

Typically these are isolated singularities. In order to classify the singularities of $f(z)$, we look at the principal part of the Laurent series. $\sum_{j=1}^{\infty}b_j(z-z_0)^{-j}$. This is the part of the Laurent series containing only negative powers of $z-z_0$.

- If $f(z)$ is bounded near z_0, then z_0 is a removable singularity.
- If there are a finite number of terms in the principal part, then one has a pole of order n if there is a term $(z-z_0)^{-n}$ and no terms of the form $(z-z_0)^{-j}$ for $j>n$.

- If there are an infinite number of terms in the principal part, then one has an essential singularity.

Example: Removable $f(z) = \dfrac{\sin z}{z}$.

At first it looks like there is a possible singularity at $z = 0$. However, we know from the first semester of calculus that $\lim_{z\to 0} \dfrac{\sin z}{z} = 1$. Furthermore, we can expand $\sin z$ about $z = 0$ and see that

$$\frac{\sin z}{z} = \frac{1}{z}\left(z - \frac{z^3}{6!} + \cdots\right) = 1 - \frac{z^2}{6!} + \cdots$$

Thus, there are only nonnegative powers in the series expansion. So, this is an example of a removable singularity.

Example: Poles $f(z) = \dfrac{e^z}{(z-1)^n}$.

For $n = 1$ we have $f(z) = \dfrac{e^z}{z-1}$.

For $n = 1$ we have $f(z) = \dfrac{e^z}{z-1}$. This function has a singularity at $z = 1$ called a simple pole. The series expansion is found by expanding ez about $z = 1$.

$$f(z) = \frac{e}{z-1}e^{z-1} = \frac{e}{z-1} + e + \frac{e}{2!}(z-1) + \cdots$$

Note that the principal part of the Laurent series expansion about $z = 1$ has only one term.

For $n = 2$ we have $f(z) = \dfrac{e^2}{(z-1)^2}$. The series expansion is found again by expanding *ez* about $z = 1$.

$$f(z) = \frac{e}{(z-1)^2}e^{z-1} = \frac{e}{(z-1)^2} + \frac{e}{z-1} + \frac{e}{2!} + \frac{e}{31}(z-1) + \cdots$$

Note that the principal part of the Laurent series has two terms involving $(z-1)^{-2}$ and $(z-1)^{-1}$. This is a pole of order 2.

Example: Essential $f(z) = \dfrac{1}{e^z}$.

In this case we have the series expansion about $z = 0$ given by

$$f(z) = e^{\frac{1}{z}} = \sum_{n=0}^{\infty} \frac{\left(\frac{1}{z}\right)^n}{n!} = \sum_{n=0}^{\infty} \frac{1}{n!} z^{-n}$$

We see that there are an infinite number of terms in the principal part of the Laurent series.

So, this function has an essential singularity at $z = 0$. In the above examples we have seen poles of order one (a simple pole) and two In general, we can define poles of order k.

Definition: $f(z)$ has a pole of order k at z_0 if and only if $(z - z_0)^k f(z)$ has a removable singularity at z_0, but $(z - z_0)^{k-1} f(z)$ for $k > 0$ does not.

Let $\phi(z) = (z - z_0)^k f(z)$ be analytic.

Then it has a Taylor series expansion about z_0. As we had seen in equation in the last section, we can write the integral representation of the $(k - 1)$st derivative of an analytic function as

$$\phi^{(k-1)}(z_0) = \frac{(k-1)!}{2\pi i} \oint_C \frac{\phi(z)}{(z-z_0)^k} dz .$$

Inserting the definition of $\phi(z)$ we then have

$$\phi^{(k-1)}(z_0) = \frac{(k-1)!}{2\pi i} \oint_C f(z) dz$$

Dividing out the factorial factor and evaluating the A(z) derivative, we have that

$$\frac{1}{2\pi i} \oint_C f(z) dz = \frac{1}{2\pi i} \oint_C f(z) dz$$

$$= \frac{1}{(k-1)!} \frac{d^{k-1}}{dz^{k-1}} \left[(z-z_0)^k f(z) \right]_{z=z_0} .$$

We note that from the integral representation of the coefficients for a Laurent series, this gives c_{-1}, or b_1.

This particular coefficient plays a role in helping to compute contour integrals surrounding poles. It is called the residue of $f(z)$ at $z = z_0$. Thus, for a pole of order k we define the residue

$$\text{Res}[f(z), z_0] = \lim_{z \to z_0} \frac{1}{(k-1)!} \frac{d^{k-1}}{dz^{k-1}} \left[(z-z_0)^k f(z) \right]$$

Again, the residue is the coefficient of the $(z - z_0)^{-1}$ term.

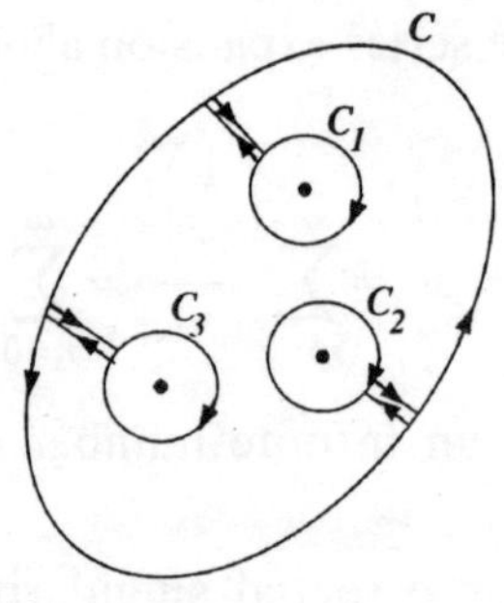

Fig. Contour for Computing $\oint_{|z|=1} \frac{dz}{\sin z}$

Referring to the last derivation, we have shown that if $f(z)$ has one pole, z_0, of order k inside a simple closed contour C, then

$$\oint_C f(z)\,dz = 2\pi i \operatorname{Res}[f(z), z_0]$$

Example: $\oint_{|z|=1} \frac{dz}{\sin z}$.

We begin by looking for the singularities of the integrand, which is when $\sin z = 0$. Thus, $z = 0, \pm\pi, \pm 2\pi, \ldots$ are the singularities. However, only $z = 0$ lies inside the contour. We note further that $z = 0$ is a simple pole, since

$$\lim_{z\to 0}(z-0)\frac{1}{\sin z} = 1.$$

Therefore, the residue is one and we have

$$\oint_{|z|=1} \frac{dz}{\sin z} = 2\pi i.$$

In general, we could have several poles of different orders. For example, we will be computing

$$\oint_{|z|=1} \frac{dz}{z^2-1}.$$

The integrand has singularities at $z^2 - 1 = 0$, or $z = \pm 1$, both poles are inside the contour. One could do a partial fraction decomposition and have two integrals with one pole each. However, in cases in which we have many poles, we can use the following theorem, known as the Residue Theorem.

Theorem: Let $f(z)$ be a function which has poles z_j, $j = 1, \ldots, N$ inside a simple closed contour C and no other singularities in this region. Then,

$$\oint_C f(z)\,dz = 2\pi i \sum_{j=1}^{N} \operatorname{Res}\left[f(z); z_j\right],$$

where the residues are computed using equation.

The proof of this theorem is based upon the contours. One constructs a new contour C' by encircling each pole. Then one connects a path from C to each circle. In the figure two paths are shown only to indicate the direction followed on the cut.

The new contour is then obtained by following C and crossing each cut as it is encountered.

Then one goes around a circle in the negative sense and returns along the cut to proceed around C. The sum of the contributions to the contour integration involve two integrals for each cut, which will cancel due to the opposing directions. Thus, we are left with

$$\oint_{C'} f(z)dz = \oint_{C} f(z)dz - \oint_{C_1} f(z)dz - \oint_{C_2} f(z)dz - \oint_{C_3} f(z)dz = 0$$

Of course, the sum is zero because $f(z)$ is analytic in the enclosed region, since all singularities have be cut out. Solving for $\oint_C f(z)dz$, one has that this integral is the sum of the integrals around the separate poles, which can be evaluated with single residue computations. Thus, the result is that $\oint_C f(z)dz$ is $2\pi i$ times the sum of the residues of $f(z)$ at each pole.

Example: $\oint_{|z|=2} \frac{dz}{z^2-1}$.

We first note that there are two poles in this integral since

$$\frac{1}{z^2-1} = \frac{1}{(z-1)(z+1)}.$$

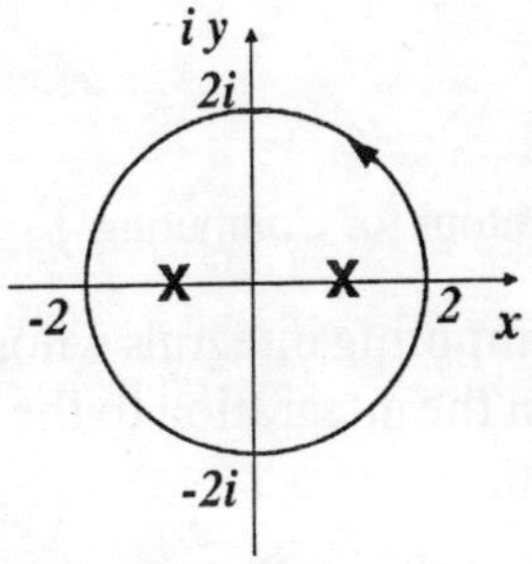

Fig. A Depiction of how one cuts out Poles to prove that the Integral Around C is the sum of the Integrals Around Circles with the poles at the Centre of Each

We plot the contour and the two poles, each denoted by an "x". Since both poles are inside the contour, we need to compute the residues for each one. They are both simple poles, so we have

$$\text{Res}\left[\frac{1}{z^2-1}; z=1\right] = \lim_{z\to 1}(z-1)\frac{1}{z^2-1}$$

$$= \lim_{z\to 1}\frac{1}{z+1} = \frac{1}{2},$$

and

$$\text{Res}\left[\frac{1}{z^2-1}; z=1\right] = \lim_{z\to 1}(z+1)\frac{1}{z^2-1}$$

$$= \lim_{z\to -1}\frac{1}{z-1} = -\frac{1}{2}.$$

Example: $\int_0^{2\pi}\frac{d\theta}{2+\cos\theta}$.

Here we have a real integral in which there are no signs of complex functions. In fact, we could apply methods from our calculus class to do this integral, attempting to write $1 + \cos\theta = 2\cos^2\frac{\theta}{2}$. We do not, however, get very far.

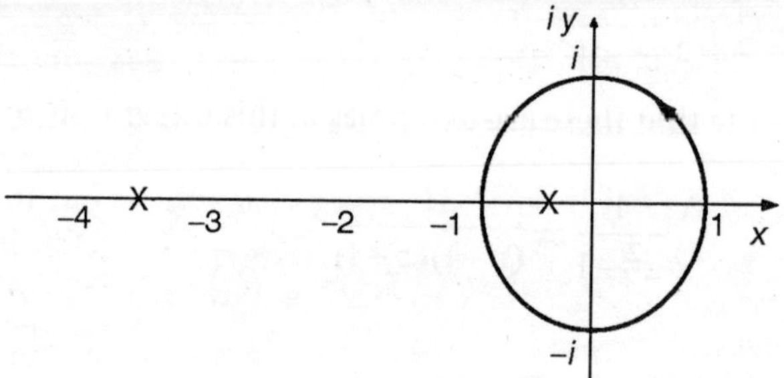

Fig. Contour for Computing $\oint_{|z|=2}\frac{dz}{z^2-1}$

One trick, useful in computing integrals whose integrand is in the form $f(\cos\theta, \sin\theta)$, is to transform the integration to the complex plane through the transformation $z = e^{i\theta}$. Then,

$$\cos\theta = \frac{e^{i\theta}+e^{-i\theta}}{2} = \frac{1}{2}\left(z+\frac{1}{z}\right),$$

$$\sin\theta = \frac{e^{i\theta}-e^{-i\theta}}{2} = \frac{i}{2}\left(z-\frac{1}{z}\right).$$

Under this transformation, $z = e^{i\theta}$, the integration now takes place around the unit circle in the complex plane. Noting that $dz = ie^{i\theta}\, d\theta = iz\, d\theta$, we have

$$\int_0^{2\pi} \frac{d\theta}{2+\cos\theta} = \oint_{|z|=1} \frac{\frac{dz}{iz}}{2+\frac{1}{2}\left(z+\frac{1}{z}\right)}$$

$$= -i\oint_{|z|=1} \frac{dz}{2z+\frac{1}{2}\left(z^2+1\right)}$$

$$= -2i\oint_{|z|=1} \frac{dz}{z^2+4z+1}.$$

We can apply the Residue Theorem to the resulting integral. The singularities occur for $z^2 + 4z + 1 = 0$. Using the quadratic formula, we have the roots $z = -2 \pm \sqrt{3}$. Only $z = -2 + \sqrt{3}$ lies inside the integration contour. We will therefore need the residue of $f(z) = \frac{-2i}{z^2+4z+1}$ simple at this pole:

$$\text{Res}[f(z), z = -2+\sqrt{3}\,] = \lim_{z\to -2+\sqrt{3}} \left(z-(-2+\sqrt{3})\right)\frac{-2i}{z^2+4z+1}$$

$$= -2i \lim_{z\to -2+\sqrt{3}} \frac{z-(-2+\sqrt{3})}{(z-(-2+\sqrt{3}))(z-(-2-\sqrt{3}))}$$

$$= -2i \lim_{z\to -2+\sqrt{3}} \frac{1}{(z-(-2-\sqrt{3}))}$$

$$= \frac{-2i}{-2+\sqrt{3}-(-2-\sqrt{3})}$$

$$= \frac{-i}{\sqrt{3}}$$

$$= \frac{-i\sqrt{3}}{3}$$

Therefore, we have

$$\int_0^{2\pi} \frac{d\theta}{2+\cos\theta} = -2i\oint_{|z|=1} \frac{dz}{z^2+4z+1} = 2\pi i\left(\frac{-i\sqrt{3}}{3}\right) = \frac{2\pi\sqrt{3}}{3}.$$

Example: $\oint_{|z|=3} \frac{z^2+1}{(z-1)^2(z+2)}dz$.

In this example there are two poles $z = 1, -2$ inside the contour. $z = 1$ is a second order pole and $z = 2$ is a simple pole. Therefore, we need the residues at each pole for $f(z) = \frac{z^2+1}{(z-1)^2(z+2)}$:

$$\text{Res}[f(z); z = 1] = \lim_{z\to 1}\frac{1}{1!}\frac{d}{dz}\left[(z-1)^2\frac{z^2+1}{(z-1)^2(z+2)}\right]$$

$$= \lim_{z\to 1}\left(\frac{z^2+4z-1}{(z-2)^2}\right)$$

$$= \frac{4}{9}.$$

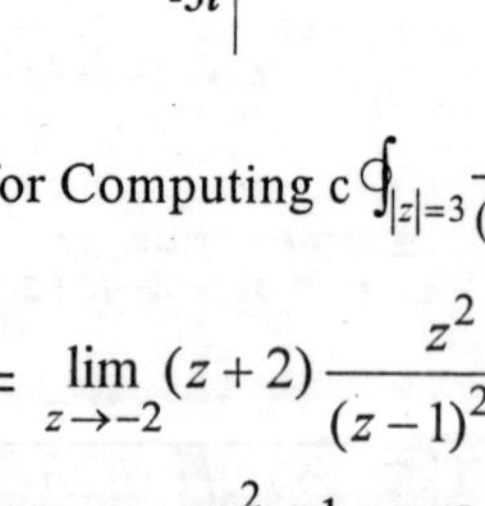

Fig. Contour for Computing c $\oint_{|z|=3}\frac{z^2+1}{(z-1)^2(z+2)}dz$

$$\text{Res}[f(z), z = -2] = \lim_{z\to -2}(z+2)\frac{z^2+1}{(z-1)^2(z+2)}$$

$$= \lim_{z\to -2}\frac{z^2+1}{(z-2)^2} = \frac{5}{9}.$$

The evaluation of the integral is now $2\pi i$ times the sum of the residues.

$$\oint_{|z|=3}\frac{z^2+1}{(z-1)^2(z+2)}dz = 2\pi i\left(\frac{4}{9}+\frac{6}{9}\right) = 2\pi i.$$

COMPUTING REAL INTEGRALS

As our final application of complex integration techniques, we will turn to the evaluation of infinite integrals of the form $\int_{-\infty}^{\infty} f(x)dx$. These types of integrals will appear later in the text and will help to tie in what seems to be a digression in our study of mathematical physics. In this section we will see that such integrals may be computed by extending the integration to a contour in the complex plane.

Recall that such integrals are improper integrals and you had seen them in your calculus classes. The way that one determines if such integrals exist, or converge, is to compute the integral using a limit:

$$\int_{-\infty}^{\infty} f(x)dx = \lim_{R\to\infty} \int_{-R}^{R} f(x)dx$$

For example,

$$\int_{-\infty}^{\infty} \frac{1}{x^2}dx = \lim_{R\to\infty} \int_{-R}^{R} \frac{1}{x^2}dx = \lim_{R\to\infty} \left(-\frac{2}{R}\right) = 0.$$

Similarly,

$$\int_{-\infty}^{\infty} x\,dx = \lim_{R\to\infty} \int_{-R}^{R} x\,dx = \lim_{R\to\infty} \left(\frac{R^2}{2} - \frac{(-R^2)}{2}\right) = 0.$$

However, the integrals $\int_{0}^{\infty} x\,dx$ and $\int_{-\infty}^{0} x\,dx$ do not exist. Note that

$$\int_{0}^{\infty} x\,dx = \lim_{R\to\infty} \int_{0}^{R} x\,dx = \lim_{R\to\infty} \left(\frac{R^2}{2}\right) = \infty.$$

Therefore,

$$\int_{-\infty}^{\infty} f(x)dx = \int_{-\infty}^{0} f(x)dx + \int_{0}^{\infty} f(x)dx.$$

does not exist while $\lim_{R\to\infty} \int_{-R}^{R} f(x)dx$ does exist. We will be interested in computing the latter type of integral. Such an integral is called the Cauchy Principal Value Integral and is denoted with either a P or PV prefix:

$$P\int_{-\infty}^{\infty} f(x)dx = \lim_{R\to\infty} \int_{-R}^{R} f(x)dx.$$

In our discussions we will be computing integrals over the real line in the Cauchy principal value sense.

We now proceed to the evaluation of such principal value integrals using complex integration methods. We want to evaluate the integral

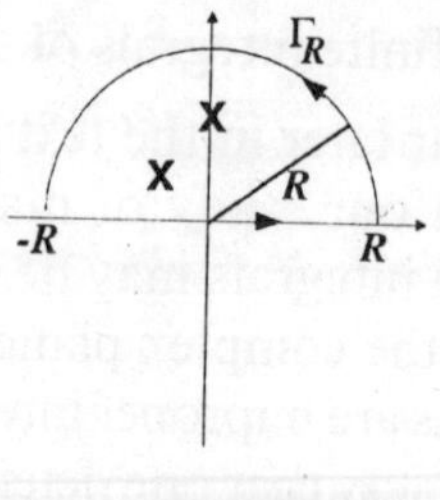

Fig. Contours for Computing $P\int_{-\infty}^{\infty} f(x)\,dx$

$\int_{-\infty}^{\infty} f(x)\,dx$. We will extend this into an integration in the complex plane.

We extend f (x) to $f(z)$ and assume that $f(z)$ is analytic in the upper half plane (Im(z) > 0). We then consider the integral $\int_{-R}^{R} f(x)\,dx$ dx as an integral over the interval ($-R$, R).

We view this interval as a piece of a contour CR obtained by completing the contour with a semicircle $-R$ of radius R extending into the upper half plane.

Note, a similar construction is sometimes needed to extend the integration into the lower half plane (Im(z) < 0) when $f(z)$ is analytic there. The integral around the entire contour CR can be computed using the Residue Theorem and is related to integrations over the pieces of the contour by

$$\oint_{C_R} f(z)\,dz = \int_{\Gamma_R} f(z)\,dz + \int_{-R}^{R} f(z)\,dz.$$

Taking the limit $R \to \infty$ and noting that the integral over ($-R$, R) is the desired integral, we have

$$P\int_{-\infty}^{\infty} f(x)\,dx = \oint_C f(z)\,dz - \int_{\Gamma_R} f(z)\,dz,$$

where we have identified C as the limiting contour as R gets large,

$$\oint_C f(z)\,dz = \lim_{R\to\infty} \oint_{C_R} f(z)\,dz.$$

Now the key to carrying out the integration is that the integral over Γ_R vanishes in the limit. This is true if $R|f(z)| \to 0$ along Γ_R as $R \to 1$.

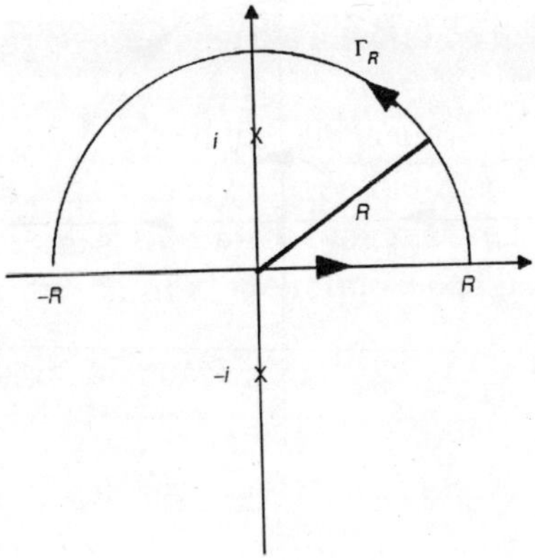

Fig. Contour for Computing $\int_{-\infty}^{\infty} \frac{dx}{1+x^2}$

This can be seen by the following argument. We can parametrize the contour Γ_R using $z = Re^{i\theta}$. We assume that the function is bounded by some function of R. Denote this functon by $M(R)$. Thus, for $|f(z)| < M(R)$, we have

$$\left| \int_{\Gamma_R} f(z)\,dz \right| = \left| \int_0^{2\pi} f(Re^{i\theta}) Re^{i\theta} d\theta \right|$$

$$\leq \int_0^{2\pi} \left| f(Re^{i\theta}) \right| d\theta$$

$$< RM(R) \int_0^{2\pi} d\theta$$

$$= 2\pi RM(R).$$

So, if $\lim_{R\to\infty} RM(R) = 0$, then $\lim_{R\to\infty} \int_{\Gamma_R} f(z)\,dz = 0$.

We show how this applies some examples.

Example: Evaluate $\int_{-\infty}^{\infty} \frac{dx}{1+x^2}$.

We already know how to do this integral from our calculus classes. We have that

$$\int_{-\infty}^{\infty} \frac{dx}{1+x^2} = \lim_{R\to\infty} \left(2\tan^{-1} R\right) = 2\left(\frac{\pi}{2}\right) = \pi.$$

We will apply the methods of this section and confirm this result. The poles of the integrand are at $z = \pm i$.

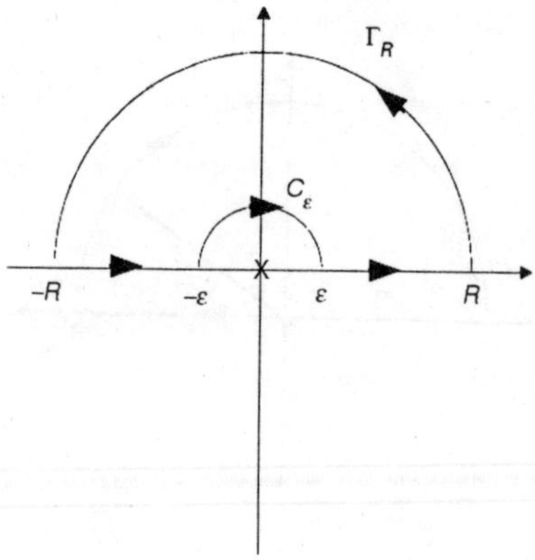

Fig. Contour for Computing $P\int_{-\infty}^{\infty}\frac{\sin x}{x}dx$

We first note that $f(z)=\frac{1}{1+z^2}$ goes to zero fast enough on Γ_R as R gets large.

$$R|f(z)|=\frac{R}{\left|1+R^2e^{2i\theta}\right|}=\frac{R}{\sqrt{1+R^2\cos\theta+R^4}}$$

Thus, as $R\to\infty$ 1, $Rf(z)\to 0$. So,

$$\int_{-\infty}^{\infty}\frac{dx}{1+x^2}=\oint_C\frac{dz}{1+z^2}.$$

We need only compute the residue at the enclosed pole, z = i.

$$\text{Res}[f(z),\ z=i]=\lim_{z\to i}(z-i)\frac{1}{1+z^2}=\lim_{z\to i}\frac{1}{z+i}=\frac{1}{2i}$$

Then, using the Residue Theorem, we have

$$\int_{-\infty}^{\infty}\frac{dx}{1+x^2}=2\pi i\left(\frac{1}{2i}\right)=\pi$$

Example: $P\int_{-\infty}^{\infty}\frac{\sin x}{x}dx$.

There are several new techniques that have to be introduced in order to carry out this integration.

We need to handle the pole at z = 0 in a special way and we need something called Jordan's Lemma to guarantee that the contour on the contour Γ_R vanishes, since the integrand does not satisfy our previous condition on the bound.

For this example the integral is unbounded at $z = 0$. Constructing the contours as before, we are faced for the first time with a pole lying on the contour. We cannot ignore this fact.

We can proceed with our computation by carefully going around this pole with a small semicircle of radius $\in$. Then our principal value integral computation becomes:

$$P\int_{-\infty}^{\infty}\frac{\sin x}{x}dx = \lim_{\in\to 0}\left(\int_{-\infty}^{-\in}\frac{\sin x}{x}dx + \int_{\in}^{\infty}\frac{\sin x}{x}dx\right).$$

We will also need to rewrite the sine function in terms of exponentials in this integral.

$$P\int_{-\infty}^{\infty}\frac{\sin x}{x}dx = \frac{1}{2i}\left(P\int_{-\infty}^{\infty}\frac{e^{ix}}{x}dx - \int_{-\infty}^{\infty}\frac{e^{-ix}}{x}dx\right).$$

We now employ Jordan's Lemma. If $f(z)$ converges uniformly to zero as $z \to \infty$, then

$$\lim_{R\to\infty}\int_{C_R} f(z)e^{ikz}dz = 0$$

where $k > 0$ and C_R is the upper half of the circle $|z| = R$. A similar result applies for $k < 0$, but one closes the contour in the lower half plane.

We now put these ideas together to compute the given integral. According to Jordan's lemma, we will need to compute the above exponential integrals using two different contours. We first consider $P\int_{-\infty}^{\infty}\frac{e^{ix}}{z}dz$. Then we have

$$\int_{C_R}\frac{e^{iz}}{z}dz = \int_{\Gamma_R}\frac{e^{iz}}{z}dz + \int_{-R}^{-\in}\frac{e^{iz}}{z}dz + \int_{C_\in}\frac{e^{iz}}{z}dz + \int_{\in}^{R}\frac{e^{iz}}{z}dz.$$

The integral $\oint_{C_R}\frac{e^{iz}}{z}dz$ vanishes, since there are no poles enclosed in the contour!

The integral over Γ_R will vanish as R gets large, according to Jordan's Lemma. The sum of the second and fourth integrals is the integral we seek as $\in \to 0$ and $R \to \infty$.

The remaining integral around the small circle has to be done separately. We have

$$\oint_{C_R}\frac{e^{iz}}{z}dz = \int_{\Gamma_R}\frac{e^{iz}}{z}dz + \int_{-R}^{-\in}\frac{e^{iz}}{z}dz + \int_{C_\in}\frac{e^{iz}}{z}dz + \int_{\in}^{R}\frac{e^{iz}}{z}dz$$

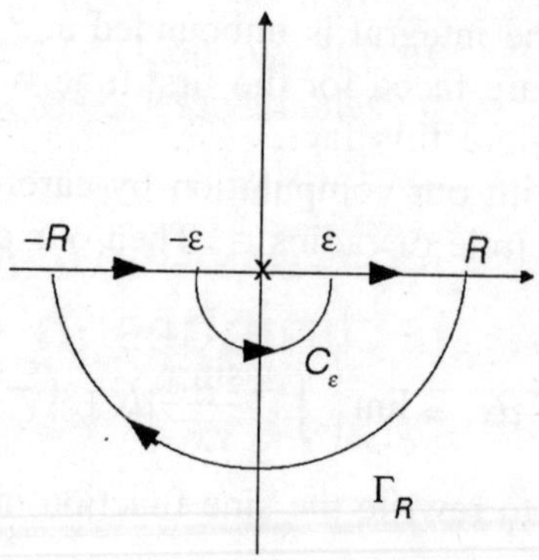

Fig. Contour in the Lower half Plane for Computing $P\int_{-\infty}^{\infty}\frac{e^{-ix}}{x}dx$

Taking the limit as $\in$ goes to zero, the integrand goes to i and we have

$$\int_{C_\in}\frac{e^{iz}}{z}dz = -\pi i.$$

So far, we have that

We can compute $P\int_{-\infty}^{\infty}\frac{e^{-ix}}{x}dx$ in a similar manner, being careful with the sign changes due to the orientations of the contours. In this case, we find the same value

$$P\int_{-\infty}^{\infty}\frac{e^{-ix}}{x}dx = \pi i.$$

Finally, we can compute the original integral as

$$P\int_{-\infty}^{\infty}\frac{\sin x}{x}dx = \frac{1}{2i}\left(P\int_{-\infty}^{\infty}\frac{e^{ix}}{x}dx - P\int_{-\infty}^{\infty}\frac{e^{-ix}}{x}dx\right)$$

$$= \frac{1}{2i}(\pi i + \pi i)$$

$$= \pi.$$

Chapter 6

Transform Techniques in Physics

INTRODUCTION

Some of the most powerful tools for solving problems in physics are transform methods. The idea is that one can transform the problem at hand to a new problem in a different space, hoping that the problem in the new space is easier to solve. Such transforms appear in many forms.

The Linearized KdV Equation

As a relatively simple example, we consider the linearized Kortweg-deVries (KdV) equation.

$$u_t + cu_x + \beta u_{xxx} = 0, -\infty < x < \infty.$$

This equation governs the propagation of some small amplitude water waves. Its nonlinear counterpart has been at the centre of attention in the last 40 years as a generic nonlinear wave equation.

We seek solutions that oscillate in space. So, we assume a solution of the form

$$u(x,\ t) = A(t)e^{ikx}$$

In that case, we found plane wave solutions of the form $e^{ik(x - ct)}$, which we could write as $e^{i(kx-\omega t)}$ by defining $\omega = kc$. We further note that one often seeks complex solutions of this form and then takes the real part in order to obtain a real physical solutions.

Inserting the guess into the linearized KdV equation, we find that

$$\frac{dA}{dt} + i(ck - \beta k^3)A = 0.$$

Thus, we have converted our problem of seeking a solution of the partial differential equation into seeking a solution to an ordinary differential equation. This new problem is easier to solve. In fact, we have

$$A(t) = A(0)e^{-i(ck-k3)t}.$$

Therefore, the solution of the partial differential equation is

$$u(x,\ t) = A(0)e^{ik(x-(c-\beta k^2)t)}.$$

We note that this takes the form $e^{i(kx-\omega t)}$, where

$$\omega = ck - \beta k^3.$$

In general, the equation $\omega = \omega(k)$ gives the angular frequency as a function of the wave number, k, and is called a dispersion relation. For $\beta = 0$, we see that c in nothing but the wave speed. For $\beta \neq 0$, the wave speed is given as

$$v = \frac{\omega}{k} = c - \beta k^2 .$$

This suggests that waves with different wave numbers will travel at different speeds. Recalling that wave numbers are related to wavelengths, $k = \frac{2\pi}{\lambda}$, this means that waves with different wavelengths will travel at different speeds. So, a linear combination of such solutions, will not maintain its shape. It is said to disperse, as the waves of differing wavelengths tend to part company.

For a general initial condition, we need to write the solutions to the linearized KdV as a superposition of waves. We can do this as the equation is linear. This should remind you of what we had done when using separation of variables.

We first sought product solutions and then took a linear combination of the product solutions to give the general solution.

In this case, we need to sum over all wave numbers. The wave numbers are not restricted to discrete values, so we have a continuous range of values. Thus, summing over k means that we have to integrate over the wave numbers. Thus, we have the general solution

$$u(x, t) = \int_{-\infty}^{\infty} A(k,0) e^{ik(x-(c-\beta k^2)t)} dk .$$

Note that we have now made A a function of k. This is similar to introducing the A_n's and B_n's in the series solution for waves on a string.

How do we determine the $A(k, 0)$'s? We introduce an initial condition. Let $u(x, 0) = f(x)$. Then, we have

$$f(x) = u(x, 0) = \int_{-\infty}^{\infty} A(k,0) e^{ikx}\, dk$$

Thus, given $f(x)$, we seek $A(k, 0)$. This involves what is called the Fourier transform of $f(x)$. This is just one of the so-called integral transforms that we will consider in this section.

The Free Particle Wave Function

A more familiar example in physics comes from quantum mechanics. The Schroodiger equation gives the wave function $\Psi(x, t)$ for a particle under the influence of forces, represented through the corresponding potential function V. The one dimensional time dependent Schrodinger equation is given by

$$i\hbar\Psi t = \frac{\hbar^2}{2m}\Psi_{xx} + V\Psi.$$

We consider the case of a free particle in which there are no forces, $V = 0$.

Then we have

$$i\hbar\Psi t = \frac{\hbar^2}{2m}\Psi_{xx}$$

Taking a hint from the study of the linearized KdV equation, take the form

$$\Psi(x, t) = \int_{-\infty}^{\infty}\phi(k,t)e^{ikx}dk.$$

[Here we have opted to use the notation, $\phi(k, t)$ instead of $A(k, t)$ as above.]

Inserting this expression into equation, we have

$$i\hbar\int_{-\infty}^{\infty}\frac{d\phi(k,t)}{dt}e^{ikx}dk = -\frac{\hbar^2}{2m}\int_{-\infty}^{\infty}\phi(k,t)(ik)^2 e^{ikx}dk.$$

Since this is true for all t, we can equate integrands, giving

$$i\hbar\frac{d\phi(k,t)}{dt} = \frac{\hbar^2}{2m}k^2\phi(k,t).$$

This is easily solved. We obtain

$$\phi(k, t) = \phi(k,0)e^{-i\frac{\hbar k^2}{2m}t}.$$

Therefore, we have found the general solution to the time dependent problem for a free particle. It is given as

$$\Psi(x, t) = \int_{-\infty}^{\infty}\phi(k,0)e^{ik\left(x-\frac{\hbar k}{2m}t\right)}dk$$

We note that this takes the familiar form

$$\Psi(x, t) = \int_{-\infty}^{\infty}\phi(k,0)e^{i(kx-\omega t)}dk$$

where

$$\omega = \frac{\hbar k^2}{2m}.$$

The wave speed is given as

$$v = \frac{\omega}{k} = \frac{\hbar k}{2m}.$$

As a special note, we see that this is not the particle velocityω Recall that the momentum is given as $p = \hbar k$. So, this wave speed is $v = \frac{p}{2m}$, which is only half the classical particle velocityω A simple manipulation of our result will clarify this "problem". We assume that particles can be represented by a localized wave function. This is the case if the major contributions to the integral are centered about a central wave number, k_0. Thus, we can expand $\omega(k)$ about k_0.

$$\omega(k) = \omega_0 + \omega_0' (k - k_0) + \ldots$$

Here $\omega_0 = \omega(k_0)$ and $\omega_0' = \omega'(k_0)$. Inserting this expression into our integral representation for $\Psi(x, t)$, we have

$$\Psi(x, t) = \int_{-\infty}^{\infty} \phi(k_0 + s, 0) e^{i((k_0+s)x-(\omega_0+\omega_0's))} ds$$

We make a change of variables, $s = k - k_0$ and rearrange the factors to find

$$\Psi(x, t) \approx \int_{-\infty}^{\infty} \phi(k_0 + s, 0) e^{i((k_0+s)x-(\omega_0+\omega_0's))} ds$$

$$= e^{i(-\omega_0 t + k_0\omega_0' t)} \int_{-\infty}^{\infty} \phi(k_0 + s, 0) e^{i((k_0+s)(x-\omega_0' t))} ds$$

$$= e^{i(-\omega_0 t + k_0\omega_0' t)} \Psi(x - \omega_0' t, 0).$$

What we have found is that for a localized wave packet with wave numbers grouped around k0 the wave function is a translated version of the initial wave function, up to a phase factor. The velocity of the wave packet is seen to be $\omega'_0 = \frac{\hbar k}{m}$. This corresponds to the classical velocity of the particle. Thus, one usually defines this to be the group velocity,

$$v_g = \frac{d\omega}{dk}$$

and the former velocity as the phase velocity,

$$v_p = \frac{\omega}{k}.$$

Transform Schemes

These examples have illustrated one of the features of transform theory. Given a partial differential equation, we can transform the equation from spatial variables to wave number space, or time variables to frequency space.

In the new space the time evolution is simpler. In these cases, the evolution is governed by an ordinary differential equation. One solves the problem in the new space and then transforms back to the original space.

This is similar to the solution of the system of ordinary differential equations $x = Ax$. In that case we diagonalized the system using the transformation $x = Sy$. This lead to a simpler system $y = \Lambda y$.

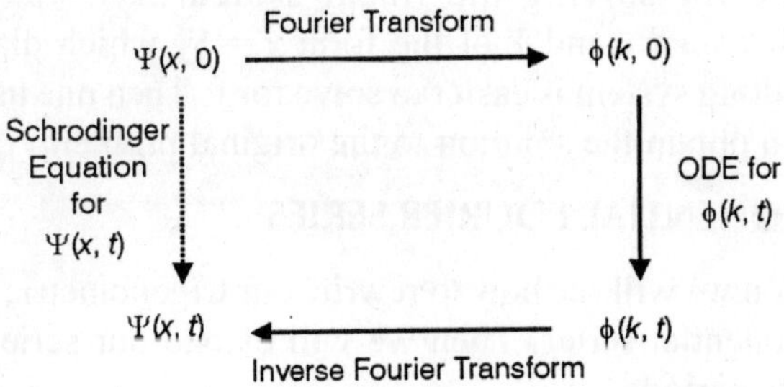

The scheme for solving the Schroodiger equation using Fourier transforms.

The goal is to solve for $\Psi(x, t)$ given $\Psi(x, 0)$. Instead of a direct solution in coordinate space (on the left side), one can first transform the initial condition obtaining $\phi(k, 0)$ in wave number space. The governing equation in the new space is found by transforming the PDE to get an ODE.

This simpler equation is solved to obtain $\phi(k, t)$. Then an inverse transform yields the solution of the original equation.

Solving for y, we inverted the solution to obtain x. Similarly, one can apply this diagonalization to the solution of linear algebraic systems of equations.

Similar transform constructions occur for many other type of problems. We will end this chapter with a study of Laplace transforms, which are useful in the study of initial value problems, particularly for linear ordinary differential equations with constant coefficients.

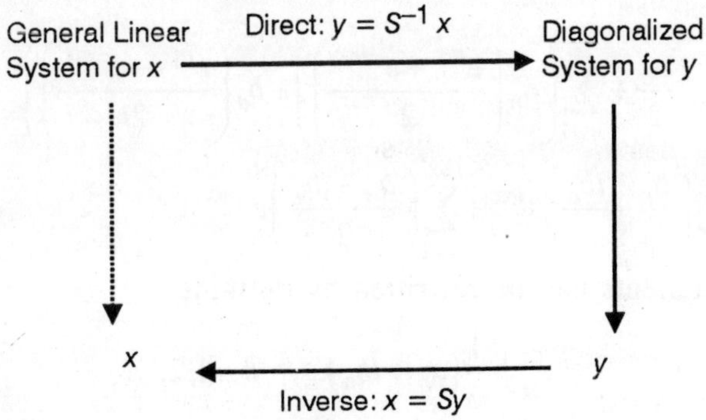

The study of Fourier transforms, these will provide an integral representation of functions defined on the real line. Such functions can also represent analog signals. Analog signals are continuous signals which may be sums over a continuous set of frequencies, as opposed to the sum over discrete frequencies.

We will then investigate a related transform, the Laplace transform, which is useful in solving initial value problems such as those encountered in ordinary differential equations.

The scheme for solving the linear system $Ax = b$. One finds a transformation between x and Y of the form $x = Sy$ which diagonalizes the system. The resulting system is easier to solve for y. Then one uses the inverse transformation to obtain the solution to the original problem.

COMPLEX EXPONENTIAL FOURIER SERIES

In this section we will see how to rewrite our trigonometric Fourier series as complex exponential series. Then we will extend our series to problems involving infinite periods.

We first recall the trigonometric Fourier series representation of a function defined on $[-\pi, \pi]$ with period 2π. The Fourier series is given by

$$f(x) \sim \frac{a_0}{2} + \sum_{n=1}^{\infty} (a_n \cos nx + b_n \sin nx),$$

where the Fourier coefficients were found as

$$a_n = \frac{1}{\pi} \int_{-\pi}^{\pi} f(x) \cos nx \, dx, \qquad n = 0, 1, \ldots,$$

$$b_n = \frac{1}{\pi} \int_{-\pi}^{\pi} f(x) \sin nx \, dx, \qquad n = 1, 2, \ldots$$

In order to derive the exponential Fourier series, we replace the trigonometric functions with exponential functions and collect like terms. This gives

$$f(x) \sim \frac{a_0}{2} + \sum_{n=1}^{\infty} \left(a_n \left(\frac{e^{inx} + e^{-inx}}{2} \right) \right) + b_n \left(\frac{e^{inx} - e^{-inx}}{2i} \right) \Bigg)$$

$$= \frac{a_0}{2} + \sum_{n=1}^{\infty} \left(\frac{a_n - ib_n}{2} \right) e^{inx} + \sum_{n=1}^{\infty} \left(\frac{a_n - ib_n}{2} \right) e^{-inx}$$

The coefficients can be rewritten by defining

$$c_n = \frac{1}{2}(a_n + ib_n), \qquad n = 1, 2, \ldots.$$

Then, we also have that

$$\bar{c}_n = \frac{1}{2}(a_n - ib_n), \qquad n = 1, 2, \ldots$$

This gives our representation as

$$f(x) \sim \frac{a_0}{2} + \sum_{n=1}^{\infty} \overline{c}_n e^{inx} + \sum_{n=1}^{\infty} c_n e^{-inx}$$

Reindexing the first sum, by letting $k = -n$, we can write

$$f(x) \sim \frac{a_0}{2} + \sum_{k=-1}^{-\infty} \overline{c}_{-k} e^{-ikx} + \sum_{n=1}^{\infty} c_n e^{-inx}$$

Now, we define

$$c_n = \overline{c}_{-n}, \quad n = -1, -2, \ldots$$

Finally, we note that we can take $c_0 = \frac{a_0}{2}$. So, we can write the complex exponential Fourier series representation as

$$f(x) \sim \sum_{n=-\infty}^{\infty} c_n e^{-inx}$$

where

$$c_n = \frac{1}{2}\left(a_n + ib_n\right), \qquad n = 1, 2, \ldots$$

$$c_n = \frac{1}{2}\left(a_n + ib_n\right), \qquad n = 1, 2, \ldots$$

$$c_0 = \frac{a_0}{2}$$

Given such a representation, we would like to write out the integral forms of the coefficients, c_n. So, we replace the a_n's and b_n's with their integral representations and replace the trigonometric functions with a complex exponential function using Euler's formula. Doing this, we have for $n = 1, 2, \ldots$.

$$\begin{aligned}
c_n &= \frac{1}{2}\left(a_n + ib_n\right) \\
&= \frac{1}{2}\left[\frac{1}{\pi}\int_{-\pi}^{\pi} f(x)\cos nx\, dx + \frac{i}{\pi}\int_{-\pi}^{\pi} f(x)\sin nx\, dx\right] \\
&= \frac{1}{2\pi}\int_{-\pi}^{\pi} f(x)(\cos nx\, dx + i\sin nx)\, dx \\
&= \frac{1}{2\pi}\int_{-\pi}^{\pi} f(x) e^{inx}\, dx
\end{aligned}$$

It is a simple matter to determine the c_n's for other values of n. For $n = 0$, we have that

$$c_0 = \frac{a_0}{2} = \frac{1}{2\pi}\int_{-\pi}^{\pi} f(x)\,dx.$$

For n = –1, –2,..., we find that

$$c_n = \bar{c}_{-n} = \frac{1}{2\pi}\int_{-\pi}^{\pi} \overline{f(x)e{-inx}}\,dx = \frac{1}{2\pi}\int_{-\pi}^{\pi} f(x)e^{inx}\,dx.$$

Therefore, for all n we have shown that

$$c_n = \frac{1}{2\pi}\int_{-\pi}^{\pi} f(x)e^{inx}\,dx$$

We have converted our trigonometric series for functions defined on $[-\pi, \pi]$ to the complex exponential series in equation with Fourier coefficients. Thus, we have obtained the Complex

Fourier Series representation

$$f(x) \sim \sum_{n=-\infty}^{\infty} c_n e^{-inx}$$

where the complex Fourier coefficients are given by

$$c_n = \frac{1}{2\pi}\int_{-\pi}^{\pi} f(x)e^{inx}\,dx$$

We can easily extend the above analysis to other intervals. For example, for $x \in [-L, L]$ the Fourier trigonometric series is

$$f(x) \sim \frac{a_0}{2} + \sum_{n=1}^{\infty}\left(a_n \cos\frac{n\pi x}{L} + b_n \sin\frac{n\pi x}{L}\right)$$

with Fourier coefficients

$$a_n = \frac{1}{L}\int_{-L}^{L} f(x)\cos\frac{n\pi x}{L}\,dx \qquad n = 0,1,\ldots,$$

$$b_n = \frac{1}{L}\int_{-L}^{L} f(x)\sin\frac{n\pi x}{L}\,dx \qquad n = 1,2,\ldots,$$

This can be rewritten in an exponential Fourier series of the form.

$$f(x) \sim \sum_{n=-\infty}^{\infty} c_n e^{-in\pi x/L}$$

with

$$c_n = \frac{1}{2L}\int_{-L}^{L} f(x)e^{in\pi x/L}\,dx.$$

Finally, we note that these expressions can be put into the form.

$$f(x) \sim \sum_{n=-\infty}^{\infty} c_n e^{-ik_n x}$$

with

$$c_n = \frac{1}{2L}\int_{-L}^{L} f(x) e^{ik_n x}\, dx$$

where we have introduced the discrete set of wave numbers

$$k_n = \frac{n\pi}{L}.$$

At times, we will also be interested in functions of time. In this case we will have a function $g(t)$ defined on a time interval $[-T, T]$. The exponential Fourier series will then take the form

$$g(t) \sim \sum_{n=-\infty}^{\infty} c_n e^{-i\omega_n t}$$

with

$$c_n = \frac{1}{2T}\int_{-T}^{T} g(x) e^{i\omega_n t}\, dt.$$

Here we have introduced the discrete set of angular frequencies, which can be related to the corresponding discrete set of frequencies by

$$\omega_n = 2\pi f_n = \frac{n\pi}{T}.$$

with

$$f_n = \frac{n}{2T}.$$

EXPONENTIAL FOURIER TRANSFORM

Both the trigonometric and complex exponential Fourier series provide us with representations of a class of functions in term of sums over a discrete set of wave numbers for functions of finite wavelength. On intervals $[-L, L]$ the wavelength is $2L$. Writing the arguments in terms of wavelengths, we have $k_n = \frac{2\pi}{\lambda_n} = \frac{2\pi}{\lambda_n}$, or the sums are over wavelengths $\lambda_n = \frac{2\pi}{n}$. This is a discrete, or countable, set of wavelengths. A similar argument can be made for time series, or functions of time, which occur more often in signal analysis.

We would now like to extend our interval to $x \in (-\infty, \infty)$ and to extend the discrete set of wave numbers to a continuous set of wave numbers. One can do this rigorously, but it amounts to letting L and n get large and keeping $\frac{n}{L}$ fixed. We define $k = \frac{2\pi}{\lambda}$ and the sum over the continuous set of wave numbers becomes an integral. Formally, we arrive at the Fourier transform

$$F[f] = \hat{f}(k) = \int_{-\infty}^{\infty} f(x)e^{ikx}\,dx$$

–1 is a generalization of the Fourier coefficients. Once we know the Fourier transform, then we can reconstruct our function using the inverse Fourier transform, which is given by

$$F^{-1}[\hat{f}] = f(x) = \frac{1}{2\pi}\int_{-\infty}^{\infty} \hat{f}(k)e^{-ikx}\,dk.$$

We note that it can be proven that the Fourier transform exists when $f(x)$ is absolutely integrable, i.e.,

$$\int_{-\infty}^{\infty} |f(k)|\,dx < \infty.$$

Such functions are said to be L_1.

The Fourier transform and inverse Fourier transform are inverse operations. This means that

$$F^{-1}[F[f]] = f(x)$$

and

$$F[F^{-1}[\hat{f}]] = \hat{f}(k).$$

We will now prove the first of these equations. The second follows in a similar way. This is done by inserting the definition of the Fourier transform into the inverse transform definition and then interchanging the orders of integration. Thus, we have

$$\begin{aligned} F^{-1}[F[f]] &= \frac{1}{2\pi}\int_{-\infty}^{\infty} F[f]e^{-ikx}\,dk \\ &= \frac{1}{2\pi}\int_{-\infty}^{\infty}\left[\int_{-\infty}^{\infty} f(\xi)e^{ik\xi}d\xi\right]e^{-ikx}\,dk \\ &= \frac{1}{2\pi}\int_{-\infty}^{\infty}\int_{-\infty}^{\infty} f(\xi)e^{ik(\xi-x)}\,d\xi\,dk \\ &= \frac{1}{2\pi}\int_{-\infty}^{\infty}\left[\int_{-\infty}^{\infty} e^{ik(\xi-x)}\,dk\right]f(\xi)\,d\xi \end{aligned}$$

In order to complete the proof, we need to evaluate the inside integral, which does not depend upon $f(x)$. This is an improper integral, so we will define

$$D_L(x) = \int_{-L}^{L} e^{ikx}\,dk$$

and compute the inner integral as

$$\int_{-\infty}^{\infty} e^{ik(\xi - x)} dk = \lim_{L\to\infty} D_L(\xi - x).$$

We can compute $D_L(x)$. A simple evaluation yields

$$D_L(x) = \int_{-L}^{L} e^{ikx} dk$$

$$= \left. \frac{e^{ikx}}{ix} \right|_{-L}^{L}$$

$$= \frac{e^{ixL} - e^{-ixL}}{2ix}$$

$$= \frac{2\sin xL}{x}.$$

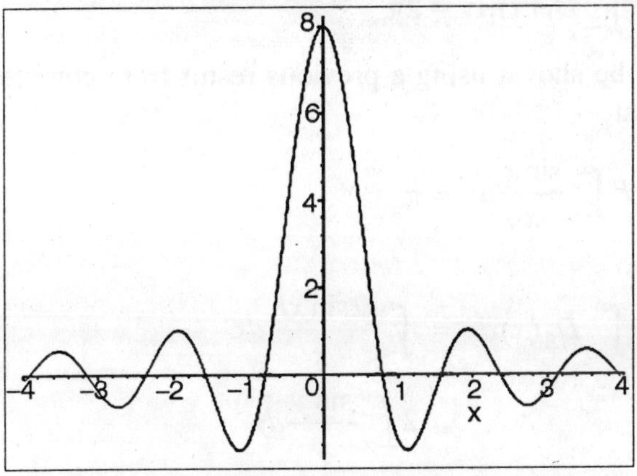

Fig. A Plot of the Function $D_L(x)$ for $L = 4$

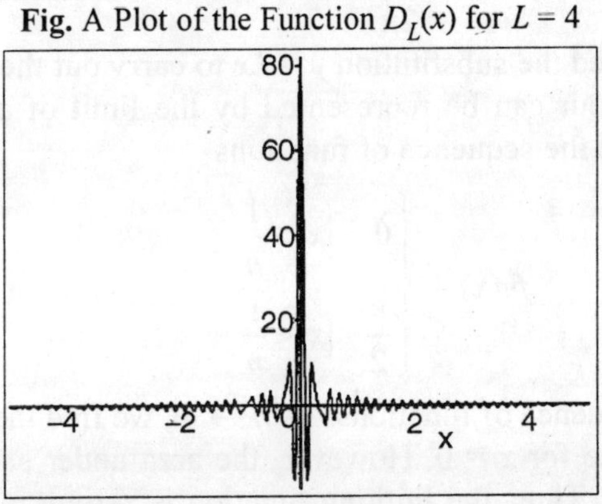

Fig. A Plot of the Function $D_L(x)$ for $L = 40$

We can graph this function. For large x, the function tends to zero. For large L the peak grows and the values of $D_L(x)$ for $x \neq 0$ tend to zero. In fact, we can show that as $x \to 0$, $D_L(x) \to 2L$:

$$\begin{aligned}\lim_{x\to 0} D_L(x) &= \lim_{x\to 0} \frac{2\sin xL}{x} \\ &= \lim_{x\to 0} 2L\frac{\sin xL}{xL} \\ &= 2L\left(\lim_{y\to 0}\frac{\sin y}{y}\right) \\ &= 2L.\end{aligned}$$

We note that in the limit $L \to 1$, $D_L(x) = 0$ for $x \neq 0$ and it is infinite at $x = 0$. However, the area is constant for each L. In fact,

$$\int_{-\infty}^{\infty} D_L(x)dx = 2\pi.$$

This can be shown using a previous result from complex analysis. We had shown that

$$P\int_{-\infty}^{\infty} \frac{\sin x}{x}dx = \pi.$$

So

$$\begin{aligned}\int_{-\infty}^{\infty} D_L(x)dx &= \int_{-\infty}^{\infty} \frac{2\sin xL}{x}dx \\ &= 2\int_{-\infty}^{\infty} \frac{\sin y}{y}dy \\ &= 2\pi,\end{aligned}$$

where we had used the substitution $y = Lx$ to carry out the integration.

This behaviour can be represented by the limit of other sequences of functions. Define the sequence of functions

$$fn(x) = \begin{cases} 0, & |x| > \dfrac{1}{n} \\ \dfrac{n}{2} & |x| < \dfrac{1}{n} \end{cases}$$

This is a sequence of functions. As $n \to \infty$, we find the limit is zero for $x \neq 0$ and is infinite for $x = 0$. However, the area under each member of the sequences is one. Thus, the limiting function is zero at most points but has area one.

The limit is not really a function. It is a generalized function. It is called the Dirac delta function, which is defined by

- $\delta(x) = 0$ for $x \neq 0$.
- $\int_{-\infty}^{\infty} \delta(x)\, dx$.

As a further note, we could have considered the sequence of functions

$$g_n(x) = \begin{cases} 0, & |x| > \frac{1}{n} \\ 2n, & |x| < \frac{1}{n} \end{cases}$$

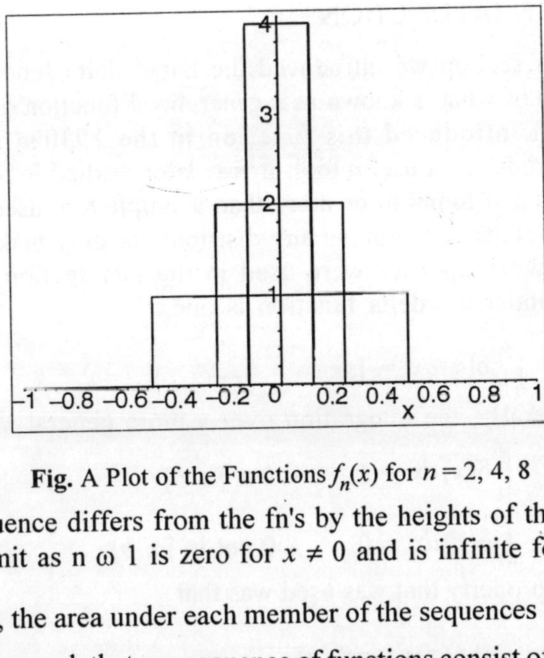

Fig. A Plot of the Functions $f_n(x)$ for $n = 2, 4, 8$

This sequence differs from the fn's by the heights of the functions. As before, the limit as n ω 1 is zero for $x \neq 0$ and is infinite for $x = 0$.

However, the area under each member of the sequences is now $2n \times \frac{2}{n} =$ 4. So, it is not enough that our sequence of functions consist of nonzero values at just one point. The height might be infinite, but the areas can vary! In this case $\lim_{n\to\infty} gn(x) = 4\delta(x)$.

Before returning to the proof, we state one more property of the Dirac delta function, which we will prove in the next section. We have that

$$\int_{-\infty}^{\infty} \delta(x-a) f(x)\, dx = f(a).$$

This is called the sifting property because it sifts out a value of the function $f(x)$.

Returning to the proof, we now have that

$$\int_{-\infty}^{\infty} e^{ik}(\xi - a)\,dk = \lim_{L\to\infty} D_L(\xi - x) = 2\pi\delta(\xi - x)$$

Inserting this into equation, we have

$$F^{-1}[F[f]] = \frac{1}{2\pi}\int_{-\infty}^{\infty}\left[\int_{-\infty}^{\infty} e^{ik(\xi-x)}dk\right] f(\xi)\,d\xi$$

$$= \frac{1}{2\pi}\int_{-\infty}^{\infty} 2\pi\delta(\xi - x)\,d\xi$$

$$= f(x)$$

Thus, we have proven that the inverse transform of the Fourier transform of f is f.

THE DIRAC DELTA FUNCTION

In the last section we introduced the Dirac delta function, $\delta(x)$. This is one example of what is known as a generalized function or a distribution.

Dirac had introduced this function in the 1930's in his study of quantum mechanics as a useful tool. It was later studied in a general theory of distributions and found to be more than a simple tool used by physicists.

The Dirac delta function, as any distribution, only makes sense under an integral. Two properties were used in the last section. First one has that the area under the delta function is one,

$$\int_{-\infty}^{\infty} \delta(x)\,dx = 1.$$

More generally, the integration over a more general interval gives

$$\int_a^b \delta(x)\,dx = 1, \qquad 0 \in [a, b]$$

$$\int_a^b \delta(x)\,dx = 0, \qquad 0 \text{ not in } [a, b]$$

The other property that was used was that

$$\int_{-\infty}^{\infty} \delta(x - a) f(x)\,dx = f(a).$$

This can be seen by noting that the delta function is zero everywhere except at $x = a$. Therefore, the integrand is zero everywhere and the only contribution from $f(x)$ will be from $x = a$. So, we can replace $f(x)$ with $f(a)$ under the integral. Since $f(a)$ is a constant, we have that

$$\int_{-\infty}^{\infty} \delta(x - a) f(x)\,dx = \text{f}$$

Other occurrences of the delta function are integrals of the form $\int_{-\infty}^{\infty} \delta(f(x))\,dx$. Such integrals can be converted into a useful form depending upon the number of zeros of $f(x)$. If there is only one zero, $f(x_1) = 0$, then one has that

$$\int_{-\infty}^{\infty} \delta(f(x))\,dx = \int_{-\infty}^{\infty} \frac{1}{|f'(x)|}\delta(x-x_1)\,dx$$

This can be proven using the substitution $y = f(x)$ and is left as an exercise for the reader. This result is often written as

$$\delta(f(x)) = \frac{1}{|f'(x_j)|}\delta(x-x_j)$$

More generally, one can show that when $f(x_j) = 0$ for x_j, $j = 1, 2,\ldots,n$, then

$$\delta(f(x)) = \sum_{j=1}^{n} \frac{1}{|f'(x_j)|}\delta(x-x_j).$$

Finally, one can show that there is a relationship between the Heaviside, or step, function and the Dirac delta function. We define the Heaviside function as

$$H(x) = \begin{cases} 0, & x<0 \\ 1, & x>0 \end{cases}$$

Then, it is easy to see that $H'(x) = \delta(x)$. For $x \neq 0$, $H'(x) = 0$. It has an infinite slope at $x = 0$. We need only check that the area is one. Thus,

$$\int_{-\infty}^{\infty} H'(x)\,dx = \lim_{L\to\infty} [H(L) - H(-L)] = 1$$

In some texts the notation $\theta(x)$ is used for the step function, $H(x)$.

Example: Evaluate $\int_{-\infty}^{\infty} \delta(3x-2)x^2\,dx$.

This is not a simple $\delta(x-a)$. So, we need to find the zeros of $f(x) = 3x - 2$. There is only one, $x = \frac{2}{3}$. Also, $|f'(x)| = 3$. Therefore, we have

$$\int_{-\infty}^{\infty} \delta(3x-2)x^2\,dx = \int_{-\infty}^{\infty} \frac{1}{3}\delta\left(x-\frac{2}{3}\right)x^2\,dx = \frac{1}{3}\left(\frac{2}{3}\right)^2 = \frac{4}{27}$$

Example: Evaluate $\int_{-\infty}^{\infty} \delta(3x)(x^2+4)\,dx$.

This problem is deceiving. One cannot just plug in $x = 0$ into the function $x^2 + 4$. One has to use the fact that the derivative of $2x$ is 2. So, $\delta(2x) = \frac{1}{2}\delta(x)$. So,

$$\int_{-\infty}^{\infty}\delta(2x)(x^2+4)\,dx = \frac{1}{2}\int_{-\infty}^{\infty}\delta(x)(x^2+4)\,dx = 2.$$

PROPERTIES OF THE FOURIER TRANSFORM

We now return to the Fourier transform. Before actually computing the Fourier transform of some functions, we prove a few of the properties of the Fourier transform.

First we recall that there are several forms that one may encounter for the Fourier transform. In applications our functions can either be functions of time, $f(t)$, or space, $f(x)$. The corresponding Fourier transforms are then written as

$$\hat{f}(\omega) = \int_{-\infty}^{\infty} f(t)e^{i\omega t}\,dt$$

$$\hat{f}(k) = \int_{-\infty}^{\infty} f(x)e^{ikx}\,dx$$

ω is called the angular frequency and is related to the frequency ν by $\omega = 2\pi\nu$. The units of frequency are typically given in Hertz (Hz). Sometimes the frequency is denoted by f when there is no confusion. Recall that k is called the wavenumber. It has units of inverse length and is related to the wavelength, λ, by $k = \frac{2\pi}{\lambda}$.

- *Linearity:* For any functions $f(x)$ and $g(x)$ for which the Fourier transform exists and constant a, we have

  $$F[f+g] = F[f] + F[g]$$

  and

  $$F[af] = aF[f].$$

  These simply follow from the properties of integration and establish the linearity of the Fourier transform.

- $F\left[\frac{df}{dx}\right] = -ik\hat{f}(k)$

  This property can be shown using integration by parts.

  $$F\left[\frac{df}{dx}\right] = \int_{-\infty}^{\infty}\frac{df}{dx}e^{ikx}\,dx$$

  $$= \lim_{L\to\infty}\left(f(x)e^{ikx}\right)\Big|_{-L}^{L} - ik\int_{-\infty}^{\infty} f(x)e^{ikx}\,dx$$

  The limit will vanish if we assume that $\lim_{x\to\pm\infty} f(x) = 0$. The integral is recognized as the Fourier transform of f, proving the given property.

- $F\left[\dfrac{d^n f}{dx^2}\right] = (-ik)^n \hat{f}(k)$

  The proof of this property follows from the last result, or doing several integration by parts. We will consider the case when $n = 2$. Noting that the second derivative is the derivative of $f'(x)$ and applying the last result, we have

$$F\left[\frac{d^2 f}{dx^2}\right] = F\left[\frac{d}{dx} f'\right]$$

$$= -kF\left[\frac{df}{dx}\right] = (-ik)^2 \hat{f}(k)$$

  This result will be true if both $\lim_{x\to\pm\infty} f(x) = 0$ and $\lim_{x\to\pm\infty} f'(x) = 0$. Generalizations to the transform of the nth derivative easily follows.

- $F\,[xf(x)] = -i\left[\dfrac{d}{dk}\right]\hat{f}(k)$

  This property can be shown by using the fact that $\dfrac{d}{dk} e^{ikx} = ixe^{ikx}$ and being able to differentiate an integral with respect to a parameter.

$$F[xf(x)] = \int_{-\infty}^{\infty} xf(x) e^{ikx}\, dx$$

$$= \int_{-\infty}^{\infty} f(x) \frac{d}{dk}\left(\frac{1}{i} e^{ikx}\right) dx$$

$$= -i \frac{d}{dk} \int_{-\infty}^{\infty} f(x) e^{ikx}\, dx$$

$$= -i \frac{d}{dk} \hat{f}(k)$$

- *Shifting Properties:* For constant a, we have the following shifting properties.

$$f(x-a) \leftrightarrow e^{ika} \hat{f}(k),$$

$$f(x) e^{-iax} \leftrightarrow \hat{f}(k-a).$$

Here we have denoted the Fourier transform pairs as $f(x) \leftrightarrow \hat{f}(k)$. These are easily proven by inserting the desired forms into the definition of the Fourier transform, or inverse Fourier transform. The first shift property is shown by the following argument. We evaluate the Fourier transform.

$$F[f(x-a)] = \int_{-\infty}^{\infty} f(x-a)e^{ikx}\,dx$$

Now perform the substitution $y = x - a$. This will make f a function on asingle variable. Then,

$$F[f(x-a)] = \int_{-\infty}^{\infty} f(y)e^{ik(y+a)}dy = e^{ika}\int_{-\infty}^{\infty} f(y)e^{iky}dy = e^{ika}\hat{f}(k)$$

The second shift property follows in a similar way.

- ***Convolution:*** We define the convolution of two functions $f(x)$ and $g(x)$ as

$$(f * g)(x) = \int_{-\infty}^{\infty} f(t)g(x-t)\,dx.$$

Then

$$F[f * g] = \hat{f}(k)\hat{g}(k)$$

Fourier Transform Examples

In this section we will compute some Fourier transforms of several functions.

Example: $f(x) = e^{-ax2/2}$.

This function is called the Gaussian function. It has many applications in areas such as quantum mechanics, molecular theory, probability and heat diffusion.

We will compute the Fourier transform of this function and show that the Fourier transform of a Gaussian is a Gaussian. In the derivation we will introduce classic techniques for computing such integrals.

We begin by applying the definition of the Fourier transform,

$$\hat{f}(k) = \int_{-\infty}^{\infty} f(x)e^{ikx}\,dx = \int_{-\infty}^{\infty} e^{-ax^2/2+ikx}\,dx.$$

The first step in computing this integral is to complete the square in the argument of the exponential. Our goal is to rewrite this integral so that a simple substitution will lead to a classic integral of the form $\int_{-\infty}^{\infty} e^{\beta y^2}\,dy$, which we can integrate. The completion of the square follows as usual.

$$-\frac{a}{2}x^2 + ikx = -\frac{a}{2}\left[x^2 - \frac{2ik}{a}x\right]$$

$$= -\frac{a}{2}\left[x^2 - \frac{2ik}{a}x + \left(-\frac{ik}{a}\right)^2 - \left(-\frac{ik}{a}\right)^2\right]$$

$$= -\frac{a}{2}\left(x - \frac{ik}{a}\right)^2 - \frac{k^2}{2a}$$

Using this result in the integral and making the substitution $y = x - \frac{ik}{a}$, we have

$$\hat{f}(k) = e^{-\frac{k^2}{2a}} \int_{-\infty}^{\infty} e^{-\frac{a}{2}\left(x - \frac{ik}{a}\right)^2} dx$$

$$= e^{-\frac{k^2}{2a}} \int_{-\infty - \frac{ik}{a}}^{\infty - \frac{ik}{a}} e^{-\beta y^2} dy, \ \beta = \frac{a}{2}.$$

One would be tempted to absorb the $-\frac{ik}{a}$ terms in the limits of integration. In fact, this is what is usually done in texts. However, we need to be careful. We know from our previous study that the integration takes place over a contour in the complex plane. We can deform this horizontal contour to a contour along the real axis since we will not cross any singularities of the integrand. So, we can safely write

$$\hat{f}(k) = e^{-\frac{k^2}{2a}} \int_{-\infty}^{\infty} e^{-\beta y^2} dy.$$

The resulting integral is a classic integral and can be performed using a standard trick. Let I be given by

$$I = \int_{-\infty}^{\infty} e^{-\beta y^2} dy.$$

Then,

$$I^2 = \int_{-\infty}^{\infty} e^{-\beta y^2} dy \int_{-\infty}^{\infty} e^{-\beta x^2} dx.$$

Note that we needed to introduce a second integration variable. We can now write this product as a double integral:

$$I^2 = \int_{-\infty}^{\infty} \int_{-\infty}^{\infty} e^{-\beta(x^2+y^2)} dx dy$$

This is an integral over the entire xy-plane. Since this is a function of $x^2 + y^2$, it is natural to transform to polar coordinates. We have that $r^2 = x^2 + y^2$ and the area element is given by $dxdy = rdrd\theta$. Therefore, we have that

$$I^2 = \int_0^{2\pi} \int_0^{\infty} e^{-\beta r^2} r dr d\theta.$$

This integral is doable. Letting $z = r^2$,

$$\int_0^\infty e^{-\beta r^2} r dr = \frac{1}{2}\int_0^\infty e^{-\beta z} dz$$

This gives $I^2 = \frac{\pi}{\beta}$. So, the final result is found by taking the square root of both sides.

$$I = \sqrt{\frac{\pi}{\beta}}.$$

We can now insert this result into equation to give the Fourier transform of the Gaussian function.

$$\hat{f}(k) = \sqrt{\frac{2\pi}{a}} e^{-k^2/2a}.$$

Example: $f(x) = \begin{cases} b, & |x| \le a \\ 0, & |x| > a \end{cases}.$

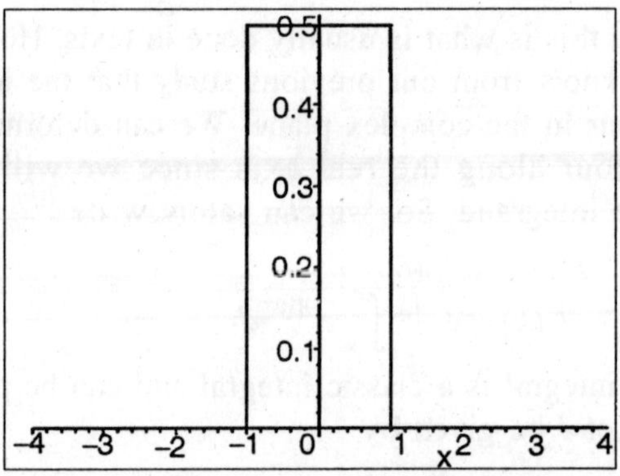

Fig. A Plot of the Box Function in Example 2

This function is called the box function, or gate function. The Fourier transform of the box function is relatively easy to compute. It is given by

$$\begin{aligned}\hat{f}(k) &= \int_{-\infty}^{\infty} f(x) e^{ikx} dx \\ &= \int_{-a}^{a} b e^{ikx} dx \\ &= \frac{b}{ik} e^{ikx} \Big|_{-a}^{a} \\ &= \frac{2b}{k} \sin ka.\end{aligned}$$

We can rewrite this using the sinc function, $\operatorname{sinc} x \equiv \frac{\sin x}{x}$, as

$$f(k) = 2ab\frac{\sin ka}{ka} = 2ab\operatorname{sinc} ka\ .$$

The sinc function appears often in signal analysis. We will consider special limiting values for the box function and its transform.

$a \to \infty$ and b fixed.

In this case, as a gets large the box function approaches the constant function $f(x) = b$. At the same time, we see that the Fourier transform approaches a Dirac delta function. We had seen this function earlier when we first defined the Dirac delta function. In fact,

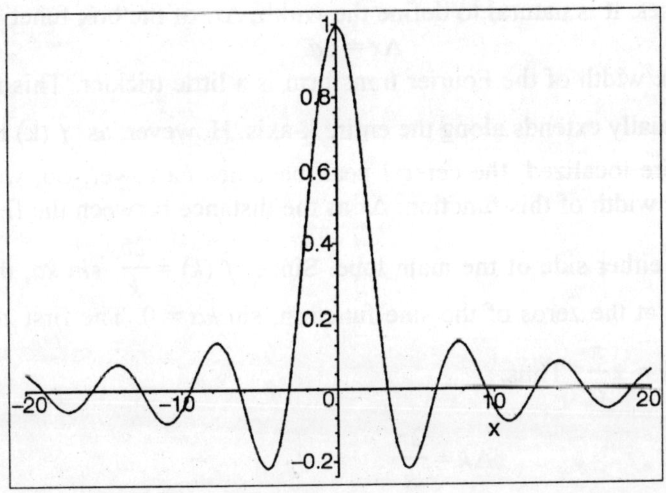

Fig. A Plot of the Fourier Transform of the Box Function in Example 2

$\hat{f}(k) = bD_a(k)$. So, in the limit we obtain $\hat{f}(k) = 2\pi b\delta(k)$. This limit gives us the fact that the Fourier transform of $f(x) = 1$ is $\hat{f}(k) = 2\pi\delta(k)$.

As the width of the box becomes wider, the Fourier transform becomes more localized. Namely, we have arrived at the result that

$$\int_{-\infty}^{\infty} e^{ikx}dx = 2\pi\delta(k).$$

$b \to \infty$, $a \to 0$, and $2ab = 1$.

In this case our box narrows and becomes steeper while maintaining a constant area of one.

This is the way we had found a representation of the Dirac delta function previously. The Fourier transform approaches a constant in this limit.

As a approaches zero, the sinc function approaches one, leaving $\hat{f}(k) \to 2ab = 1$. Thus, the Fourier transform of the Dirac delta function is one. Namely, we have

$$\int_{-\infty}^{\infty} \delta(x) e^{ikx}\, dx = 1.$$

In this case we have that the more localized the function $f(x)$ is, the more spread out the Fourier transform is.

The Uncertainty Principle: The widths of the box function and its Fourier transform are related as we have seen in the last two limiting cases. It is natural to define the width, Δx, of the box function as

$$\Delta x = 2a.$$

The width of the Fourier transform is a little trickier. This function actually extends along the entire k-axis. However, as $\hat{f}$ (k) becomes more localized, the central peak becomes narrower. So, we define the width of this function, Δk as the distance between the first zeros on either side of the main lobe. Since $\hat{f}(k) = \frac{2b}{k} \sin ka$, the zeros are at the zeros of the sine function, $\sin ka = 0$. The first zeros are at $k = \pm\frac{\pi}{a}$. Thus,

$$\Delta k = \frac{2\pi}{a}.$$

Combining the expressions for the two widths, we find that

$$\Delta x \Delta k = 4\pi.$$

Thus, the more localized a signal (smaller δx), the less localized its transform (larger δk).

This notion is referred to as the Uncertainty Principle. For more general signals, one needs to define the effective widths more carefully, but the main idea holds.

$$\Delta x \Delta k \geq 1.$$

While this is a result of Fourier transforms, the uncertainty principle arises in other forms elsewhere. In particular, it appears in quantum mechanics, where is it most known. In quantum mechanics, one finds that the momentum is given in terms of the wave number, $p = \hbar k$, where $\hbar$ is Planck's constant divided by 2π. Inserting this into the above condition, one obtains

$$\Delta x \Delta p \geq \hbar.$$

This gives the famous uncertainty relation between the uncertainties in position and momentum.

Example: $f(x) = \begin{cases} e^{-ax} & x \geq 0 \\ 0, & x < 0 \end{cases}, \quad a > 0$

The Fourier transform of this function is

$$\hat{f}(k) = \int_{-\infty}^{\infty} f(x) e^{ikx}\, dx$$

$$= \int_{0}^{\infty} e^{ikx-ax}\, dx$$

$$= \frac{1}{a - ik}.$$

Next, we will compute the inverse Fourier transform of this result and recover the original function.

Example: $\hat{f}(k) = \dfrac{1}{a - ik}$.

The inverse Fourier transform of this function is

$$f(x) = \frac{1}{2\pi} \int_{-\infty}^{\infty} \hat{f}(k) e^{-ikx}\, dk = \frac{1}{2\pi} \int_{-\infty}^{\infty} \frac{e^{-ikx}}{a - ik}\, dk.$$

This integral can be evaluated using contour integral methods.

If $f(z)$ converges uniformly to zero as $z \to 1$, then

$$\lim_{R \to \infty} \int_{C_R} f(z) e^{ikz}\, dz = 0$$

where $k > 0$ and C_R is the upper half of the circle $|z| = R$.A similar result applies for $k < 0$, but one closes the contour in the lower half plane.

In this example, we have to evaluate the integral

$$I = \int_{-\infty}^{\infty} \frac{e^{-ixz}}{a - iz}\, dz.$$

According to Jordan's Lemma, we need to enclose the contour with a semicircle in the upper half plane for $x < 0$ and in the lower half plane for $x > 0$. The integrations along the semicircles will vanish and we will have

$$f(x) = \frac{1}{2\pi} \int_{-\infty}^{\infty} \frac{e^{-ixx}}{a - ik}\, dk$$

$$= \pm\frac{1}{2\pi}\oint_C \frac{e^{-ixz}}{a-iz}dz$$

$$= \begin{cases} 0, & x<0 \\ -\dfrac{1}{2\pi}2\pi i\,\mathrm{Res}[z=-ia], & x>0 \end{cases}$$

$$= \begin{cases} 0, & x<0 \\ e^{-ax}, & x>0 \end{cases}.$$

Example: $\hat{f}(\omega) = \pi\delta(\omega + \omega_0) + \pi\delta(\omega - \omega_0)$.

We would like to find the inverse Fourier transform of this function. Instead of carrying out any integration, we will make use of the properties of Fourier transforms.

Since the transforms of sums are the sums of transforms, we can look at each term individually.

Consider $\delta(\omega - \omega_0)$. This is a shifted function. From the Shift Theorems we have

$$e^{i\omega_0 t} f(t) \leftrightarrow \hat{f}(\omega - \omega_0).$$

Recalling from a previous example that

$$\int_{-\infty}^{\infty} 1e^{i\omega t}dt = 2\pi\delta(\omega),$$

we have

$$F^{-1}[\delta(\omega-\omega_0)] = \frac{1}{2\pi}e^{-i\omega_0 t},$$

The other term can be transformed similarly. Therefore, we have

$$F^{-1}[\pi\delta(\omega + \omega_0) + \pi\delta(\omega - \omega_0] = \frac{1}{2}e^{i\omega_0 t} + \frac{1}{2}e^{-i\omega_0 t} = \cos\omega_0 t$$

Example: The Finite Wave Train $f(x) = \begin{cases} \cos\omega_0 t, & |t| \le a \\ 0 & |t| > a \end{cases}$

For our last example, we consider the finite wave train, which often appears in signal analysis.

A straight forward computation gives

$$\hat{f}(\omega) = \int_{-\infty}^{\infty} f(t)e^{i\omega t}dt$$

$$= \int_{-a}^{a} \cos\omega_0 t e^{i\omega t}dt$$

$$= \int_{-a}^{a} \cos\omega_0 t \cos\omega t\, dt$$

$$= \frac{1}{2}\int_{-a}^{a}\left[\cos(\omega_0 + \omega_0)t + \cos(\omega_0 - \omega)t\, dt\right]$$

$$= \frac{\sin(\omega_0 + \omega)a}{\omega + \omega_0} + \frac{\sin(\omega_0 - \omega)a}{\omega - \omega_0}.$$

THE CONVOLUTION THEOREM -OPTIONAL

In our list of properties, we defined the convolution of two functions, $f(x)$ and $g(x)$ to be the integral

$$(f * g)(x) = \int_{-\infty}^{\infty} f(\xi)g(x-\xi)d\xi$$

In some sense one is looking at a sum of the overlaps of one of the functions and all of the shifted versions of the other function. The German word for convolution is faltung, which means 'folding'. First, we note that the convolution is commutative. $f * g = g * f$. This is easily shown by replacing $x - \xi$ with a new variable, y.

$$(g * f)(x) = \int_{-\infty}^{\infty} g(\xi)f(x-\xi)d\xi$$

$$= -\int_{\infty}^{-\infty} g(x-y)f(y)dy$$

$$= -\int_{-\infty}^{\infty} f(y)g(x-y)dy$$

$$= (f * g)(x).$$

Example: Graphical Convolution.

In order to understand the convolution operation, we need to apply it to several functions.

We will do this graphically for the box function

$$f(x) = \begin{cases} 1, & |x| < 1 \\ 0, & |x| > 1 \end{cases}$$

and the triangular function

$$g(x) = \begin{cases} x, & |x| < 1 \\ 0, & |x| > 1 \end{cases}.$$

In order to determine the contributions to the integrand, we look at the shifted and reflected function $g(\xi - x)$ for various values of ξ. For $\xi = 0$, we

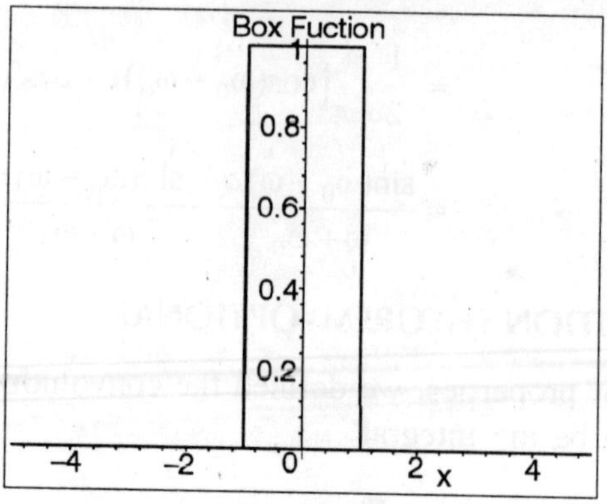

Fig. A Plot of the Box Function $f(x)$

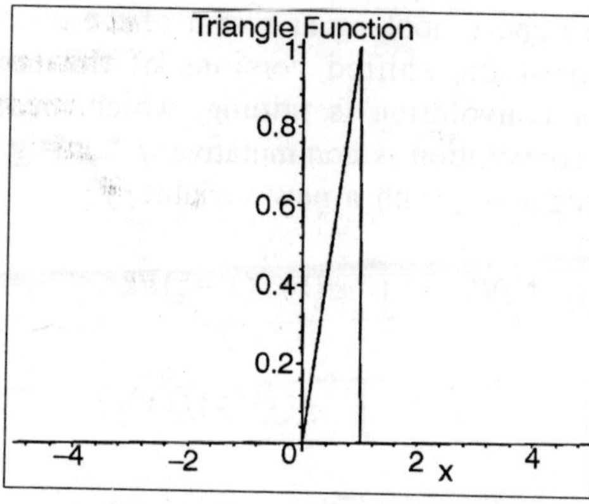

Fig. A Plot of the Triangle Function

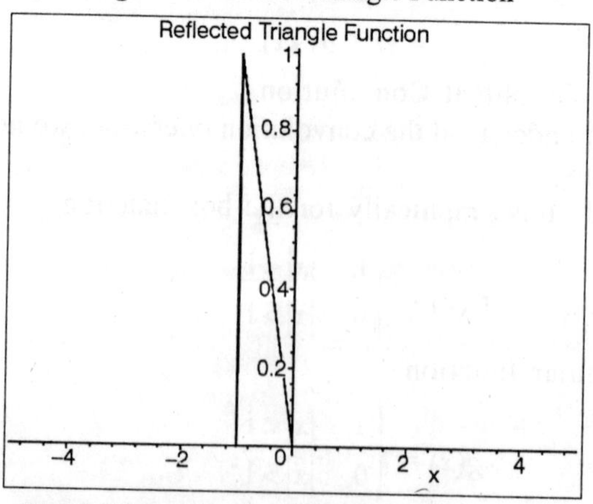

Fig. A Plot of the Reflected Triangle Function

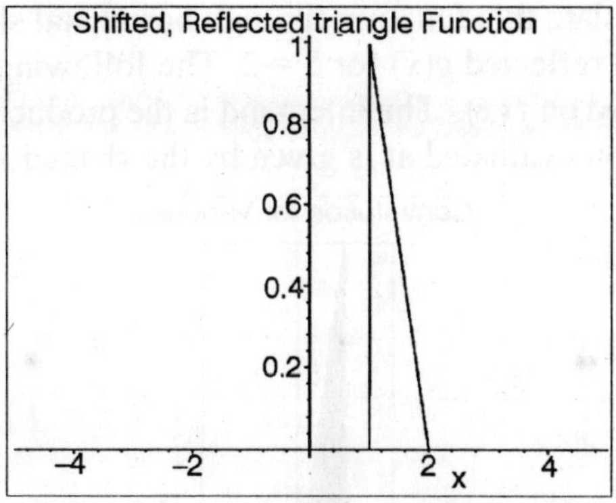

Fig. A Plot of the Reflected Triangle Function Shifted by 2 Units

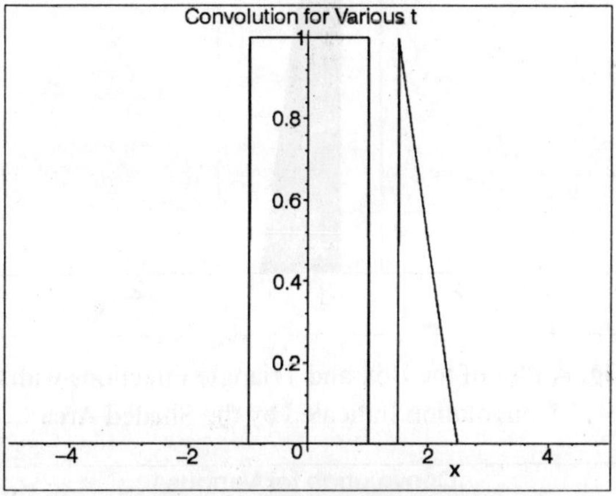

Fig. A Plot of the Box and Triangle Functions with the Convolution Indicated by the Shaded Area have $g(-x)$

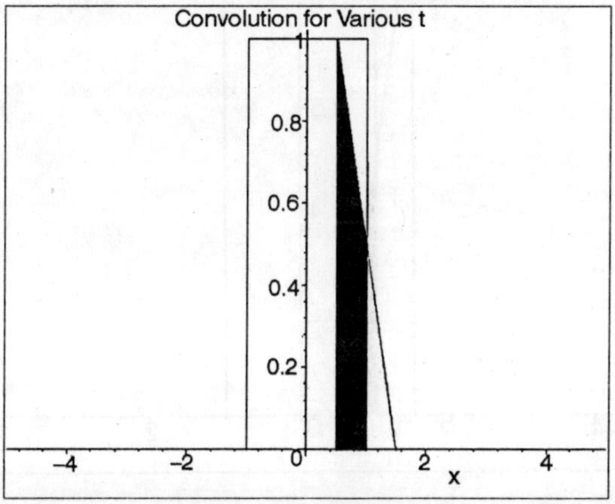

Fig. A Plot of the Box and Triangle Functions with the Convolution Indicated by the Shaded Area

We then translate this function through horizontal shifts by ξ. We show such a shifted and reflected $g(x)$ for $\xi = 2$. The following figures show other shifts superimposed on $f(x)$. The integrand is the product of $f(x)$ and $g(\xi - x)$ and the convolution evaluated at is given by the shaded areas.

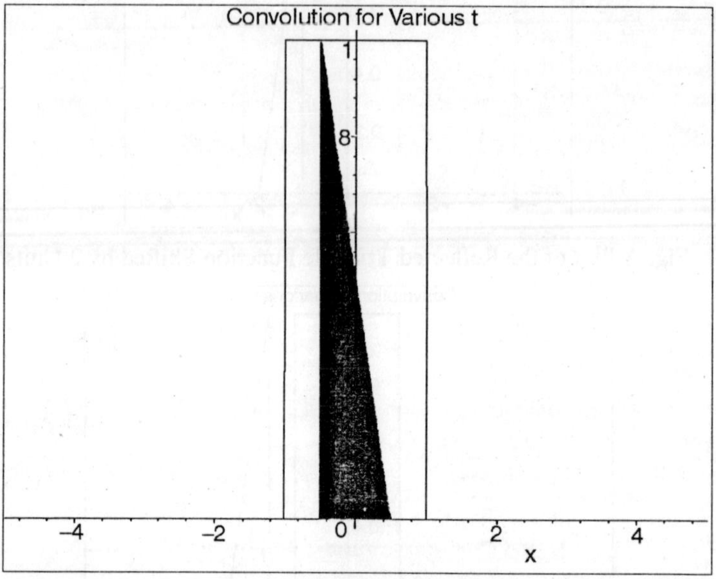

Fig. A Plot of the Box and Triangle Functions with the Convolution Indicated by the Shaded Area

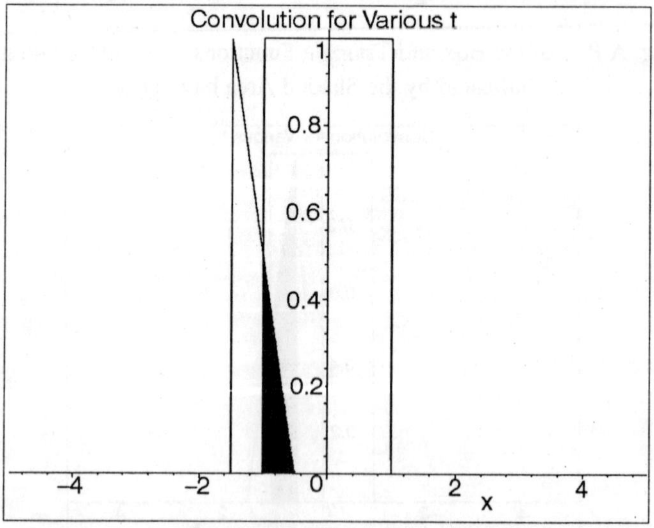

Fig. A Plot of the Box and Triangle Functions with the Convolution Indicated by the Shaded Area

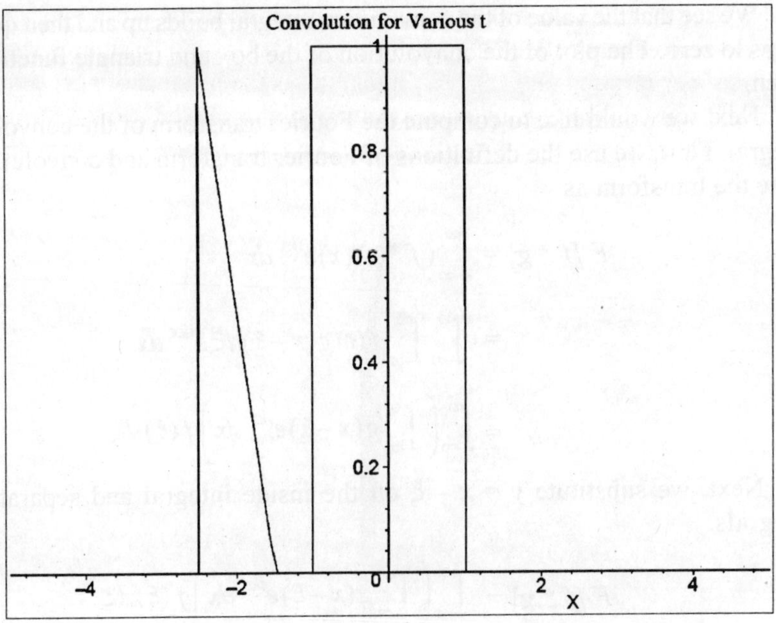

Fig. A Plot of the Box and Triangle Functions
with the Convolution Indicated by the Shaded Area

The area is zero, as there is no overlap of the functions. Intermediate shift values are displayed in figures and the convolution is shown by the area under the product of the two functions.

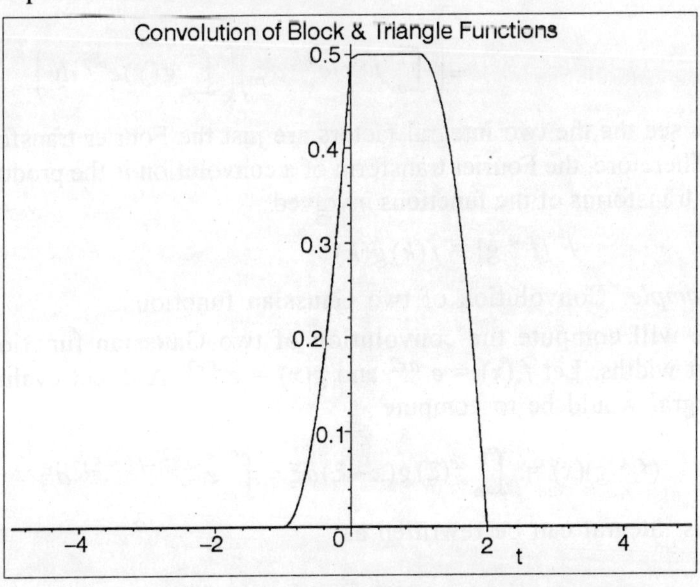

Fig. A Plot of the Convolution of the Box and Triangle Functions

We see that the value of the convolution integral builds up and then quickly drops to zero. The plot of the convolution of the box and triangle functions is given.

Next we would like to compute the Fourier transform of the convolution integral. First, we use the definitions of Fourier transform and convolution to write the transform as

$$F[f*g] = \int_{-\infty}^{\infty} (f*g)(x) e^{ikx}\, dx$$

$$= \int_{-\infty}^{\infty}\int_{-\infty}^{\infty} f(t) g(x-\xi) d\xi\, e^{ikx}\, dx$$

$$= \int_{-\infty}^{\infty}\left(\int_{-\infty}^{\infty} g(x-\xi) e^{ikx}\, dx\right) f(\xi)\, d\xi .$$

Next, we substitute $y = x - \xi$ on the inside integral and separate the integrals.

$$F[f*g] = \int_{-\infty}^{\infty}\left(\int_{-\infty}^{\infty} g(x-\xi) e^{ikx}\, dx\right) f(\xi)\, d\xi$$

$$= \int_{-\infty}^{\infty}\left(\int_{-\infty}^{\infty} g(y) e^{ik(y+\xi)}\, dy\right) f(\xi)\, d\xi$$

$$= \int_{-\infty}^{\infty}\left(\int_{-\infty}^{\infty} g(y) e^{iky}\, dy\right) f(\xi) e^{ik\xi} d\xi$$

$$= \left(\int_{-\infty}^{\infty} f(x) e^{ik\xi}\, d\xi\right)\left(\int_{-\infty}^{\infty} g(y) e^{iky}\, dy\right).$$

We see the the two integral factors are just the Fourier transforms of f and g. Therefore, the Fourier transform of a convolution is the product of the Fourier transforms of the functions involved.

$$F[f*g] = \hat{f}(k)\hat{g}(k).$$

Example: Convolution of two Gaussian functions.

We will compute the convolution of two Gaussian functions with different widths. Let $f(x) = e^{-ax2}$ and $g(x) = e^{-bx2}$. A direct evaluation of the integral would be to compute

$$(f*g)(x) = \int_{-\infty}^{\infty} f(\xi) g(x-\xi) d\xi = \int_{-\infty}^{\infty} e^{-a\xi^2 - b(x-\xi)^2} d\xi$$

This integral can be rewritten as

$$(f*g)(x) = e^{-bx^2}\int_{-\infty}^{\infty} e^{-(a+b)\xi^2 + 2bx\xi} d\xi .$$

One could proceed to complete the square and finish carrying out the integration. However, we will use the Convolution Theorem to evaluate the convolution. Recalling the Fourier transform of a Gaussian, we have

$$\hat{f}(k) = F[e^{-ax^2}] = \sqrt{\frac{2\pi}{a}}\, e^{-k^2/2a}$$

and

$$\hat{g}(k) = F[e^{-bx^2}] = \sqrt{\frac{2\pi}{b}}\, e^{-k^2/2b}$$

Denoting the convolution function by $h(x) = (f * g)(x)$, the Convolution Theorem gives

$$\hat{h}(k) = \hat{f}(k)\,\hat{g}(k) = \frac{2\pi}{\sqrt{ab}} e^{-k^2/2b} e^{-k^2/2b}$$

This is another Gaussian function, as seen by rewriting the Fourier transform of $h(x)$ as

$$\hat{h}(k) = \hat{h}(k) = \frac{2\pi}{\sqrt{ab}} e^{-\frac{1}{2}\left(\frac{1}{a}+\frac{1}{b}\right)k^2} = \frac{2\pi}{\sqrt{ab}} e^{-\frac{a+b}{2ab}k^2}$$

To complete the evaluation of the convolution of these two Gaussian functions, we need to find the inverse transform of the Gaussian in equation. We have first that

$$F^{-1}\left[\sqrt{\frac{2\pi}{a}}\, e^{-k^2/2a}\right] = e^{-ax^2}.$$

Moving the constants, we then obtain

$$F^{-1}[e^{-k^2/2a}] = \sqrt{\frac{a}{2\pi}}\, e^{-ax^2}.$$

We now make the substitution $\alpha = \dfrac{1}{2a}$,

$$F^{-1}[e^{-\alpha k^2}] = \sqrt{\frac{1}{4\pi\alpha}}\, e^{-x^2/2\alpha}.$$

This is in the form needed to invert. Thus, for $\alpha = \dfrac{a+b}{2ab}$ we and

$$(f * g)(x) = h(x) = \sqrt{\frac{2\pi}{a+b}}\, e^{-\frac{ab}{a+b}x^2}.$$

APPLICATIONS OF THE CONVOLUTION THEOREM

There are many applications of the convolution operation. In this section we will describe a few of the applications.

The first application is filtering signals. For a given signal there might be some noise in the signal, some undesirable high frequencies, or the device used for recording an analog signal might naturally not be able to record high frequencies.

Let $f(t)$ denote the amplitude of a given analog signal and $\hat{f}(\omega)$ be the Fourier transform of this signal.

Recall that the Fourier transform gives the frequency content of the signal and that $\omega = 2\pi\nu$, where ν is the frequency in Hertz, or cycles per second (cps).

There are many ways to filter out unwanted frequencies. The simplest would be to just drop all of the high frequencies, $|\omega| > \omega_0$ for some cutoff frequency ω_0.

The Fourier transform of the filtered signal would then be zero for $|\omega| > \omega_0$. This could be accomplished by multiplying the Fourier

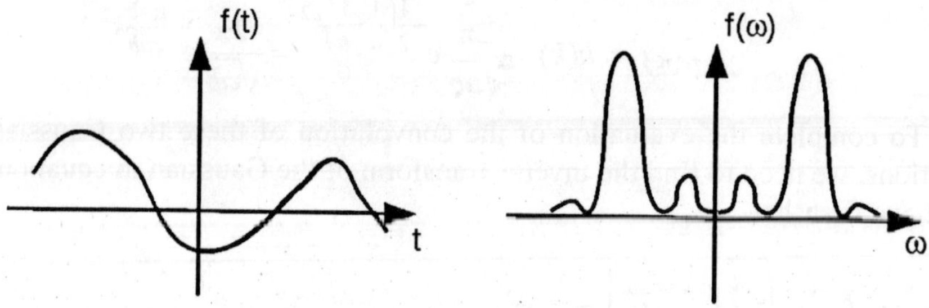

Fig. Plot of a Signal $f(t)$ and its Fourier Transform $\hat{f}(\omega)$

transform of the signal by a function that vanishes for $|\omega| > \omega_0$. For example, we could consider the gate function

$$p\omega_0(\omega) = \begin{cases} 1, & |\omega| \le \omega_0 \\ 0, & |\omega| > \omega_0 \end{cases}$$

Shows how the Gate Function is Used to Filter the Signal. In general, we multiply the Fourier transform of the signal by some filtering function $\hat{h}(t)$ to get the Fourier transform of the filtered signal,

$$\hat{g}(\omega) = \hat{f}(\omega)\hat{h}(\omega).$$

The new signal, $g(t)$ is then the inverse Fourier transform of this product, giving the new signal as a convolution.

$$g(t) = F^{-1}\left[\hat{f}(\omega)\hat{h}(\omega)\right] = \int_{-\infty}^{\infty} h(t-\tau)f(\tau)\,d\tau.$$

Such processes occur often in systems theory as well. One thinks of $f(t)$ as the input signal into some filtering device which in turn produces the output, $g(t)$. The function $h(t)$ is called the impulse response. This is because it is a response to the impulse function, $\delta(t)$. In this case, one has

$$\int_{-\infty}^{\infty} h(t-\tau)\delta(\tau)\,d\tau = h(t)$$

Another application of the convolution is in windowing. This represents what happens when one measures a real signal. Real signals cannot be recorded for all values of time. Instead data is collected over a finite time

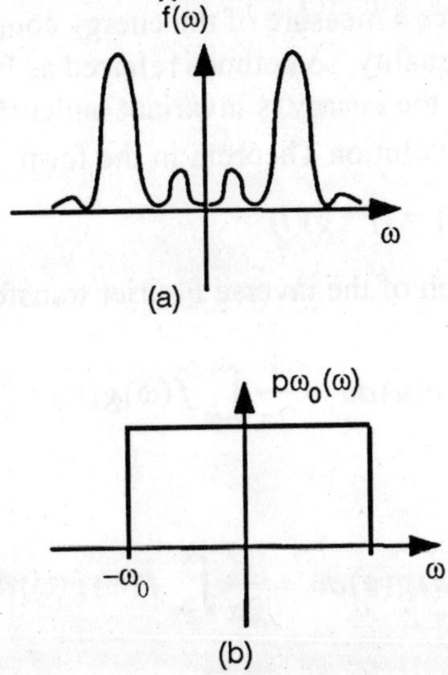

Fig. (a) Plot of the Fourier transform $\hat{f}(\omega)$ of a signal. (b) The gatefunction $p_{\omega 0}(\omega)$ used to filter out high frequencies.

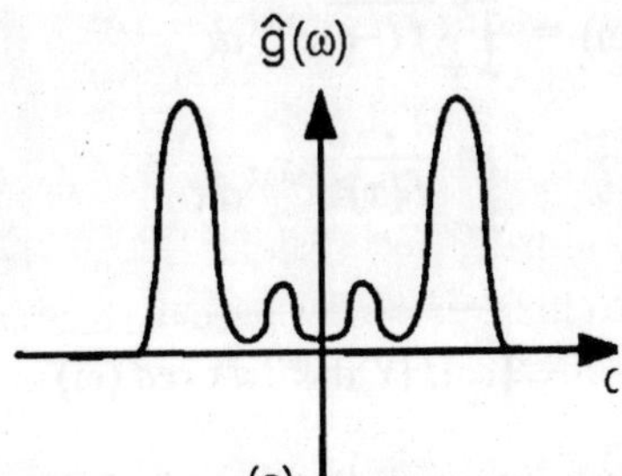

Fig. (c) The product of the functions, $\hat{g}(\omega) = \hat{f}(\omega)p_{\omega 0}(\omega)$, in (a) and (b)

interval. If the length of time the data is collected is T, then the resulting signal is zero outside this time interval. This can be modeled in the same way as with filtering, except the new signal will be the product of the old signal with the windowing function.

The resulting Fourier transform of the new signal will be a convolution of the Fourier transforms of the original signal and the windowing function. We can also use the convolution theorem to derive Parseval's Equality.

$$\int_{-\infty}^{\infty} |f(t)|^2\, dt = \frac{1}{2\pi}\int_{-\infty}^{\infty} \left|\hat{f}(\omega)\right|^2 d\omega$$

This equality has a physical meaning for signals. The integral on the left side is a measure of the energy content of the signal in the time domain.

The right side provides a measure of the energy content of the transform of the signal. Parseval's equality, sometimes referred as Plancherel's formula, is simply a statement that the energy is invariant under the transform.

Let's rewrite the Convolution Theorem in the form

$$F^{-1}[\hat{f}(k)\hat{g}(k)] = (f * g)(t).$$

Then, by the definition of the inverse Fourier transform, we have

$$\int_{-\infty}^{\infty} f(t-u)\, g(u)\, du = \frac{1}{2\pi}\int_{-\infty}^{\infty} \hat{f}(\omega)\,\hat{g}(\omega)\, e^{-i\omega t}\, d\omega\,.$$

Setting $t = 0$,

$$\int_{-\infty}^{\infty} f(t-u)\, g(u)\, du = \frac{1}{2\pi}\int_{-\infty}^{\infty} \hat{f}(\omega)\,\hat{g}(\omega)\, d\omega\,.$$

Now, let $g(t) = \overline{f(-t)}$, or $f(-t) = \overline{g(-t)}$. Then, the Fourier transform of $g(t)$ is related to the Fourier transform of $f(t)$:

$$\begin{aligned}\hat{g}(\omega) &= \int_{-\infty}^{\infty} \overline{f(-t)}\, e^{i\omega t}\, dt \\ &= \int_{-\infty}^{\infty} \overline{f(\tau)}\, e^{-i\omega\tau}\, d\tau \\ &= \overline{\int_{-\infty}^{\infty} f(\tau) e^{-i\omega\tau}\, d\tau} = \overline{\hat{f}(\omega)}.\end{aligned}$$

So, inserting this result into equation, we find that

$$\int_{-\infty}^{\infty} f(-u)\,\overline{f(-u)}\, du = \frac{1}{2\pi}\int_{-\infty}^{\infty} \left|\hat{f}(\omega)\right|^2 d\omega$$

which implies Parseval's Equality.

THE LAPLACE TRANSFORM

Up until this point we have only explored Fourier exponential transforms as one type of integral transform. The Fourier transform is useful on infinite domains.

However, students are often introduced to another integral transform, called the Laplace transform, in their introductory differential equations class. These transforms are defined over semi-infinite domains and are useful for solving ordinary differential equations. They also have proven useful in engineering for solving circuit problems and doing systems analysis.

The Laplace transform of a function $f(t)$ is defined as

$$F(s) = \mathcal{L}[f](s) = \int_0^{\infty} f(t)e^{-st}\,dt, \quad s > 0.$$

This is an improper integral and one needs

$$\lim_{t \to \infty} f(t)e^{-st} = 0$$

to guarantee convergence.

It is typical that one makes use of Laplace transforms by referring to a Table of transform pairs.

Combining some of these simple Laplace transforms with the properties of the Laplace transform, we can deal with many applications of the Laplace transform. We will first prove a few of the given Laplace transforms and show how they can be used to obtain new transform pairs. In the next section we will show how these can be used to solve ordinary differential equations.

We begin with some simple transforms. These are found by simply using the definition of the Laplace transform.

Example: $\mathcal{L}[1]$

For this example, we insert $f(t) = 1$ into our integral.

$$\mathcal{L}[1] = \int_0^{\infty} e^{-st}\,dt.$$

This is an improper integral and the computation is understood by introducing an upper limit of a and then letting $a \to \infty$. We will not always write this limit, but it will be understood that this is how one computes such improper integrals. Thus, we have

$$\mathcal{L}[1] = \int_0^{\infty} e^{-st}\,dt$$

$$= \lim_{a \to \infty} \int_0^{\infty} e^{-st}\,dt$$

$$= \lim_{a\to\infty}\left(-\frac{1}{s}e^{-st}\right)_0^a$$

$$= \lim_{a\to\infty}\left(-\frac{1}{s}e^{-st} + \frac{1}{s}\right) = \frac{1}{s}.$$

Table: Table of Selected Laplace Transform Pairs

$f(t)$	$F(s)$	$f(t)$	$F(s)$
c	$\frac{c}{s}$	e^{at}	$\frac{1}{s-a}$, $s > a$
t^n	$\frac{n!}{s^n+1}$, $s > 0$	$t_n e^{at}$	$\frac{n!}{(s-a)^{n+1}}$
$\sin \omega t$	$\frac{\omega}{s^2+\omega^2}$	$e^{at} \sin \omega t$	$\frac{\omega}{(s-a)^2+\omega^2}$
$\cos \omega t$	$\frac{s}{s^2+\omega^2}$	$e^{at} \sin \omega t$	$\frac{s-a}{(s-a)^2+\omega^2}$
$t \sin \omega t$	$\frac{2\omega s}{(s^2+\omega)^2}$	$t \cos \omega t$	$\frac{s^2-\omega^2}{(s^2+\omega^2)^2}$
$\sinh at$	$\frac{a}{s^2+a^2}$	$\cosh at$	$\frac{s}{s^2-a^2}$

$H(t-a)\frac{e^{-as}}{s}$, $s > 0$ $\delta(t-a)$ e^{-as}, $a \geq 0$, $s > 0$

Thus, we have that the Laplace transform of 1 is $\frac{1}{s}$. This can be extended to any constant c, using the property of linearity of the transform. The Laplace transform is simply an integral.

So, $\mathcal{L}[c] = c\mathcal{L}[1]$, since we can pull a constant factor out from under the integral. Therefore, we have

$$\mathcal{L}[c] = \frac{c}{s}.$$

Example: $\mathcal{L}[e^{at}]$,

For this example, we can easily compute the transform. It again is simply the integral of an exponential function.

$$\mathcal{L}[e^{at}] = \int_0^\infty e^{at}e^{-st}\,dt$$

$$= \int_0^\infty e^{(a-s)t}\,dt$$

$$= \left(\frac{1}{a-s} e^{(a-s)t} \right)_0^\infty$$

$$= \lim_{t\to\infty} \frac{1}{a-s} e^{(a-s)t} - \frac{1}{a-s} = \frac{1}{s-a}.$$

Note that the last limit was computed as $\lim_{t\to\infty} e^{(a-s)t} = 0$. This is only true if $a - s < 0$, or $s > a$. [Actually, a could be complex. In this case we would only need that s is greater than the real part of a.]

Example: $\mathcal{L}[\cos at]$ and $\mathcal{L}[\sin at]$

In these cases, we could again insert the functions directly into the transform. For example,

$$\mathcal{L}[\cos at] = \int_0^\infty e^{-st} \cos at \, dt.$$

Recall how one does such integrals involving both the trigonometric function and the exponential function.

One integrates by parts two times and then obtains an integral of the form with which one started. Rearranging the result the answer can be obtained.

However, there is a much simpler way to compute these transforms.

Recall that $e^{iat} = \cos at + i \sin at$. Making use of the linearity of the Laplace transform, we have

$$\mathcal{L}[e^{iat}] = \mathcal{L}[\cos at] + i\mathcal{L}[\sin at].$$

Thus, transforming this complex exponential and looking at the real and imaginary parts of the results will give both transforms at the same time! The transform is simply computed as

$$\mathcal{L}[e^{iat}] = \int_0^\infty e^{iat} e^{-st} \, dt = \int_0^\infty e^{-(s-ia)t} \, dt = \frac{1}{s-ia}.$$

Note that we could easily have used the result for the transform of an exponential, which was already proven. In this case $s > Re(ia) = 0$.

We now extract the real and imaginary parts of the result using the complex conjugate of the denominator.

$$\frac{1}{s-ia} = \frac{1}{s-ia} \frac{s+ia}{s+ia} = \frac{s+ia}{s^2+a^2}.$$

Reading off the real and imaginary parts gives

$$\mathcal{L}[\cos at] = \frac{s}{s^2+a^2}$$

$$\mathcal{L}[\sin at] = \frac{a}{s^2+a^2}.$$

Example: $\mathcal{L}[t]$

For this example we need to evaluate

$$\mathcal{L}[t] = \int_0^\infty te^{-st}\,dt\,.$$

This integration can be done using integration by parts. (Pick $u = t$ and $dv = e^{-st}\,dt$. Then, $du = dt$ and $v = -\frac{1}{s}e^{-st}$.)

$$\int_0^\infty te^{-st}\,dt = -t\frac{1}{s}e^{-st}\Big|_0^\infty \frac{1}{s}\int_0^\infty te^{-st}\,dt$$

$$= \frac{1}{s^2}\,.$$

Example: $\mathcal{L}[t^n]$

We can generalize the last example to powers greater than $n = 1$. In this case we have to do the integral

$$\mathcal{L}[t^n] = \int_0^\infty t^n e^{-st}\,dt.$$

Following the previous example, we integrate by parts.

$$\int_0^\infty t^n e^{-st}\,dt = -t^n\frac{1}{s}e^{-st}\Big|_0^\infty + n\frac{1}{s}\int_0^\infty t^{-n}e^{-st}\,dt$$

$$= n\frac{1}{s}\int_0^\infty t^n e^{-st}\,dt.$$

We could continue to integrate by parts until the final integral can be computed. However, look at the integral that resulted after one integration by parts. It is just the Laplace transform of t^{n-1}. So, we can write the result as

$$\mathcal{L}[t^n] = \frac{n}{s}\mathcal{L}[t^{n-1}].$$

This is an example of an recursive definition of a sequence, in this case a sequence of integrals. Denoting

$$I_n = \mathcal{L}[t^n] = \int_0^\infty t^n e^{-st}\,dt$$

and noting that $I\,[0] = \mathcal{L}[1] = \frac{1}{s}$, we have the following.

$$I_n = \frac{n}{s}I_{n-1}, \quad I_0 = \frac{1}{s}.$$

This is also what is called a difference equation. It is a first order difference equation with an initial condition", I_0. There is a whole theory of difference equations, which we will not get into here.

Our goal is to solve the above difference equation. It is easy to do by simple iteration. Note that replacing n with $n - 1$, we have

$$I_{n-1} = \frac{n-1}{s} I_{n-2}.$$

So, repeating the process we find

$$\begin{aligned} I_n &= \frac{n}{s} I_{n-1} \\ &= \frac{n}{s}\left(\frac{n-1}{s}\right) I_{n-2} \\ &= \frac{n(n-1)}{s^2} I_{n-2}. \end{aligned}$$

We can repeat this process until we get to I_0, which we know. In some cases you need to be careful so that you can count the number of iterations of the process. So, we first ask what the result is after k steps. This can be seen by watching for patterns. Continuing the iteration process, we have

$$\begin{aligned} I_n &= \frac{n}{s} I_{n-1} \\ &= \frac{n(n-1)}{s^2} I_{n-2} \\ &= \frac{n(n-1)(n-2)}{s^3} I_{n-3} \\ &= \ldots \\ &= \frac{n(n-1)(n-2)\ldots(n-k+1)}{s^k} I_{n-k} \end{aligned}$$

Since we know I_0, we choose to stop at $k = n$ obtaining

$$I_n = \frac{n(n-1)(n-2)\ldots(2)(1)}{s^n} I_0 = \frac{n!}{s^{n+1}}$$

Therefore, we have shown that $\mathcal{L}[t^n] = \frac{n!}{s^{n+1}}$. [Such iterative techniques are useful in obtaining a variety of of integrals, such as $I_n = \int_{-\infty}^{\infty} x^{2n} e^{-x^2} dx$.]

As a final note, one can extend this result to cases when n is not an integer. To do this, one introduces what is called the Gamma function. This function is defined as

$$\Gamma(x) = \int_0^\infty t^{x-1} e^{-t} dt$$

Note the similarity to the Laplace transform of t^{x-1}:

$$\mathcal{L}[t^{x-1}] = \int_0^\infty t^{x-1} e^{-st} dt.$$

For $x - 1$ an integer and $s = 1$, we have that

$$\Gamma(x) = (x - 1)!.$$

Thus, the Gamma function seems to be a generalization of the factorial.

$$\mathcal{L}[t^p] = \frac{\Gamma(p+1)}{s^{p+1}}$$

for values of $p > -1$.

Example: $\mathcal{L}\left[\frac{df}{dt}\right]$.

We have to compute

$$\mathcal{L}\left[\frac{df}{dt}\right] = \int_0^\infty \frac{df}{dt} e^{-st} dt.$$

Laplace Transform Properties

$$\mathcal{L}[af(t) + bg(t)] = aF(s) + bG(s)$$

$$\mathcal{L}[tf(t)] = \frac{d}{ds} F(s)$$

$$\mathcal{L}\left[\frac{dy}{dt}\right] = sY(s) - y(0)$$

$$\mathcal{L}\left[\frac{d^2y}{dt^2}\right] = s^2Y(s) - sy(0) - y'(0)$$

$$\mathcal{L}[e^{at}f(t)] = F(s - a)$$

$$\mathcal{L}[H(t - a)f(t - a)] = e^{-as}F(s)$$

$$\mathcal{L}[(f * g)(t)] = \mathcal{L}[\textstyle\int_0^t f(t - u)g(u)\, du] = F(s)G(s)$$

We can move the derivative off of f by integrating by parts. This is similar to what we had done when finding the Fourier transform of the derivative of a function.

Thus letting $u = e^{-st}$ and $v = f(t)$, we have

$$\mathcal{L}\left[\frac{df}{dt}\right] = \int_0^\infty \frac{df}{dt} e^{-st}\,dt$$

$$= f(t)e^{-st}\Big|_0^\infty + s\int_0^\infty f(t)e^{-st}\,dt$$

$$= -f(0) + sF(s).$$

Here we have assumed that $f(t)e^{-st}$ vanishes for large t. The final result is that

$$\mathcal{L}\left[\frac{df}{dt}\right] = sF(s) - f(0).$$

Example: $\mathcal{L}\left[\frac{d^2 f}{dt^2}\right]$.

We can compute this using two integrations by parts, or we could make use of the last result.

Letting $g(t) = \frac{df(t)}{dt}$, we have

$$\mathcal{L}\left[\frac{d^2 f}{dt^2}\right] = \mathcal{L}\left[\frac{dg}{dt}\right] = sG(s) - g(0) = sG(s) - f'(0).$$

But,

$$G(s) = \mathcal{L}\left[\frac{df}{dt}\right] = sF(s) - f(0).$$

So,

$$\mathcal{L}\left[\frac{d^2 f}{dt^2}\right] = sG(s) - f'(0)$$

$$= s(sF(s) - f(0)) - f'(0)$$

$$= s^2F(s) - sf(0) - f'(0).$$

SOLUTION OF ODES USING LAPLACE TRANSFORMS

One of the typical applications of Laplace transforms is the solution of nonhomogeneous linear constant coefficient differential equations. In the following examples we will show how this works.

The general idea is that one transforms the equation for an unknown function y(t) into an algebraic equation for its transform, $Y(t)$. Typically, the algebraic equation is easy to solve for $Y(s)$ as a function of s. Then one transforms back into t-space using Laplace transform tables and the properties of Laplace transforms.

There is an integral form for the inverse transform. This is typically not covered in introductory differential equations classes as one needs carry out integrations in the complex plane.

Example: Solve the initial value problem $y' + 3y = e^{2t}$, $y(0) = 1$. The first step is to perform a Laplace transform of the initial value problem. The transform of the left side of the equation is

$$L[y' + 3y] = sY - y(0) + 3Y = (s + 3)Y - 1.$$

Transforming the right hand side, we have

$$L[e^{2t}] = \frac{1}{s-2}.$$

Combining these, we obtain

$$(s+3)Y - 1 = \frac{1}{s-2}.$$

The next step is to solve for $Y(s)$.

$$Y(s) = \frac{1}{s+3} + \frac{1}{(s-2)(s+3)}.$$

One transforms the initial value problem for $y(t)$ and obtains an algebraic equation for $Y(s)$. Solve for $Y(s)$ and the inverse transform give the solution to the initial valuc problem. Now, we need to find the inverse Laplace transform. Namely, we need to figure out what function has a Laplace transform of the above form. It is easy to do if we only had the first term. The inverse transform of the first term is e^{-3t}.

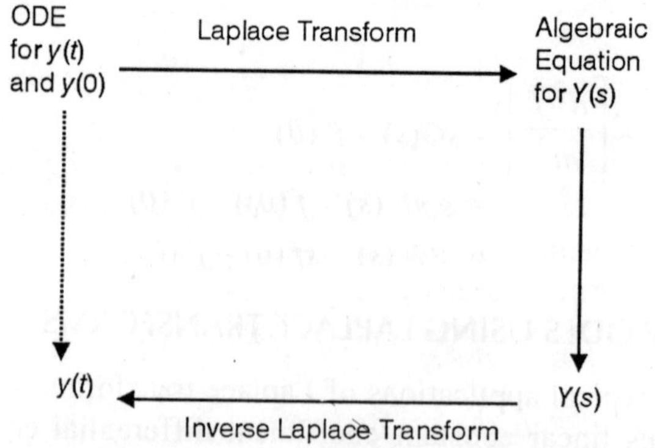

Fig. The Scheme for Solving an Ordinary Differential Equation using Laplace Transforms

We have not seen anything that looks like the second form in the table of transforms that we have compiled so far. However, we are not stuck. We know that we can rewrite the second term by using a partial fraction decomposition.

Let's recall how to do this. The goal is to find constants, A and B, such that

$$\frac{1}{(s-2)(s+3)}=\frac{A}{s-2}+\frac{B}{s+3}.$$

We picked this form because we know that recombining the two terms into one term will have the same denominator. We just need to make sure the numerators agree afterwards. So, adding the two terms, we have

$$\frac{1}{(s-2)(s+3)}=\frac{A(s+3)+B(s-2)}{(s-2)(s+3)}.$$

Equating numerators,

$$1 = A(s + 3) + B(s - 2).$$

This has to be true for all s. Rewriting the equation by gathering terms with common powers of s, we have

$$(A + B)s + 3A - 2B = 1.$$

The only way that this can be true for all s is that the coefficients of the different powers of s agree on both sides. This leads to two equations for A and B:

$$A + B = 0$$
$$3A - 2B = 1.$$

The first equation gives $A = -B$, so the second equation becomes $-5B = 1$. The solution is then $A = -B = \frac{1}{5}$.

Returning to the problem, we have found that

$$Y(s)=\frac{1}{s+3}\frac{1}{5}\left(\frac{1}{s-2}-\frac{1}{s+3}\right).$$

[Of course, we could have tried to guess the form of the partial fraction decomposition as we had done earlier when talking about Laurent series.] In order to finish the problem at hand, we find a function whose Laplace transform is of this form. We easily see that

$$y(t) = e^{-3t} + \frac{1}{5}\left(e^{2t}-e^{-3t}\right)$$

works. Simplifying, we have the solution of the initial value problem

$$y(t)=\frac{1}{5}e^{2t}+\frac{4}{5}e^{-3t}$$

Example: Solve the initial value problem $y'' + 4y = 0, y(0) = 1, y'(0) = 3$.

We can probably solve this without Laplace transforms, but it is a simple exercise. Transforming the equation, we have

$$\begin{aligned} 0 &= s^2Y - sy(0) - y'(0) + 4Y \\ &= (s^2 + 4)Y - s - 3. \end{aligned}$$

Solving for Y, we have

$$Y(s) = \frac{s+3}{s^2+4}.$$

We now ask if we recognize the transform pair needed. The denominator looks like the type needed for the transform of a sine or cosine. We just need to play with the numerator. Splitting the expression into two terms, we have

$$Y(s) = \frac{s}{s^2+4} + \frac{3}{s^2+4}.$$

The first term is now recognizable as the transform of $\cos 2t$. The second term is not the transform of $\sin 2t$.

It would be if the numerator were a 2. This can be corrected by multiplying and dividing by 2:

$$\frac{s}{s^2+4} = \frac{3}{2}\frac{s}{s^2+4}.$$

So, out solution is then found as

$$y(t) = \mathcal{L}\left[\frac{s}{s^2+4} + \frac{3}{2}\frac{3}{s^2+4}\right] = \cos 2t + \frac{3}{2}\sin 2t.$$

STEP AND IMPULSE FUNCTIONS

The initial value problems that we have solved so far can be solved using the Method of Undetermined Coefficients or the Method of Variation of Parameters.

However, using variation of parameters can be messy and involves some skill with integration. Many circuit designs can be modelled with systems of differential equations using Kirchoff's Rules.

Such systems can get fairly complicated. However, Laplace transforms can be used to solve such systems and electrical engineers have long used such methods in circuit analysis.

In this section we add a couple of more transform pairs and transform properties that are useful in accounting for things like turning on a driving force, using periodic functions like a square wave, or introducing an impulse force. We first recall the Heaviside step function, given by

$$H(t) = \begin{cases} 0, & t < 0 \\ 1, & t > 0 \end{cases}.$$

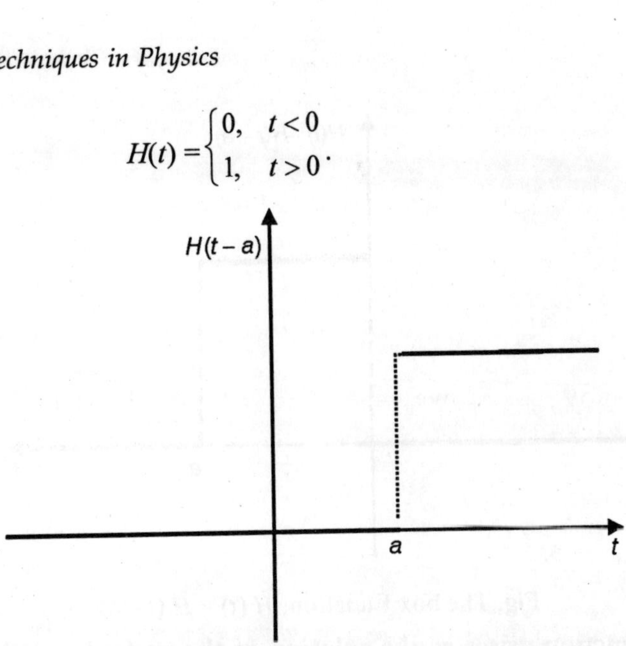

Fig. A Shifted Heaviside Function, $H(t - a)$

A more general version of the step function is the horizontally shifted step function, H (t – a). The Laplace transform of this function is found for $a > 0$ as

$$\mathcal{L}[H(t-a)] = \int_0^\infty H(t-a)e^{-st}\,dt$$

$$= \int_a^\infty H(t-a)e^{-st}\,dt$$

$$= \left.\frac{e^{-st}}{s}\right|_a^\infty = \frac{e^{-as}}{s}.$$

Just like the Fourier transform, the Laplace transform has two shift theorems involving multiplication of $f(t)$ or $F(s)$ by exponentials. These are given by

$$\mathcal{L}[e^{at}f(t)] = F(s-a)$$
$$\mathcal{L}[f(t-a)H(t-a)] = e^{-as}F(s).$$

We prove the first shift theorem and leave the other proof as an exercise for the reader. Namely,

$$\mathcal{L}[e^{at}f(t)] = \int_0^\infty e^{at} f(t)e^{-st}\,dt$$

$$= \int_0^\infty f(t)e^{-(s-a)t}\,dt = F(s-a).$$

Example: Compute the Laplace transform of $e^{-at} \sin \omega t$.

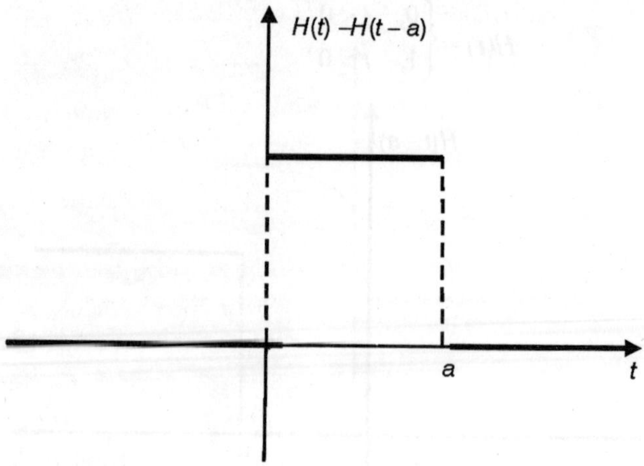

Fig. The box Function, $H(t) - H(t-a)$

This function arises as the solution of the underdamped harmonic oscillator. We first note that the exponential multiplies a sine function. The shift theorem tells us that we need the transform of this function. So,

$$F(s) = \frac{\omega}{s^2 + \omega^2}.$$

Knowing this, we can write the solution as

$$\mathcal{L}[e^{-at} \sin \omega t] = F(s + a) = \frac{\omega}{(s + a)^2 + \omega^2}.$$

More interesting examples can be found in piecewise functions. First we consider the function H $(t) - H(t - a)$. For $t < 0$ both terms are zero. In the interval [0, a] the function H $(t) = 1$ and H $(t - a) = 0$. Therefore, $H(t) - H(t - a) = 1$ for $t \in [0, a]$. Finally, for $t > a$, both functions are one and therefore the difference is zero. We now consider the piecewise defined function

$$g(t) = \begin{cases} f(t), & 0 \le t \le a, \\ 0, & t < 0, t > a \end{cases}.$$

This function can be rewritten in terms of step functions. We only need to multiply $f(t)$ by the above box function, $g(t) = f(t)[H(t) - H(t - a)]$.

Even more complicated functions can be written out in terms of step functions. We only need to look at sums of functions of the form $f(t)[H(t - a) - H(t - b)]$ for $b > a$. This is just a box between a and b of height $f(t)$. It can be represented as a sum of an infinite number of boxes,

$$f(t) = \sum_{n=-\infty}^{\infty} [H(t - 2na) - H(t - (2n + 1)a)].$$

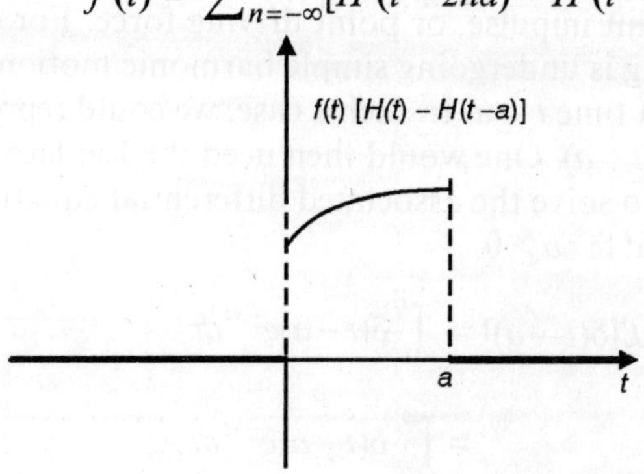

Fig. Formation of a Piecewise Function, $f(t)[H(t) - H(t - a)]$

Example: Laplace Transform of a square wave turned on at $t = 0$,

$$f(t) = \sum_{n=0}^{\infty} [H(t-2na) - H(t-(2n+1)a)]$$

Using the properties of the Heaviside function, we have

$$\mathcal{L}[\delta(t - a)] = \sum_{n=0}^{\infty} [\mathcal{L}[H(t-2na) - \mathcal{L}[H(t-(2n+1)a)]]$$

$$= \sum_{n=0}^{\infty} \left[\frac{e^{-2nas}}{s} - \frac{e^{-2(n+1)as}}{s} \right]$$

$$= \frac{1-e^{-as}}{s} \left(\frac{1}{1-e^{-2as}} \right) = \frac{1-e^{-as}}{s(1-e^{-2as})}.$$

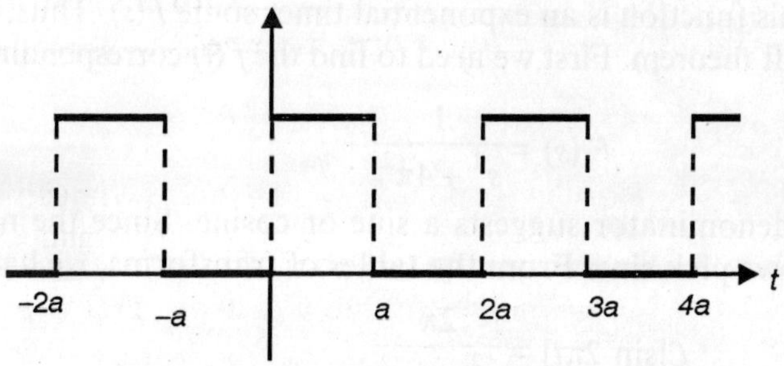

Fig. A Square Wave, $f(t) = \sum_{n=-\infty}^{\infty} [H(t - 2na) - H(t - (2n + 1)a)]$

Note that the third line in the derivation is a geometric series. We summed this series to get our answer in a compact form.

Another interesting example is the delta function. The delta function represents a point impulse, or point driving force. For example, while a mass on a spring is undergoing simple harmonic motion, one could hit it for an instant at time t = a. In such a case, we could represent the force as a multiple of $\delta(t - a)$. One would then need the Laplace transform of the delta function to solve the associated differential equation.

We find that for $a > 0$.

$$\mathcal{L}[\delta(t-a)] = \int_0^\infty \delta(t-a)e^{-st}\,dt$$

$$= \int_{-\infty}^\infty \delta(t-a)e^{-st}\,dt.$$

$$= e^{-as}.$$

Example. Solve the initial value problem

$$y'' + 4\pi^2 y = \delta(t-2),\ y(0) = y'(0) = 0.$$

In this case we see that we have a nonhomogeneous spring problem. Without the forcing term, given by the delta function, this spring is initially at rest and not stretched. The delta function models a unit impulse at $t = 2$. Of course, we anticipate that at this time the spring will begin to oscillate. We will solve this problem using Laplace transforms.

First, transform the differential equation.

$$s^2 Y - sy(0) - y'(0) + 4\pi^2 Y = e^{-2s}.$$

Inserting the initial conditions, we have

$$(s^2 + 4\pi^2)Y = e^{-2s}.$$

Solve for $Y(s)$.

$$Y(s) = \frac{e^{-2s}}{s^2 + 4\pi^2}.$$

We now seek the function for which this is the Laplace transform. The form of this function is an exponential times some $F(s)$. Thus, we need the second shift theorem. First we need to find the $f(t)$ corresponding to

$$F(s) = \frac{1}{s^2 + 4\pi^2}.$$

The denominator suggests a sine or cosine. Since the numerator is constant, we pick sine. From the tables of transforms, we have

$$\mathcal{L}[\sin 2\pi t] = \frac{2\pi}{s^2 + 4\pi^2}.$$

So, we write

$$F(s) = \frac{1}{2\pi}\frac{2\pi}{s^2 + 4\pi^2}.$$

This gives $f(t) = (2\pi)^{-1} \sin 2\pi t$.

We now apply the second shift theorem,

$\mathcal{L}[f(t-a)H(t-a)] = e^{-as} F(s)$.

$$y(t) = H(t-2)f(t-2)$$

$$= \frac{1}{2\pi} H(t-2)\sin 2\pi(t-2).$$

This solution tells us that the mass is at rest until $t = 2$ and then begins to oscillate at its natural frequency.

Finally, we consider the convolution of two functions. Often we are faced with having the product of two Laplace transforms that we know and we seek the inverse transform of the product. For example, let's say you end up with $Y(s) = \frac{1}{(s-1)(s-2)}$ while trying to solve a differential equation. We know how to do this if we only have one of the denominators present.

Of course, we could do a partial fraction decomposition. But, there is another way to find the inverse transform, especially if we cannot perform a partial fraction decomposition.

We define the convolution of two functions defined on $[0, \infty)$ much the same way as we had done for the Fourier transform. We define

$$(f * g)(t) = \int_0^t f(u)g(t-u)\,du.$$

The convolution operation has two important properties.

- The convolution is commutative. $f * g = g * f$

Proof: The key is to make a substitution $y = t - u$ int the integral to make f a simple function of the integration variable.

$$(g * f)(t) = \int_0^t f(u)g(t-u)\,du$$

$$= -\int_t^0 g(t-y)f(y)\,dy$$

$$= \int_t^0 f(y)g(t-y)\,dy$$

$$= (f * g)(t)$$

***The Convolution** Theorem:* The Laplace transform of a convolution is the product of the Laplace transforms of the individual functions:

$$\mathcal{L}[f * g] = F(s)G(s)$$

Proving this theorem takes a bit more work. We will make some assumptions that will work in many cases. First, we assume that our functions are causal, $f(t) = 0$ and $g(t) = 0$ for $t < 0$. Secondly, we will assume that we can interchange integrals, which needs more rigorous attention than will be provided here.

The first assumption will allow us to write the finite integral as an infinite integral. Then a change of variables will allow us to split the integral into the product of two integrals that are recognized as a product of two Laplace transforms.

$$\mathcal{L}[f*g] = \int_0^\infty\left(\int_0^t f(u)g(t-u)\,du\right)e^{-st}\,dt$$

$$= \int_0^\infty\left(\int_0^\infty f(u)g(t-u)\,du\right)e^{-st}\,dt$$

$$= \int_0^\infty f(u)\left(\int_0^\infty g(t-u)e^{-st}\,dt\right)du$$

$$= \int_0^\infty f(u)\left(\int_0^\infty g(\tau)e^{-s(\tau+u)}\,d\tau\right)du$$

$$= \int_0^\infty f(u)e^{-su}\left(\int_0^\infty g(\tau)e^{-s\tau}\,d\tau\right)du$$

$$= \left(\int_0^\infty f(u)e^{-su}\,du\right)\left(\int_0^\infty g(\tau)e^{-s\tau}\,d\tau\right)$$

$$= F(s)G(s).$$

We make use of the Convolution theorem to do the following example.

Example: $y(t) = \mathcal{L}^{-1}\left[\dfrac{1}{(s-1)(s-2)}\right].$

We note that this is a product of two functions

$$Y(s) = \frac{1}{(s-1)(s-2)} = \frac{1}{s-1}\frac{1}{s-2}$$

$$= F(s)\,G(s).$$

We know the inverse transforms of the factors. $f(t) = e^t$ and $g(t) = e^{2t}$.

Using the Convolution Theorem, we find that $y(t) = (f*g)(t)$. We compute the convolution:

$$y(t) = \int_0^t f(u)g(t-u)\,du$$

$$= \int_0^t e^u e^{2(t-u)}\,du$$

$$= e^{2t}\int_0^t e^{-u}\,du$$

$$= e^{2t}[-e^t + 1] = e^{2t} - e^t.$$

You can confirm this by carrying out the partial fraction decomposition.

The Inverse Laplace Transform

Up until this point we have seen that the inverse Laplace transform can be found by making use of Laplace transform tables and properties of Laplace transforms. This is typically the way Laplace transforms are taught and used. One can do the same for Fourier transforms. However, in that case we introduced an inverse transform in the form of an integral. does such an inverse exist for the Laplace transform? Yes, it does. In this section we will introduce the inverse Laplace transform integral and show how it is used.

We begin by considering a function $f(t)$ which vanishes for $t < 0$. We define the function $g(t) = f(t)e^{-ct}$. For $g(t)$ absolutely integrable,

$$\int_{-\infty}^{\infty} |g(t)|\, dt = \int_0^{\infty} |f(t)|\, e^{-ct} dt < \infty,$$

we can write the Fourier transform,

$$\hat{g}(\omega) = \int_{-\infty}^{\infty} g(t) e^{-i\omega t} dt = \int_0^{\infty} f(t) e^{i\omega t - ct} dt$$

and the inverse Fourier transform,

$$g(t) = f(t)e^{-ct} = \frac{1}{2\pi} \int_{-\infty}^{\infty} \hat{g}(\omega) e^{-i\omega t} d\omega.$$

Multiplying by e^{ct} and inserting $\hat{g}(\omega)$ into the integral for $g(t)$, we find

$$f(t) = \frac{1}{2\pi} \int_{-\infty}^{\infty} \int_{-\infty}^{\infty} f(\tau) e^{(i\omega - c)\tau} d\tau e^{-(i\omega - c)t} d\omega.$$

Letting $s = c - i\omega$ (so $d\omega = ids$), we have

$$f(t) = \frac{i}{2\pi} \int_{c+i\infty}^{c-i\infty} \int_0^{\infty} f(\tau) e^{-s\tau} d\tau e^{-st} ds$$

Note that the inside integral is simply $F(s)$. So, we have

$$f(t) = \frac{1}{2\pi i} \int_{c-i\infty}^{c+i\infty} F(s) e^{st} ds.$$

This is the inverse Laplace transform, called the Bromwich integral. This integral is evaluated along a path in the complex plane. The typical way to compute this integral is to chose c so that all poles are to the left of the contour and to close the contour with a semicircle enclosing the poles. One then relies on Jordan's lemma extended into the second and third quadrants.

Example: Find the inverse Laplace transform of $F(s) = \dfrac{1}{s(s+1)}$.

The integral we have to compute is

$$f(t) = \frac{1}{2\pi i}\int_{c-i\infty}^{c+i\infty} \frac{e^{st}}{s(s+1)} ds .$$

This integral has poles at $s = 0$ and $s = -1$. We enclose the contour with a semicircle to the left of the path in the complex s-plane. One has to verify that the integral over the semicircle vanishes as the radius goes to infinity. Assuming that we have done this, then the result is simply obtained as $2\pi i$ times the sum of the residues. The residues in this case are:

$$\text{Res}\left[\frac{e^{zt}}{z(z+1)}; z=0\right] = \lim_{z\to 0}\frac{e^{zt}}{(z+1)} = 1$$

and

$$\text{Res}\left[\frac{e^{zt}}{z(z+1)}; z-1\right] = \lim_{z\to -1}\frac{e^{zt}}{z} = -e^{-t}.$$

Therefore, we have

$$f(t) = 2\pi i\left(\frac{1}{2\pi i}(1) + \frac{1}{2\pi i}(-e^{-t})\right) = 1 - e^{-t}.$$

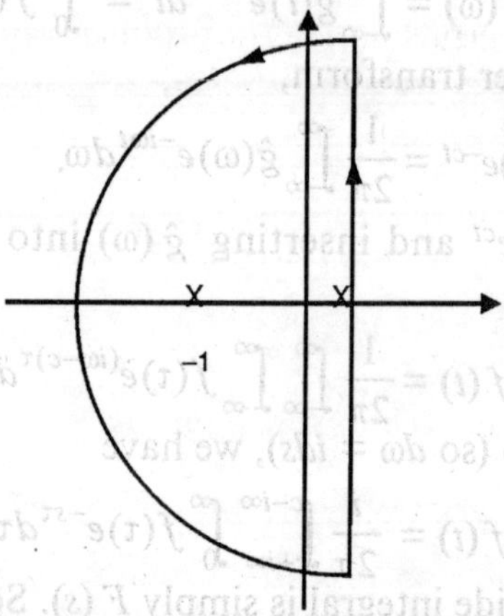

Fig. The Contour used for Applying the Bromwich Integral to $F(s) = \dfrac{1}{s(s+1)}$

We can verify this result using the Convolution Theorem or using partial fraction decomposition. The decomposition is simplest.

$$\frac{1}{s(s+1)} = \frac{1}{s} - \frac{1}{s+1}.$$

The first term leads to an inverse transform of 1 and the second term gives an e^{-t}. Thus, we have verified the result from doing a contour integration.

Chapter 7

Problems in Higher Dimensions

We will explore several generic examples of the solution of initial-boundary value problems involving higher spatial dimensions. These are described by higher dimensional partial differential equations. The method of solution will be the method of separation of variables. This will result in a system of ordinary differential equations for each problem.

Of these equations, several are boundary value problems, which are eigenvalue problems. We solve the eigenvalue problems for the eigenvalues and eigenfunctions, leading to a set of product solutions satisfying the partial differential equation and the boundary conditions. The general solution can be written as a linear superposition of the product solutions.

As you go through these examples, you will see some common features. For example, the main equation that we have seen are the heat equation and the wave equation. For higher dimensional problems these take the form

$$u_t = k\nabla^2 u,$$

$$u_{tt} = c^2\nabla^2 u.$$

One can first separate out the time dependence. Let $u(r, t) = \phi(r)T(t)$. For these two equations we have

$$T'\phi = kT\nabla^2\phi,$$

$$T''\phi = c^2 T\nabla^2\phi.$$

Separating out the time and space dependence, we find

$$\frac{1}{k}\frac{T'}{T} = \frac{\nabla^2\phi}{\phi} = -\lambda,$$

$$\frac{1}{c^2}\frac{T''}{T} = \frac{\nabla^2\phi}{\phi} = -\lambda.$$

Note that in each case we have a function of time equals a function of spatial variables. Thus, they must be a constant, $-\lambda < 0$. The sign of λ is chosen because we expect decaying solutions in time for the heat equation and oscillations in time for the wave equation.

This leads to the respective set of equations for $T(t)$.

$$T' = -\lambda kT,$$

$$T'' + c^2 \lambda T = 0.$$

These are easily solved. We have

$$T(t) = T(0)e^{-\lambda kt},$$

$$T(t) = a\cos \omega t + b \sin \omega t \quad \omega = c\sqrt{\lambda}.$$

In both cases the spatial equation becomes

$$\nabla^2\phi + \lambda\phi = 0.$$

This is called the Helmholtz equation.

For one dimensional problems, which wc havc already solved, the Helmholtz equation takes the form $\lambda'' + \lambda\phi = 0$. We had to impose the boundary conditions and found that there were a discrete set of eigenvalues, λ_n, and associated eigenfunctions, ϕ_n.

In higher dimensional problems we need to further separate out the space dependence.

We will again use the boundary conditions and find the eigenvalues and eigenfunctions, though they will be labelled with more than one index. The resulting boundary value problems belong to a class of eigenvalue problems called Sturm-Liouville problems.

Also, in the solution of the ordinary differential equations we will find solutions other than the sines and cosines that we have seen in previous problems, such as the vibrations of a string. The special functions also form a set of orthogonal functions, leading to general Fourier-type series expansions.

VIBRATIONS OF RECTANGULAR MEMBRANES

Our first example will be the study of the vibrations of a rectangular membrane.

You can think of this as a drum with a rectangular cross-section. We stretch the membrane over the drumhead and fasten the material to the boundaries of the rectangle.

The hight of the vibrating membrane is described by its hight from equilibrium, $u(x, y, t)$. This problem is a much simpler example of higher dimensional vibrations than that possed by the oscillating electric and magnetic fields.

The problem is given by a partial differential equation,

$$u_{tt} = c^2\nabla^2 u = c^2(u_{xx} + u_{yy}),\ t > 0,\ 0 < x < L,\ 0 < y < H.$$

a set of boundary conditions,

$$u(0, y, t) = 0,\ u(L, y, t) = 0,\ t > 0, \quad 0 < y < \mathrm{H},$$

$$u(x, 0, t) = 0,\ u(x, H, t) = 0,\ t > 0, \quad 0 < x < \mathrm{L},$$

and a pair of initial conditions (since the equation is second order in time),

$$u(x, y, 0) = f(x, y),\ ut(x, y, 0) = g(x, y).$$

The first step is to separate the variables. $u(x, y, t) = X(x)y(y)T(t)$. Inserting into the wave equation, we have

$$X(x)y(y)T''(t) = c^2 -X''(x)Y(y)T(t) + X(x)Y''(y)T(t)).$$

Dividing by both $u(x, y, t)$ and c^2, we obtain

$$\underbrace{\frac{1}{c^2}\frac{T''}{T}}_{\text{Function of } t} = \underbrace{\frac{X''}{X}\frac{Y''}{Y}}_{\text{Function of } t \text{ and } y} = -\lambda.$$

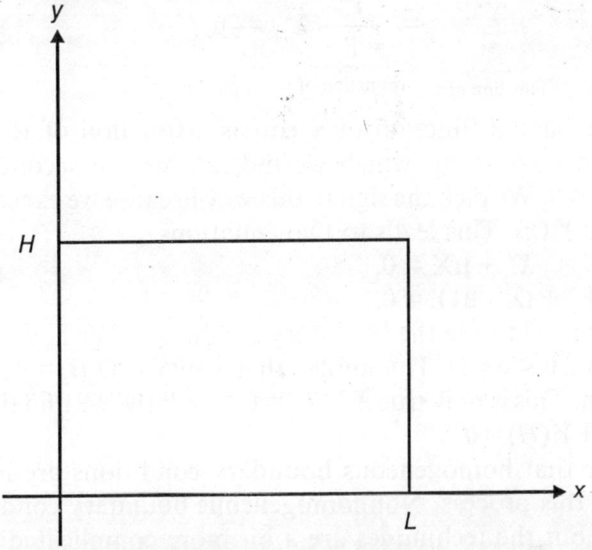

Fig. The Rectangular Membrane of Length L and Width H. There are Fixed Boundary Conditions Along the Edges

We see that we have a function of t equals a function of x and y. Thus, both expressions are constant. We expect oscillations in time, so we chose the constant to be $\lambda > 0$. (Note. As usual, the primes mean differentiation with respect to the specific dependent variable. So, there should be no ambiguity.)

These lead to two equations.

$$T'' + c^2\lambda T = 0,$$

and

$$\frac{X''}{X} + \frac{Y''}{Y} = -\lambda.$$

The first equation is easily solved. We have

$$T(t) = a \cos \omega t + b \sin \omega t,$$

where

$$\omega = c\sqrt{\lambda}.$$

This is the angular frequency in terms of the separation constant, or eigenvalue. It leads to the frequency of oscillations for the various harmonics of the vibrating membrane as

$$\nu = \frac{\omega}{2\pi} = \frac{c}{2\pi}\sqrt{\lambda}.$$

Once we know we can compute these frequencies.

Now we solve the spatial equation. Again, we need to do a separation of variables. Rearranging the spatial equation, we have

$$\underbrace{\frac{X''}{X}}_{\text{Function of } t} = \underbrace{-\frac{Y''}{Y} - \lambda}_{\text{Function of } y} = -\mu.$$

Here we have a function of x equals a function of y. So, the two expressions are constant, which we indicate with a second separation constant, $-\mu < 0$. We pick the sign in this way because we expect oscillatory solutions for $X(x)$. This leads to two equations.

$$X'' + \mu X = 0,$$
$$Y'' + (\lambda - \mu)Y = 0.$$

We now need to use the boundary conditions. We have $u(0, y, t) = 0$ for all $t > 0$ and $0 < y < H$. This implies that $X(0)Y(y)T(t) = 0$ for all t and y in the domain. This is only true if $X(0) = 0$. Similarly, we find that $X(L) = 0$, $Y(0) = 0$, and $Y(H) = 0$.

We note that homogeneous boundary conditions are important in carrying out this process. Nonhomogeneous boundary conditions could be imposed, but the techniques are a bit more complicated and we will not discuss these techniques here.

The boundary values problems we need to solve are.

$$X'' + \mu X = 0,\ X(0) = 0,\ X(L) = 0.$$
$$Y'' + (\lambda - \mu)Y = 0,\ Y(0) = 0,\ Y(H) = 0.$$

We have seen the first of these problems before, except with a λ instead of a μ. The solution is

$$X(x) = \frac{n\pi x}{L},\quad \lambda = \left(\frac{n\pi}{L}\right)^2,\quad n = 1, 2, 3,...$$

The second equation is solved in the same way. The differences are that our "eigenvalue" is $\lambda - \mu$, the independent variable is y, and the interval is $[0, H]$. Thus, we can quickly write down the solution as λ instead of a μ. The solution is

$$Y(y) = \sin\frac{m\pi x}{H},\quad \lambda - \mu = \left(\frac{m\pi}{H}\right)^2,\quad m = 1, 2, 3,....$$

We have successfully carried out the separation of variables for the wave equation for the vibrating rectangular membrane. The product solutions can be written as

$$u_{nm} = (a \cos \omega_{nm}t + b \sin \omega_{nm}t) \sin\frac{n\pi x}{L} \sin\frac{m\pi y}{H}.$$

Recall that ω is given in terms of λ. We have that

$$\lambda_{mn} - \mu_n = \left(\frac{m\pi}{H}\right)^2$$

and

$$\mu_n = \left(\frac{n\pi}{L}\right)^2.$$

Therefore,

$$\lambda_{nm} = \left(\frac{n\pi}{L}\right)^2 + \left(\frac{m\pi}{H}\right)^2.$$

So, s

$$\omega_{nm} = c\sqrt{\left(\frac{n\pi}{L}\right)^2 + \left(\frac{m\pi}{H}\right)^2}.$$

The most general solution can now be written as a linear combination of the product solutions and we can solve for the expansion coefficients that will lead to a solution satisfying he initial conditions. However, we will first concentrate on the two dimensional harmonics of this membrane.

For the vibrating string the nth harmonic corresponded to the function $\sin\frac{n\pi x}{L}$. The various harmonics corresponded to the pure tones supported by the string. These then lead to the corresponding frequencies that one would hear.

The actual shapes of the harmonics could be sketched by locating the nodes, or places on the string that did not move.

In the same way, we can explore the shapes of the harmonics of the vibrating membrane. These are given by the spatial functions

$$\phi_{nm}(x, y) = \sin\frac{n\pi x}{L} \sin\frac{m\pi y}{H}.$$

Instead of nodes, we will look for the nodal curves, or nodal lines. These are the points (x, y) at which $\phi(x, y) = 0$. Of course, these depend on the indices, n and m.

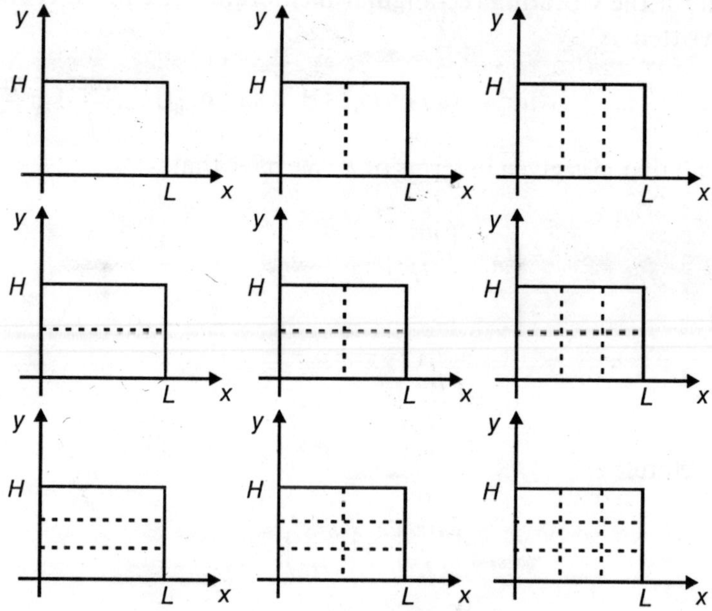

Fig. The First few modes of the Vibrating Rectangular Membrane. The Dashed Lines show the Nodal lines Indicating the Points that do not move for the Particular Mode

For example, when n = 1 and m = 1, we have

$$\sin\frac{n\pi x}{L}\sin\frac{m\pi y}{H} = 0.$$

These are zero when either

$$\sin\frac{n\pi x}{L} = 0, \quad \text{or} \quad \sin\frac{m\pi y}{H} = 0.$$

Of course, this can only happen for $x = 0, L$ and $y = 0, H$. Thus, there are no interior nodal lines.

When $n = 2$ and $m = 1$, we have $y = 0, H$ and

$$\sin\frac{2\pi x}{L} = 0,$$

or, $x = 0, \frac{L}{2}, L$. Thus, there is one interior nodal line at $x = \frac{L}{2}$. These points stay fixed during the oscillation and all other points oscillate on either side of this line.

A similar solution shape results for the (1,2)-mode, i.e., $n = 1$ and $m = 2$.

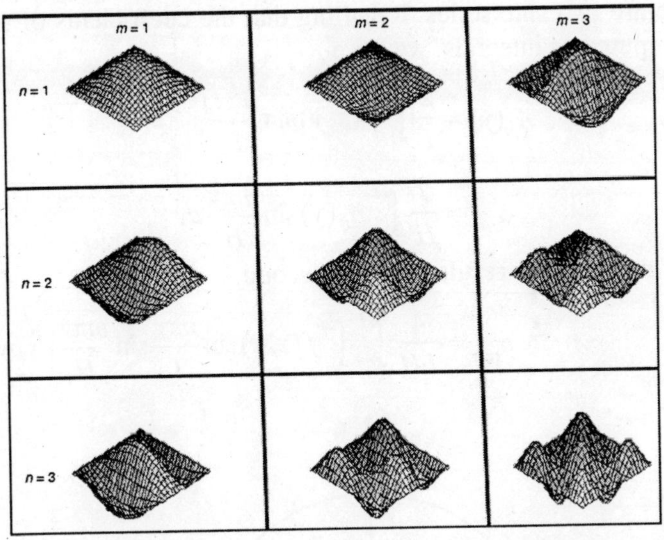

Fig. A three Dimensional view of the Vibrating Rectangular Membrane for the Lowest Modes

The nodal lines for several modes for n, $m = 1, 2, 3$ The blocked regions appear to vibrate independently. The frequencies of vibration are easily computed using the formula for ω_{nm}.

For completeness, we now see how one satisfies the initial conditions. The general solution is given by a linear superposition of the product solutions. There are two indices to sum over. Thus, the general solution is

$$u(x, y, t) = \sum_{n=1}^{\infty}\sum_{m=1}^{\infty}(a_{nm}\cos\omega_{nm}t + b_{nm}\sin\omega_{nm}t)\sin\frac{n\pi x}{L}\sin\frac{m\pi y}{H}\bigg).$$

The first initial condition is $u(x, y, 0) = f(x, y)$. Setting $t = 0$ in the general solution, we obtain

$$f(x, y) = \sum_{n=1}^{\infty}\sum_{m=1}^{\infty} a_{nm}\sin\frac{n\pi x}{L}\sin\frac{m\pi y}{H}.$$

This is a double sine series. The goal is to find the unknown coefficients a_{nm}. This can be done knowing what we already know about Fourier sine series. We can write this as the single sum

$$f(x, y) = \sum_{n=1}^{\infty} A_n(y)\sin\frac{n\pi x}{L},$$

where

$$A_n(y) = \sum_{m=1}^{\infty} a_{nm}\sin\frac{m\pi y}{H}.$$

These are two sine series. Recalling that the coefficients of sine series can be computed as integrals, we have

$$A_n(y) = \frac{2}{L}\int_0^L f(x,y)\sin\frac{n\pi x}{L},$$

$$a_{nm} = \frac{2}{H}\int_0^H A_n(y)\sin\frac{m\pi y}{H}dy.$$

Inserting the first result into the second,

$$a_{nm} = \frac{4}{LH}\int_0^H\int_0^L f(x,y)\sin\frac{n\pi y}{L}\sin\frac{m\pi y}{H}dx\,dy.$$

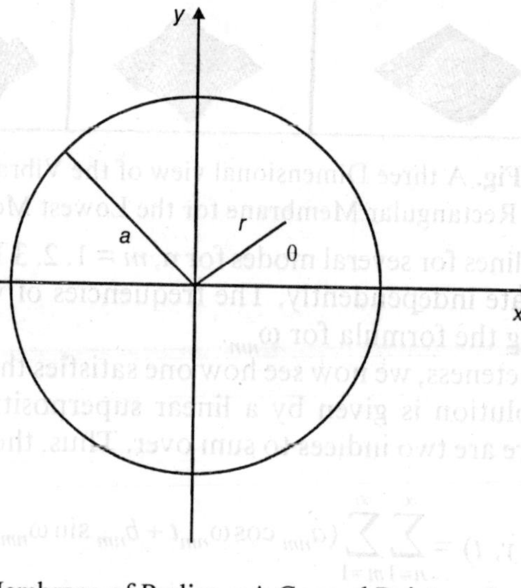

Fig. The Circular Membrane of Radius *a* A General Point on the Membrane is Given by the Distance from the centre, r, and the Angle,. There are Fixed Boundary Conditions Along the Edge at $r = a$.

We can carry out the same process for satisfying the second initial condition, *ut* (*x*, *y*, 0) = *g*(*x*, *y*) for the initial velocity of each point. Inserting this into the general solution, we have

$$g(x,\ y) = \sum_{n=1}^{\infty}\sum_{m=1}^{\infty} b_{nm}\omega_{nm}\sin\frac{n\pi x}{L}\sin\frac{m\pi y}{H}.$$

Again, we have a double sine series. But, now we can write down Fourier coefficients quickly:

$$b_{nm} = \frac{4}{\omega_{nm}LH}\int_0^H\int_0^L g(x,y)\sin\frac{n\pi x}{L}\sin\frac{m\pi y}{H}dxdy.$$

This completes the full solution of the vibrating rectangular membrane problem.

VIBRATIONS OF A KETTLE DRUM

In this section we consider the vibrations of a circular membrane of radius. Again we are looking for the harmonics of the vibrating membrane, but with the membrane fixed around the circular boundary given by $x^2 + y^2 = a^2$. However, expressing the boundary condition in Cartesian coordinates would be awkward. In this case the boundary condition would be $u = 0$ at $r = a$.

Before solving the initial-boundary value problem, we have to cast it in polar coordinates. This means that we need to rewrite the Laplacian in r and θ. To do so would require that we know how to transform derives in x and y into derivatives with respect to r and θ.

First recall that transformations

$$x = r\cos\theta,\ y = r\sin\theta$$

and

$$r = \sqrt{x^2 + y^2}, \quad \tan\theta = \frac{y}{x}.$$

Now, consider a function $f = f(x(r, \theta), y(r, \theta)) = g(r, \theta)$. (Technically, once we transform a given function of Cartesian coordinates we obtain a new function g of the polar coordinates. Many texts do not rigourously distinguish between the two which is fine when this point is clear.)

Thinking of $x = x(r, \theta)$ and $y = y(r, \theta)$, we have from the chain rule for functions of two variables.

$$\frac{\partial f}{\partial x} = \frac{\partial g}{\partial r}\frac{\partial r}{\partial x} + \frac{\partial g}{\partial \theta}\frac{\partial \theta}{\partial x}$$

$$= \frac{\partial g}{\partial r}\frac{x}{r} - \frac{\partial g}{\partial \theta}\frac{y}{r^2}$$

$$= \cos\theta\frac{\partial g}{\partial r} - \frac{\sin\theta}{r}\frac{\partial g}{\partial \theta}.$$

Here we have used

$$\frac{\partial r}{\partial x} = \frac{x}{\sqrt{x^2 + y^2}} = \frac{x}{r};$$

and

$$\frac{\partial \theta}{\partial x} = \frac{d}{dx}\left(\tan^{-1}\frac{y}{x}\right) = \frac{-y/x^2}{1 + \left(\dfrac{y}{x}\right)^2} = -\frac{y}{r^2}.$$

Similarly,

$$\frac{\partial f}{\partial y} = \frac{\partial g}{\partial r}\frac{\partial r}{\partial y} + \frac{\partial g}{\partial \theta}\frac{\partial \theta}{\partial y}$$

$$= \frac{\partial g}{\partial r}\frac{y}{r} + \frac{\partial g}{\partial \theta}\frac{x}{r^2}$$

$$= \sin\theta\frac{\partial g}{\partial r} + \frac{\cos\theta}{r}\frac{\partial g}{\partial \theta}.$$

2D Laplacian can now be computed as

$$\frac{\partial^2 f}{\partial x^2} + \frac{\partial^2 f}{\partial y^2} = \cos\theta\frac{\partial}{\partial r}\left(\frac{\partial f}{\partial x}\right) - \frac{\sin\theta}{r}\frac{\partial}{\partial \theta}\left(\frac{\partial f}{\partial x}\right) + \sin\theta\frac{\partial}{\partial r}\left(\frac{\partial f}{\partial y}\right) + \frac{\cos\theta}{r}\frac{\partial}{\partial \theta}\left(\frac{\partial f}{\partial y}\right)$$

$$= \cos\theta\frac{\partial}{\partial r}\left(\cos\theta\frac{\partial g}{\partial r} - \frac{\sin\theta}{r}\frac{\partial g}{\partial \theta}\right) - \frac{\sin\theta}{r}\frac{\partial}{\partial \theta}\left(\cos\theta\frac{\partial g}{\partial r} - \frac{\sin\theta}{r}\frac{\partial g}{\partial \theta}\right)$$

$$+ \sin\theta\frac{\partial}{\partial r}\left(\sin\theta\frac{\partial g}{\partial r} - \frac{\cos\theta}{r}\frac{\partial g}{\partial \theta}\right) + \frac{\cos\theta}{r}\frac{\partial}{\partial \theta}\left(\sin\theta\frac{\partial g}{\partial r} + \frac{\sin\theta}{r}\frac{\partial g}{\partial \theta}\right)$$

$$= \cos\theta\left(\cos\theta\frac{\partial^2 g}{\partial r^2} + \frac{\sin\theta}{r^2}\frac{\partial g}{\partial \theta} - \frac{\sin\theta}{r}\frac{\partial^2 g}{\partial r\partial \theta}\right)$$

$$-\frac{\sin\theta}{r}\left(\cos\theta\frac{\partial^2 g}{\partial r\partial \theta} - \frac{\sin\theta}{r}\frac{\partial^2 g}{\partial \theta^2} - \sin\theta\frac{\partial g}{\partial r} - \frac{\cos\theta}{r}\frac{\partial g}{\partial \theta}\right)$$

$$+ \sin\theta\left(\sin\theta\frac{\partial^2 g}{\partial r^2} + \frac{\cos\theta}{r}\frac{\partial^2 g}{\partial r\partial \theta} - \frac{\cos\theta}{r^2}\frac{\partial g}{\partial \theta}\right)$$

$$+ \frac{\cos\theta}{r}\left(\sin\theta\frac{\partial^2 g}{\partial \theta\partial r} + \frac{\cos\theta}{r}\frac{\partial^2 g}{\partial \theta^2} + \cos\theta\frac{\partial g}{\partial r} - \frac{\sin\theta}{r}\frac{\partial g}{\partial \theta}\right)$$

$$= \frac{\partial^2 g}{\partial r^2} + \frac{1}{r}\frac{\partial g}{\partial r} + \frac{1}{r^2}\frac{\partial^2 g}{\partial \theta^2}$$

$$= \frac{1}{r}\frac{\partial}{\partial r}\left(r\frac{\partial g}{\partial r}\right) + \frac{1}{r^2}\frac{\partial^2 g}{\partial \theta^2}.$$

The last form often occurs in texts because it is in the form of a Sturm-Liouville operator. Now that we have written the Laplacian in polar coordinates we can pose the problem of a vibrating circular membrane. It is given by a partial differential equation,

$$u_{tt} = c^2\left[\frac{1}{r}\frac{\partial}{\partial\theta}\left(r\frac{\partial u}{\partial r}\right)+\frac{1}{r^2}\frac{\partial^2 u}{\partial\theta^2}\right],$$

$$t > 0,\ 0 < r < a,\ -\pi < \theta < \pi,$$

the boundary condition,

$$u(a, \Theta, t) = 0,\ t > 0, \quad -\pi < < \pi,$$

and the initial conditions,

$$u(r, \theta, 0) = f(r, \theta),\ u_t(r, \theta, 0) = g(r, \theta).$$

Now were are ready to solve this problem using separation of variables. As before, we can separate out the time dependence. Let $u(r, \theta, t) = T(t)\,\phi(r, \theta)$. As usual, $T(t)$ can be written in terms of sines and cosines. This leaves the Helmhotz equation,

$$\nabla^2\phi + \lambda\phi = 0.$$

We can separate this equation by letting $\phi(r, \theta) = R(r)\Theta(\theta)$. This gives

$$\frac{1}{r}\frac{\partial}{\partial r}\left(r\frac{\partial R\Theta}{\partial r}\right)+\frac{1}{r^2}\frac{\partial^2 R\Theta}{\partial\theta^2}+\lambda R\Theta = 0.$$

Dividing by $u = R\Theta$, as usual, leads to

$$\frac{1}{rR}\frac{d}{dr}\left(r\frac{dR}{\partial r}\right)+\frac{1}{r^2\Theta}\frac{d^2\Theta}{\partial\theta^2}+\lambda = 0.$$

The last term is a constant. The first term is a function of r. However, the middle term involves both r and θ. This can be fixed by multiplying the equation by r^2. Rearranging the equation, we can separate out the θ-dependence from the radial dependence. Letting μ be the separation constant, we have

$$\frac{r}{R}\frac{d}{dr}\left(r\frac{dR}{\partial r}\right)+\lambda r^2 = -\frac{1}{\Theta}\frac{d^2\Theta}{\partial\theta^2} = \mu.$$

This gives us two ordinary differential equations.

$$\frac{d^2\Theta}{\partial\theta^2}+\mu\Theta = 0,$$

$$r\frac{d}{dr}\left(r\frac{dR}{\partial r}\right)+(\lambda r^2-\mu)R = 0.$$

Let's consider the first of these equations. It should look familiar. If $\mu > 0$, then the general solution is

$$\Theta(\theta) = a\cos\sqrt{\mu}\theta + b\sin\sqrt{\mu}\theta.$$

Now we apply the boundary condition in θ. Looking at our boundary conditions in the problem, we do not see anything involving θ. This is a case where the boundary conditions needed are implied and not stated outright. We can derive the boundary conditions by making some observations.

Let's consider the solution corresponding to the endpoints $\theta = \pm\pi$, noting that at these values for any $r < a$ we are at the same physical point. So, we would expect the solution to have the same value at $\theta = -\pi$ as it has at $\theta = -\pi$. Namely, the solution is continuous at these physical points.

Similarly, we expect the slope of the solution to be the same at these points. This tells us that

$$\Theta(\pi) = \Theta'(-\pi) \quad \Theta'(\pi) = \Theta'(-\pi).$$

Such boundary conditions are called periodic boundary conditions.

Let's apply these conditions to the general solution for $\Theta(\theta)$ First, we set $\Theta(\pi) = \Theta(-\pi)$.

$$a\cos\sqrt{\mu}\pi + b\sin\sqrt{\mu}\pi = a\cos\sqrt{\mu}\pi - b\sin\sqrt{\mu}\pi .$$

This implies that

$$\sin\sqrt{\mu}\pi = 0.$$

So, $\sqrt{\mu} = m$ for $m = 0, 1, 2, 3,\ldots$. For $\Theta'(\pi) = \Theta'(-\pi)$, we have

$$-am \sin m\pi + bm \cos m\pi = am \sin m\pi + bm \cos m\pi.$$

But, this gives no new information.

To summarize so far, we have found the general solutions to the temporal and angular equations. The product solutions will have various products of $\{\cos\omega t, \sin\omega t\}$ and $\{\cos m\theta, \sin m\theta\}_{m=0}^{\infty}$. We also know that $\mu = m^2$ and $\omega = c\sqrt{\lambda}$. That leaves us with the radial equation. Inserting $\mu = m^2$, we have

$$r\frac{d}{dr}\left(r\frac{dR}{dr}\right) + (\lambda r^2 - m^2)R = 0.$$

This is not an equation we have encountered before (unless you took a course in differential equations). We need to find solutions to this equation. It turns out that under a simple transformation it becomes an equation whose solutions are well known.

Let $z = \sqrt{\lambda}r$ and $\omega(z) = R(r)$. We have changed the name of the dependent variable since inserting the transformation into $R(r)$ leads to a function of z that is not the same function.

We need to change derivatives with respect to r into derivative with respect to z. We use the Chain Rule to obtain

$$\frac{dR}{dr} = \frac{dw}{dz}\frac{dz}{dr} = \sqrt{\lambda}\frac{dw}{dz}.$$

Thus, the derivatives transform as

$$r\frac{d}{dr} = z\frac{d}{dz}.$$

Inserting the transformation into the differential equation, we have

$$0 = r\frac{d}{dr}\left(r\frac{dR}{dr}\right) + (\lambda r^2 - m^2)R$$

$$0 = z\frac{d}{dr}\left(z\frac{dw}{dr}\right) + (z^2 - m^2)w.$$

Expanding the derivative terms, we obtain Bessel's equation.

$$z^2\frac{d^2w}{dz^2} + z\frac{dw}{dz} + (z^2 - m^2)w = 0.$$

The history of the solutions of this equation, called Bessel functions, does not originate in the study of partial differential equations. These solutions originally came up in the study of the Kepler problem, describing planetary motion. According to Watson in his Treatise on Bessel Functions, the formulation and solution of Kepler's Problem was discovered by Lagrange in 1770.

Namely, the problem was to express the radial coordinate and what is called the eccentric anomaly, E, as functions of time. Lagrange found expressions for the coefficients in the expansions of *r* and E in trigonometric functions of time. However, he only computed the fist few coefficients. In 1816 Bessel had shown that the coefficients in the expansion for r could be given an integral representation. In 1824 he presented a thorough study of these functions, which are now called Bessel functions.

There are two linearly independent solutions of this second order equation. $J_m(z)$, the Bessel function of the first kind of order m,and $N_m(z)$, the Bessel function of the second kind of order m. Sometimes the N_m's are called Neumann functions. So, we have the general solution of our transformed equation is

$$w(z) = c_1 J_m(z) + c_2 N_m(z).$$

Transforming back into r variables, this becomes

$$R(r) = c_1 J_m(\sqrt{\lambda}r) + c_2 N_m(\sqrt{\lambda}r).$$

Now we are ready to apply the boundary conditions to our last factor in our product solutions. Looking at the original problem we find only one condition. $u(a, \theta, t) = 0$ for $t > 0$ and $-\pi < < \pi$. This implies that $R(0) = 0$. But where is our second condition?

This is another unstated boundary condition. Look again at the plots of the Bessel functions. Notice that the Neumann functions are not well behaved at the origin. Do you expect that our solution will become infinite at the centre of our drum? No, the solutions should be finite at the centre. So, this is the second boundary condition. Namely, $|R(0)| < \infty$. This implies that $c^2 = 0$.

Now we are left with p

$$R(r) = J_m(\sqrt{\lambda}r).$$

We have set $c_1 = 1$ for simplicity. We can apply the vanishing condition at $r = a$. This gives

$$J_m(\sqrt{\lambda}a) = 0.$$

Looking again at the plots of $J_m(z)$, we see that there are an infinite number of zeros, but they are not as easy as π! In Table we list the nth zeros of J_p. Let's denote the nth zero of $J_m(z)$ by z_{mn}. Then our boundary condition tells us that

$$(\sqrt{\lambda}a) = z_{mn}.$$

This gives use our eigenvalue as

$$\lambda_{mn} = \left(\frac{z_{mn}}{a}\right)^2.$$

Thus, our radial function, satisfying the boundary conditions, is

$$R(r) = J_m\left(\frac{z_{mn}}{a}r\right).$$

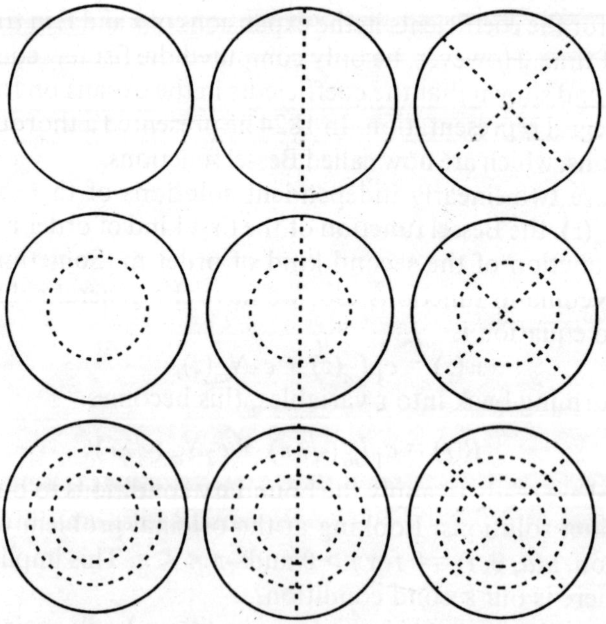

Fig. The First Few Modes of the Vibrating Circular Membrane. The Dashed Lines show the Nodal lines Indicating the Points that do not move for the Particular Mode

We are finally ready to write out the product solutions. They are given by

$$u(r, \theta, t) = \begin{Bmatrix} \cos \omega_{mn} t \\ \sin \omega_{mn} t \end{Bmatrix} \begin{Bmatrix} \cos m\theta \\ \sin m\theta \end{Bmatrix} J_m\left(\frac{z_{mn}}{a} r\right).$$

Here we have indicated choices with the braces, leading to four different types of product solutions. Also, m = 0, 1, 2,..., and

$$\omega_{mn} = \frac{z_{mn}}{a} c.$$

As with the rectangular membrane, we are interested in the shapes of the harmonics. So, we consider the spatial solution (t = 0) $\phi(r, \theta) = \cos m\theta J_m\left(\frac{z_{mn}}{a} r\right)$. Adding the solutions involving sin mθ will only rotate these modes. The nodal curves are given by A(r, θ) = 0, or cos mθ = 0, or $J_m\left(\frac{z_{mn}}{a} r\right) = 0$.

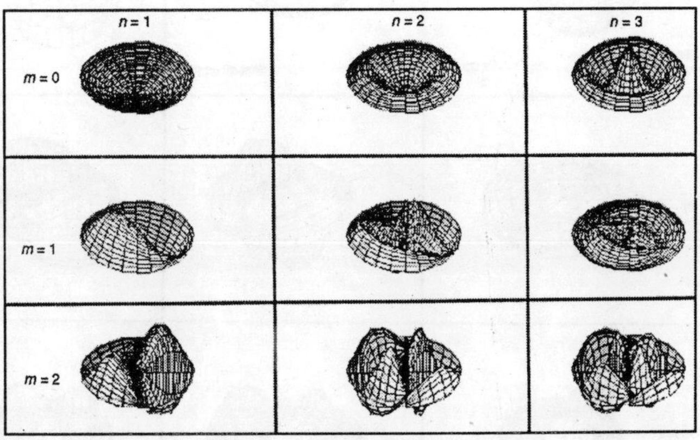

Fig. A three Dimensional View of the Vibrating Circular Membrane for the Lowest Modes.

For the angular part, we easily see that the nodal curves are radial lines. For $m = 0$, there are no solutions. For $m = 1$, we have $\cos \theta = 0$ implies that $\theta = \pm\frac{\pi}{2}$.

These values give the same line. For m = 2, cos 2θ = 0 implies that $\theta = \frac{\pi}{4}, \frac{3\pi}{4}$.

We can also consider the nodal curves defined by the Bessel functions.

We seek values of r for which $\frac{z_{mn}}{a}r$ is a zero of the Bessel function and lies in the interval $[0, a]$. Thus, we have

$$\frac{z_{mn}}{a}r = z_{mj},$$

or

$$r = \frac{z_{mj}}{z_{mn}}a.$$

These will give circles of this radius with $z_{mj} < z_{mn}$. The zeros can be found in Table. For $m = 0$ and $n = 1$, there is only one zero and $r = a$. For $m = 0$ and $n = 2$, we have two circles, $r = a$ and $r = 2\frac{2.405}{5.520}a \approx 0.436a$.

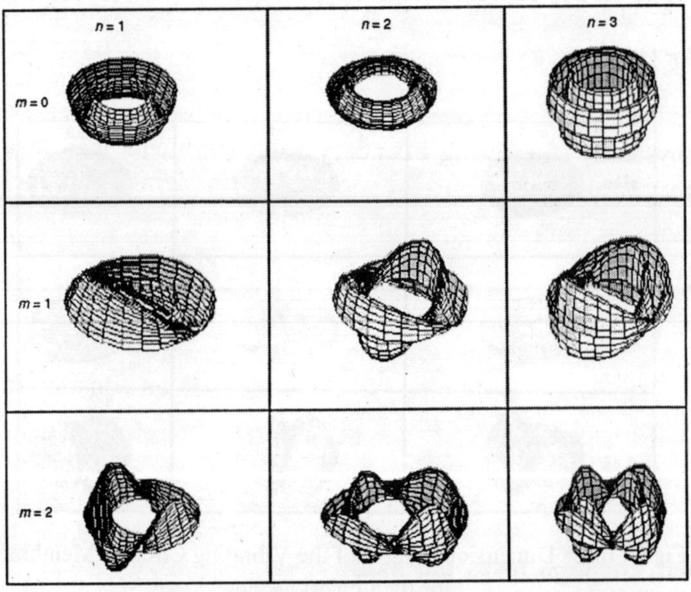

Fig. A three Dimensional View of the Vibrating Annular Membrane for the Lowest Modes

For $m = 0$ and $n = 3$ we obtain circles of radii $r = a, \frac{5.520}{8.654}a, \frac{2.405}{8.654}a$. Similar computations result for larger values of m.

Imagine that the various regions are oscillating independently and that the points on the nodal curves are not moving. More complicated vibrations can be dreamt up for this geometry. We could consider an annulus in which the drum is formed from two concentric circular cylinders and the membrane is stretch between the two with an annular cross section. The separation would follow as before except now the boundary conditions are that the membrane is fixed around the two circular boundaries. In this case we cannot toss out the Neumann functions because the origin is not part of the drum head.

In this case,specifying the radii as a and b with $b < a$, we have to satisfy the conditions.

$$R(a) = c_1 J_m(\sqrt{\lambda}a) + c_2 N_m(\sqrt{\lambda}a) = 0\,,$$

$$R(b) = c_1 J_m(\sqrt{\lambda}b) + c_2 N_m(\sqrt{\lambda}b) = 0\,.$$

This leads to two homogeneous equations for c_1 and c_2. The determinant has to vanish, giving a nice complicated condition on λ. We show various modes for a particular choice of a and b.

STURM-LIOUVILLE PROBLEMS

We have explored the solution of boundary value problems that led to trigonometric eigenfunctions. Such functions can be used to represent functions in Fourier series expansions. We would like to generalize some of these techniques in order to solve other boundary value problems. We will see that many physically interesting boundary value problems lead to a class of problems called Sturm-Liouville eigenvalue problems. This includes our previous examples of the wave and heat equations.

We begin by noting that in physics many problems arise in the form of boundary value problems involving second order ordinary differential equations. For example, we might want to solve the equation

$$a_2(x)y'' + a_1\ (x)y' + a'\ (x)y = f(x)$$

subject to boundary conditions. We can write such an equation in operator form by defining the differential operator

$$L = a_2(x)\frac{d^2}{dx^2} + a_1(x)\frac{d}{dx} + a_0(x)\,.$$

Then, equation takes the form

$$Lu = f.$$

In many of the problems we have encountered, the resulting solutions were written in term of series expansions over a set of eigenfunctions. This indicates that perhaps we might seek solutions to the eigenvalue problem $L\phi = \lambda\phi$ with homogeneous boundary conditions and then seek a solution as an expansion of the eigenfunctions.

Formally, we let $u = \sum_{n=1}^{\infty} c_n \phi_n(x)dx$. However, we are not guaranteed a nice set of eigenfunctions. We need an appropriate set to form a basis in the function space. Also, it would be nice to have orthogonality so that we can solve for the expansion coefficients as was done. It turns out that any linear second order operator can be turned into an operator that posses just the right properties for us to carry out this procedure. The resulting operator is the Sturm-Liouville operator, which we will now explore.

We define the Sturm-Liouville operator as

$$\mathcal{L} = \frac{d}{dx} p(x) \frac{d}{dx} + q(x).$$

The regular Sturm-Liouville eigenvalue problem is given by the differential equation

$$\mathcal{L}u = -\lambda\sigma(x)u,$$

or

$$\frac{d}{dx}\left(p(x)\frac{du}{dx}\right) + q(x)u + \lambda\sigma(x)u = 0,$$

for $x \in [a, b]$ and the set of homogeneous boundary conditions

$$\alpha_1 u(a) + \beta_1 u'(a) = 0,$$

$$\alpha_2 u(b) + \beta_2 u'(b) = 0.$$

The functions $p(x)$, $q(x)$ and $\sigma(x)$ are assumed to be continuous on (a, b) and $p(x) > 0$, $q(x) > 0$ on $[a, b]$.

The alpha's and β's are constants. For different values, one has special types of boundary conditions. For $\beta_i = 0$, we have what are called Dirichlet conditions. Namely, $u(a) = 0$ and $u(b) = 0$. For $\alpha_i = 0$, we have Neumann boundary conditions.

In this case, $u'(a) = 0$ and $u'(b) = 0$. In terms of the heat equation example, Dirichlet conditions correspond to maintaining a fixed temperature at the ends of the rod. The Neumann boundary conditions would correspond to no heat flow across the ends, as there would be no temperature gradient at those points. Another type of boundary condition that is often encountered is the periodic boundary condition. Consider the heated rod that has been bent to form a circle. Then the two end points are physically the same. So, we would expect that the temperature and the temperature gradient should agree at those points. For this case we write $u(a) = u(b)$ and $u'(a) = u'(b)$. *Theorem:* Any second order linear operator can be put into the form of a Sturm-Liouville operator.

The proof of this is straight forward. Consider the equation. If $a_1(x) = a_2'(x)$, then we can write the equation in the form

$$f(x) = a_2(x)y'' + a_1(x)y' + a_0(x)y$$

$$= (a_2(x)y')' + a_0(x)y.$$

This is in the correct form. We just identify $p(x) = a_2(x)$ and $q(x) = a_0(x)$.

However, consider the differential equation

$$x^2y'' + xy' + 2y = 0.$$

In this case $a_2(x) = x^2$ and $a_2'(x) = 2x \neq a_1(x)$. This equation is not of Sturm-Liouville type. But, we can change it to a Sturm Liouville type of equation.

In the Sturm Liouville operator the derivative terms are gather together into one perfect derivative. This is similar to what we saw in the first chapter when we solved linear first order equations. In that case we sought an integrating factor. We can do the same thing here. We seek a multiplicative function $\mu(x)$ that we can multiply through so that it can be written in Sturm-Liouville form. We first divide out the $a_2(x)$, giving

$$y'' + \frac{a_1(x)}{a_2(x)}y' + \frac{a_0(x)}{a_0(x)}y = \frac{f(x)}{a_2(x)}.$$

Now, we multiply the differential equation by μ:

$$\mu(x)y'' + \mu(x)\frac{a_1(x)}{a_2(x)}y' + \mu(x)\frac{a_0(x)}{a_2(x)}y = \mu(x)\frac{f(x)}{a_2(x)}.$$

The first two terms can now be combined into an exact derivative $(\mu y')'$ if $\mu(x)$ satisfies

$$\frac{d\mu}{dx} = \mu(x)\frac{a_1(x)}{a_2(x)}.$$

This is easily solved to give

$$\mu(x) = e^{\int \frac{a_1(x)}{a_2(x)}dx}$$

Thus, the original equation can be multiplied by

$$\frac{1}{a_2(x)}e^{\int \frac{a_1(x)}{a_2(x)}dx}$$

to turn it into Sturm-Liouville form.

For the example above,

$$x^2y'' + xy' + 2y = 0,$$

we need only multiply by

$$\frac{1}{x^2}e^{\int \frac{dx}{x}} = \frac{1}{x}.$$

Therefore, we obtain

$$0 = xy'' + y' + \frac{2}{x}y = (xy')' + \frac{2}{x}y.$$

PROPERTIES OF STURM-LIOUVILLE PROBLEMS

There are several properties that can be proven for the Sturm-Liouville eigenvalue problems. However, we will not prove them all here.

- The eigenvalues are real, countable, ordered and there is a smallest eigenvalue. Thus, we can write them as $\lambda_1 < \lambda_2 <\ldots$. However, there is no largest eigenvalue and $n \to 1$, $\lambda_n \to \infty$.
- For each eigenvalue λ_n there exists an eigenfunction ϕ_n with $n - 1$ zeros on (a, b).
- Eigenfunctions corresponding to different eigenvalues are orthogonal with respect to the weight function, $\sigma(x)$. Defining the inner product of $f(x)$ and $g(x)$ as

$$<f, g> = \int_a^b f(x)g(x)\sigma(x)\,dx,$$

then the orthogonality condition can be written in the form

$$<\phi_n, \phi_m> = <\phi_n, \phi_n> \delta_{nm}.$$

- The set of eigenfunctions is complete, i.e., any piecewise smooth function can be represented by a generalized Fourier series expansion of the eigenfunctions,

$$f(x) \sim \sum_{n=1}^{\infty} c_n \phi_n(x)$$

where

$$c_n = \frac{<f, \phi_n>}{<\phi_n, \phi_n>}.$$

Actually, one needs $f(x) \in L_\sigma^2\,[a, b]$, the set of square integrable functions over $[a, b]$ with weight function $\sigma(x)$. $<f, f> < \infty$.

- Multiply the eigenvalue problem

$$\mathcal{L}\phi_n = -\lambda_n \sigma(x)\phi_n$$

by ϕ_n and integrate. Solve this result for ϕ_n, to find the Rayleigh Quotient

$$\lambda_n = \frac{-p\phi_n \dfrac{d\phi_n}{dx}\Big|_a^b - \displaystyle\int_a^b \left[p\left(\dfrac{d\phi_n}{dx}\right)^2 - a\phi_n^2 \right] dx}{<\phi_n, \phi_n>}$$

The Rayleigh quotient is useful for getting estimates of eigenvalues and proving some of the other properties.

Identities and Adjoint Operators

Before turning to any proofs, we will need two more tools. For the Sturm-Liouville operator, $\mathcal{L}$, we have two identities.

Lagrange's Identity: $u\mathcal{L}v - v\mathcal{L}u = [p(uv' - vu')]'$.

The proof follows by a simple manipulations of the operator.

$$u\mathcal{L}v - v\mathcal{L}u = u\left[\frac{d}{dx}\left(p\frac{dv}{dx}\right) + qv\right] - v\left[\frac{d}{dx}\left(p\frac{du}{dx}\right) + qu\right]$$

$$= u\frac{d}{dx}\left(p\frac{dv}{dx}\right) - v\frac{d}{dx}\left(p\frac{du}{dx}\right)$$

$$= u\frac{d}{dx}\left(p\frac{dv}{dx}\right) + p\frac{du}{dx}\frac{dv}{dx} - v\frac{d}{dx}\left(p\frac{du}{dx}\right) - p\frac{du}{dx}\frac{dv}{dx}$$

$$= \frac{d}{dx}\left[pu\frac{dv}{dx} - pv\frac{du}{dx}\right].$$

Green's Identity: $\int_a^b (u\mathcal{L}v - v\mathcal{L}u)\,dx = [p(uv' - vu')]|_a^b$.

This identity is simply proven by integrating Lagrange's identity.

Next, we define the domain of an operator and introduce the notion of adjoint operators.

Definition: The domain of a differential operator L is the set of all $u \in L^2_\sigma[a, b]$ satisfying a given set of homogeneous boundary conditions.

Definition: The adjoint, $L^\dagger$, of operator L satisfies

$$< u, Lv > = < L^\dagger u, v >$$

for all v in the domain of L and u in the domain of $L^\dagger$.

Example: Find the adjoint of $L = a_2(x)\frac{d^2}{dx^2} + a_1(x)\frac{d}{dx} + a_0(x)$.

We first look at the inner product

$$< u, Lv > = \int_a^b u(a_2 v'' + a_1 v' + a_0 v)\,dx.$$

Then we have to move the operator L from v and determine what operator is acting on u. For a simple operator like $L = \frac{d}{dx}$, this is easily done using integration by parts. For the given operator, we apply several integrations by parts to individual terms. Noting that

$$\int_a^b u(x)a_1(x)v'(x)\,dx = a_1(x)u(x)v(x)|_a^b - \int_a^b (u(x)a_1(x))'v(x)\,dx$$

and

$$\int_a^b u(x)a_2(x)v''(x)\,dx = a_2(x)u(x)v'(x)|_a^b - \int_a^b (u(x)a_2(x))'v(x)'\,dx$$

$$= [a_2(x)u(x)v'(x) - (a_2(x)u(x))'v(x)]|_a^b + \int_a^b (u(x)a_2(x))''v(x)dx,$$

we have

$$< u,Lv >= \int_a^b u(a_2v'' + a_1v' + a_0v)dx$$

$$= [a_1(x)u(x)v(x) + a_2(x)u(x)v'(x) - a_2(x)u(x))'v(x)|_a^b.$$

$$+ \int_a^b [(a_2u)'' - (a_1u)' + a_0u]v\,dx.$$

Inserting the boundary conditions for v, one finds boundary conditions for u such that

$$[a_1(x)u(x)v(x) + a_2(x)u(x)v'(x) - a_2(x)u(x))'v(x)|_a^b = 0.$$

This leaves

$$< u,\ Lv > = \sum_{n=1}^{\infty} \frac{2}{n\pi} \frac{[(-1)^m - 1]}{n^2\pi^2 - 1} \sin(n\pi \ln(x)) \ < L^{\dagger}\ u,\ v >.$$

Therefore,

$$L^{\dagger} = \frac{d^2}{dx^2}a_2(x) - \frac{d}{dx}a_1(x) + a_0(x).$$

When $L^{\dagger} = L$, the operator is called formally self-adjoint. When the domain of L is the same as the domain of $L^{\dagger}$, the term self-adjoint is used.

Example: Determine $L^{\dagger}$ and its domain for operator $Lu = \frac{du}{dx}$ and u satisfying $u(0) = 2u(1)$ on [0, 1].

We need to find the adjoint operator satisfying $< v,\ Lu >=< L^{\dagger}v,\ u >$. Therefore, we rewrite the integral

$$< v,\ Lu > = \int_0^1 v\frac{du}{dx}dx = uv|_0^1 - \int_0^1 u\frac{dv}{dx}dx.$$

From this we have

- $L^{\dagger} = -\frac{d}{dx}$.
- $uv|_0^1 = 0 \Rightarrow 0 = u(1)[v(1) - 2v(0)] \Rightarrow v(1) = 2v(0).$

Orthogonality of Eigenfunctions

We are now ready to prove That the eigenvalues of a Sturm-Liouville problem are real and the corresponding eigenfunctions are orthogonal.

Theorem: The eigenvalues of the Sturm-Liouville problem are real.

Let $\phi_n(x)$ be a solution of the eigenvalue problem associated with λ_n:

$$\mathcal{L}\phi_n = -\lambda_n\sigma\phi_n.$$

The complex conjugate of this equation is

$$\mathcal{L}\bar{\phi}_n = -\bar{\lambda}_n \sigma \bar{\phi}_n .$$

Now, multiply the first equation by $\bar{\phi}_n$ and the second equation by ϕ_n and then subtract the results. We obtain

$$\bar{\phi}_n \mathcal{L}\phi_n - \phi_n \mathcal{L}\bar{\phi}_n = (\bar{\lambda}_n - \lambda_n)\sigma\phi_n\bar{\phi}_n .$$

Integrating both sides and noting that by Green's identity and the boundary conditions for a self-adjoint operator, we have

$$0 = (\bar{\lambda}_n - \lambda_n)\int_a^b \sigma \|\phi_n\|^2 \, dx$$

The integral is nonnegative, so we have $\bar{\lambda}_n = \lambda_n$. Therefore, the eigenvalues are real.

Theorem: The eigenfunctions corresponding to different eigenvalues of the Sturm-Liouville problem are orthogonal.

This is proven similar to the last proof.

Let $\phi_n(x)$ be a solution of the eigenvalue problem associated with λ_n,

$$\mathcal{L}\phi_n = -\lambda_n \sigma \phi_n,$$

and let $\phi_m(x)$ be a solution of the eigenvalue problem associated with $\lambda_m \neq \lambda_n$,

$$\mathcal{L}\phi_m = -\lambda_m \sigma \phi_m,$$

Now, multiply the first equation by ϕ_m and the second equation by ϕ_n. Subtracting the results, we obtain

$$\phi_m \mathcal{L}\phi_n - \phi_n \mathcal{L}\phi_m = (\lambda_m - \lambda_n)\, \sigma\, \phi_n \phi_m$$

Integrating both sides and using Green's identity and the boundary conditions for a self-adjoint operator, we have

$$0 = (\lambda_m - \lambda_n)\int_a^b \sigma \phi_n \phi_m dx .$$

Since the eigenvalues are different, we have that

$$\int_a^b \sigma \phi_n \phi_m dx = 0.$$

Therefore, the eigenfunctions are orthogonal with respect to the weight function $\sigma(x)$.

THE EIGENFUNCTION EXPANSION METHOD

The eigenfunctions of a differential operator to solve a nonhomogeneous boundary value problem. In this chapter we have seen that Sturm-Liouville eigenvalue problems have the requisite set of orthogonal eigenfunctions. In this section we will apply this method to solve a particular nonhomogenous eigenvalue problem. Recall that one starts with a nonhomogeneous differential equation

$$\mathcal{L}y = f$$

where $y(x)$ is to satisfy given homogeneous boundary conditions. The method makes use of the eigenfunctions satisfying the eigenvalue problem

$$\mathcal{L}\phi_n = -\lambda_n \sigma \phi_n$$

subject to the given boundary conditions. Then one assumes that $y(x)$ can be written as an expansion in the eigenfunctions,

$$y(x) = \sum_{n=1}^{\infty} c_n \phi_n(x),$$

and inserts the expansion into the nonhomogeneous equation. This gives

$$f(x) = \mathcal{L}\left(\sum_{n=1}^{\infty} c_n \phi_n(x)\right) = -\sum_{n=1}^{\infty} c_n \lambda_n \sigma(x) \phi_n(x)$$

The expansion coefficients are then found by making use of the orthogonality of the eigenfunctions. Thus,

$$c_m = \frac{\int_a^b f(x)\phi_m(x)\,dx}{\lambda_m \int_a^b \phi_m^2(x)\sigma(x)\,dx}.$$

As an example, we consider the solution of the boundary value problem

$$(xy')' + \frac{y}{x} = \frac{1}{x}, \quad x \in [1, e],$$

$$y(1) = 0 = y(e).$$

This equation is already in self-adjoint form. So, we know that the associated Sturm-Liouville eigenvalue problem has an orthogonal set of eigenfunctions. We first determine this set. Namely, we need to solve

$$(x\phi')' + \frac{\phi}{x} = -\lambda\sigma\phi, \qquad \phi(1) = 0 = \phi(e)$$

Rearranging the terms and multiplying by x, we have that

$$x^2\phi'' + x\phi' + (1 + \lambda\sigma x)\phi = 0.$$

This is almost an equation of Cauchy-Euler type. Picking the weight function $\sigma(x) = \frac{1}{x}$, we have

$$x^2\phi'' + x\phi' + (1 + \lambda)\phi = 0.$$

This is easily solved. The characteristic equation is

$$r^2 + (1 + \lambda) = 0.$$

One obtains nontrivial solutions of the eigenvalue problem satisfying the boundary conditions when $\lambda > -1$. The solutions are

$$\phi_n(x) = A \sin(n\pi \ln x), \quad , n = 1, 2, \ldots$$

where $\lambda_n = n^2\pi^2 - 1$.

It is often useful to normalize the eigenfunctions. This means that one chooses A so that the norm of each eigenfunction is one. Thus, we have

$$1 = \int_1^e \phi_n(x)^2 \sigma(x)\, dx$$

$$= A^2 \int_1^e \sin(n\pi \ln x) \frac{1}{x} dx$$

$$= A^2 \int_0^1 \sin(n\pi y)\, dy = \frac{1}{2} A^2.$$

Thus, $A = \sqrt{2}$.

We now turn towards solving the nonhomogeneous problem, $\mathcal{L}y = \frac{1}{x}$.

We first expand the unknown solution in terms of the eigenfunctions,

$$y(x) = \sum n = 1^\infty c_n \sqrt{2} \sin(n\pi \ln x).$$

Inserting this solution into the differential equation, we have

$$\frac{1}{x} = \sum_{n=1}^{\infty} c_n \lambda_n \sqrt{2} \sin(n\pi \ln x) \frac{1}{x}.$$

Next, we make use of orthogonality. Multiplying both sides by $\phi_m(x) = \sqrt{2} \sin(m\pi \ln x)$ and integrating, gives

$$\lambda_m c_m = \int_1^e \sqrt{2} \sin(m\pi \ln x) \frac{1}{x} dx = \frac{\sqrt{2}}{m\pi} [(-1)^m - 1].$$

Solving for c_m, we have

$$c_m = \frac{\sqrt{2}}{m\pi} \frac{[(-1)^m - 1]}{m^2\pi^2 - 1}.$$

Finally, we insert our coefficients into the expansion for $y(x)$. The solution is then

$$y(x) = \sum_{n=1}^{\infty} \frac{2}{n\pi} \frac{[(-1)^m - 1]}{n^2\pi^2 - 1} \sin(n\pi \ln(x)).$$

Chapter 8

Special Functions

Some additional functions which arise often in physical applications and are eigenfunctions for some Sturm-Liouville boundary value problem. We begin with a collection of special functions, called the classical orthogonal polynomials. These include such polynomial functions as the Legendre polynomials, the Hermitc polynomials, the Tchebychef and the Gegenbauer polynomials. Also, Bessel functions occur quite often. We will spend most of our time exploring the Legendre and Bessel functions. These functions are typically found as solutions of differential equations using power series methods in a first course in differential equations.

CLASSICAL ORTHOGONAL POLYNOMIALS

We begin by noting that the sequence of functions $\{1, x, x^2,...\}$ is a basis of linearly independent functions. In fact, by the Stone-Weierstrass Approximation Theorem in analysis this set is a basis of L_σ^2 (a, b), the space of square integrable functions over the interval $[a, b]$ relative to weight $\sigma(x)$. We are familiar with being able to expand functions over this basis, since

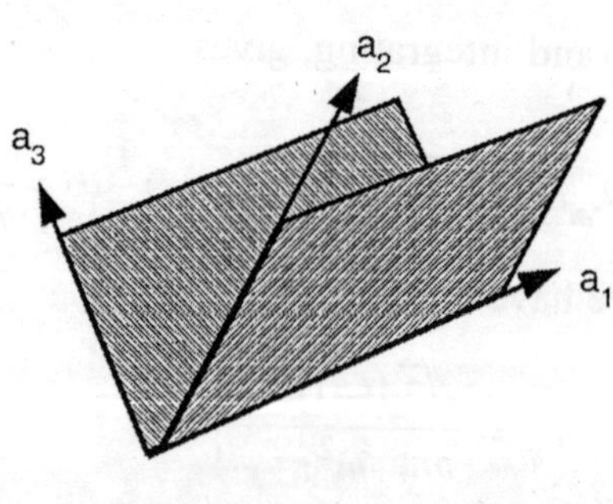

Fig. The Basis a_1, a_2, and a_3, of R^3 Considered in the Text.

the expansions are just Maclaurin series representations of the functions,

$$f(x) \sim \sum_{n=0}^{\infty} c_n x^n.$$

However, this basis is not an orthogonal set of basis functions. One can easily see this by integrating the product of two even, or two odd, basis functions with $\sigma(x) = 1$ and $(a, b) = (-1, 1)$. For example,

$$\int_{-1}^{1} x^0 x^2 dx = \frac{2}{3}.$$

Since we have found that orthogonal bases have been useful in determining the coefficients s for expansions of given functions, we might ask if it is possible to obtain an orthogonal basis involving these powers of x. Finite combinations of these basis element are just polynomials.

"Given a set of linearly independent basis vectors, can one find an orthogonal basis of the given space?" The answer is yes. We recall from introductory linear algebra, which mostly covers finite dimensional vector spaces, that there is a method for carrying this out called the Gram-Schmidt Orthogonalization Process. We will recall this process for finite dimensional vectors and then generalize to function spaces.

Let's assume that we have three vectors that span R^3, given by a_1, a_2, and a_3. We seek an orthogonal basis e_1, e_2, and e_3, beginning one vector at a time.

First we take one of the original basis vectors, say a_1, and define

$$e_1 = a_1.$$

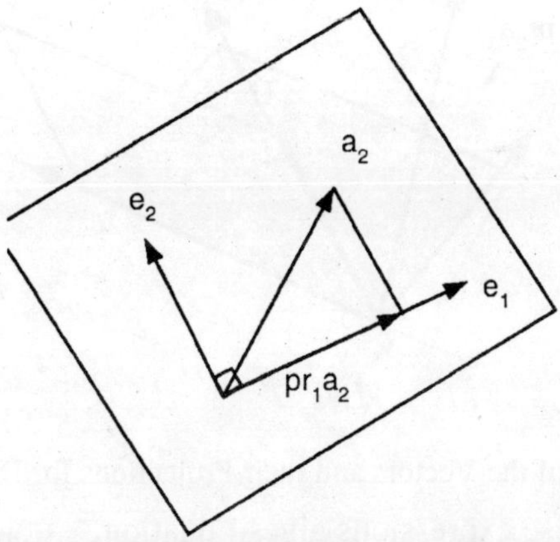

Fig. A Plot of the Vectors e_1, a_2, and e_2 Needed to find the Projection of a_2, on e_1 Illustrating the Gram-Schmidt Orthogonalization Process

Of course, we might want to normalize our new basis vectors, so we would denote such a normalized vector with a 'hat'.

$$\hat{e}_1 = \frac{e_1}{e_1},$$

where $e_1 = \sqrt{e_1 \cdot e_1}$.

Next, we want to determine an e_2 that is orthogonal to e_1. We take the next element of the original basis, a_2. We see the orientation of the vectors. Note that the desired orthogonal vector is e_2. a_2 can be written as a sum of e_2 and the projection of a_2 on e_1. Denoting this projection by pr_1a_2, we then have

$$e_2 = a_2 - pr_1a_2.$$

We recall from our vector calculus class the projection of one vector onto another,

$$pr_1a_2 = \frac{a_2 \cdot e_1}{e_1^2} e_1 .$$

Note that this is easily proven. First write the projection as a vector of length $a_2 \cos\theta$ in direction $\hat{e}_1$, where θ is the angle between e_1 and a_2,

$$pr_1a_2 = a_2 \cos\theta \frac{e_1}{e_1} .$$

Recall that the angle between e_1 and a_2 is obtained from

$$\cos\theta = \frac{a_2 \cdot e_1}{a_2 e_1} .$$

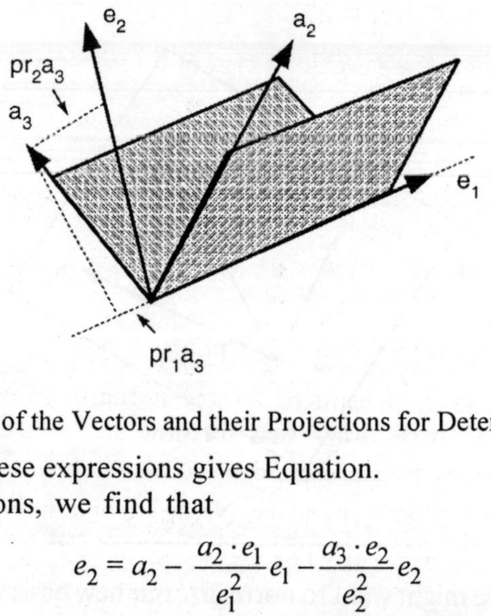

Fig. A Plot of the Vectors and their Projections for Determining e_3

Combining these expressions gives Equation.

From Equations, we find that

$$e_2 = a_2 - \frac{a_2 \cdot e_1}{e_1^2} e_1 - \frac{a_3 \cdot e_2}{e_2^2} e_2$$

It is a simple matter to verify that e_2 is orthogonal to e_1.

Now, we seek a third vector e_3 that is orthogonal to both e_1 and e_2. Pictorially, we can write the given vector a_3 as a combination of vector projections along e_1 and e_2 and the new vector e_3. Then we have,

$$e_3 = a_3 - \frac{a_3 \cdot e_1}{e_1^2} e_1 - \frac{a_3 \cdot e_2}{e_2^2} e_2.$$

Again, it is a simple matter to compute the scalar products with e_1 and e_2 to verify orthogonality.

We can generalize the procedure to the N -dimensional case. Let a_n, $n = 1,..., N$ be a set of linearly independent vectors in R^N. Then, an orthogonal basis can be found by setting $e_1 = a_1$ and for $n > 1$,

$$e_n = a_n - \sum_{j=1}^{n-1} \frac{a_n \cdot e_j}{e_j^2} e_j.$$

Now, we can generalize this idea to function spaces. Let $f_n(x)$, $n \in N$ and $x \in [a, b]$ be a linearly independent sequence of continuous functions. Then, an orthogonal basis of functions, $\phi_n(x)$, $n \in N$ can be found and is given by $\phi_0(x) = f_0(x)$ and

$$\phi_n(x) = f_n(x) - \sum_{j=1}^{n-1} \frac{< f_n, \phi_j >}{\|\phi_j\|^2} \phi_j(x), \quad n = 1, 2, \ldots$$

Here we are using inner products of real valued functions relative to weight $\sigma(x)$,

$$< f, g > = \int_a^b f(x) g(x) \sigma(x) dx$$

and $\|f\|^2 = < f, f >$. Note the similarity between this expression and the expression for the finite dimensional case in equation.

Example: Apply the Gram-Schmidt Orthogonalization process to the set $f_n(x) = x^n$, $n \in N$, when $x \in (-1, 1)$ and $\sigma(x) = 1$.

First, we have $\phi_0(x) = f_0(x) = 1$. Note that

$$\int_{-1}^{1} \phi_0^2(x) dx = \frac{1}{2}.$$

We could use this result to fix the noramlization of our new basis, but we will hold off on doing that for now.

Now, we compute the second basis element.

$$\phi_1(x) = f_1(x) - \frac{< f_1, \phi_0 >}{\|\phi_0\|^2} \phi_0(x)$$

$$= x - \frac{< x, 1 >}{\|1\|^2} 1 = x,$$

since $< x, 1 >$ is the integral of an odd function over a symmetric interval.

For $\phi_2(x)$, we have

$$\phi_2(x) = f_2(x) - \frac{< f_2, \phi_0 >}{\|\phi_0\|^2}\phi_0(x) - \frac{< f_2, \phi_1 >}{\|\phi_1\|^2}\phi_1(x)$$

$$= x^2 - \frac{< x^2, 1 >}{\|1\|^2}1 - \frac{< x^2, x >}{\|x\|^2}x$$

$$= x^2 - \frac{\int_{-1}^{1} x^2 dx}{\int_{-1}^{1} dx}$$

$$= x^2 - \frac{1}{3}.$$

Table: Common classical orthogonal polynomials with the interval and weight function used to define them

Polynomial	*Symbol*	*Interval*	$\sigma(x)$
Hermite	$H_n(x)$	$(-\infty, \infty)$	e^{-x2}
Laguerre	$L_n^{\alpha}(x)$	$[0, \infty)$	e^{-x}
Legendre	$Pn(x)$	$(-1, 1)$	1
Gegenbauer	$C_n^{\lambda}(x)$	$(-1, 1)$	$(1-x^2)^{\lambda-1/2}$
Tchebychef of the 1st kind $T_n(x)$	$(-1, 1)$	$(1-x^2)^{-1/2}$	
Tchebychef of the 2nd kind $U_n(x)$	$(-1, 1)$	$(1-x^2)^{-1/2}$	
Jacobi	$P_n^{(\nu,\mu)}(x)$	$(-1, 1)$	$(1-x)^{\nu}(1-x)^{\mu}$

So far, we have the orthogonal set $\left\{1, x, x^2 - \frac{1}{3}\right\}$. If one chooses to normalize these by forcing $\phi_n(1) = 1$, then one obtains the classical Legendre polynomials, $P_n(x) = \phi_n(x)$. This is not the typical normalization. Also, it might not be clear where the normalization constant is.

The ϕ_n's can be multiplied by any constant and this will only affect the "length", $\|\phi_n\|^2$. Thus, we have so far that $\phi_2(x) = C\left(x^2 - \frac{1}{3}\right)$. Setting $x = 1$, $\phi_2(1) = \frac{2}{3}$. Therefore, one obtains

$$P_2(x) = \phi_2(x) = \frac{1}{2}(3x^2 - 1).$$

The set of Legendre polynomials is just one set of classical orthogonal polynomials that can be obtained in this way. Many had originally appeared as solutions of important boundary value problems in physics.

Table: Tabular Computation of the Legendre Polynomials Using the Rodrigues Formula

n	$(x^2 - 1)^n$	$\frac{d^n}{dx^n}(x^2 - 1)^n$	$\frac{1}{2^n n!}$	$P_n(x)$
0	1	1	1	1
1	$x^2 - 1$	$2x$	$\frac{1}{2}$	x
2	$x^4 - 2x^2 + 1$	$12x^2 - 4$	$\frac{1}{8}$	$\frac{1}{2}(3x^2 - 1)$
3	$x_6 - 3x^4 + 3x^2 + 1$	$12x^2 - 4$	$120x^3 - 72x$	$\frac{1}{2}(5x^3 - 3x)$

LEGENDRE POLYNOMIALS

In the last section we saw the Legendre polynomials in the context of orthogonal bases for a set of square integrable functions in $L^2(-1, 1)$. In your first course in differential equations, you saw these polynomials as one of the solutions of the differential equation

$$(1 - x^2)y'' - 2xy' + n(n + 1)y = 0, \; n \in N.$$

Recall that these were obtained by using power series expansion methods.

A few of the properties of these functions. First, there is the Rodrigues formula.

$$P_n(x) = \frac{1}{2^n n!} \frac{d^n}{dx^n}(x^2 - 1)^n, \quad n \in N.$$

From this, one can see that $P_n(x)$ is an nth degree polynomial. Also, for n odd, the polynomial is an odd function and for n even, the polynomial is an even function.

One can systematically generate the Legendre polynomials in tabular form. Note that we get the same result as we found in the last section using orthogonalization.

The classical orthogonal polynomials also satisfy three term recursion formulae. In the case of the Legendre polynomials, we have

$$(2n + 1)xP_n(x) = (n + 1)P_{n+1}(x) + nP_{n-1}(x), \; n = 1, 2, \ldots.$$

This can also be rewritten by replacing n with $n - 1$ as

$$(2n - 1)xP_{n-1}(x) = nP_n(x) + (n - 1)P_{n-2}(x), \; n = 1, 2, \ldots.$$

We will prove this two different ways.

First, we use the orthogonality properties of Legendre polynomials and the fact that the coefficients of x^n in $P_n(x)$ is $\frac{1}{2^n n!}\frac{(2n)!}{n!}$. This last fact can be obtained from Rodrigues formula. We see this by focussing on the leading coefficients of $(x^2 - 1)^n$, which is x^{2n}. The first derivative is $2nx^{2n-1}$. The second derivative is $2n(2n - 1)x^{2n-2}$.

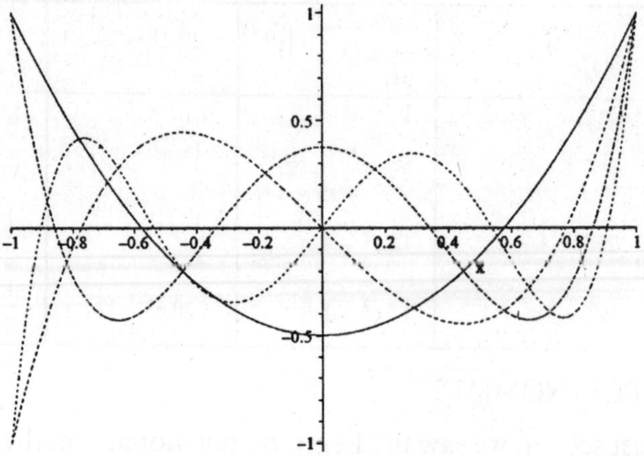

Fig. Plots of the Legendre Polynomials $P_2(x)$, $P_3(x)$, $P_4(x)$, and $P_5(x)$

The jth derivative is $[2n(2n - 1)... (2n - j + 1)]x^{2n-j}$. Thus, the nth derivative is $[2n(2n - 1)... (n + 1)]x^n$. This proves that $P_n(x)$ has degree n. The leading coefficients of $P_n(x)$ can now be written as

$$\frac{1}{2^n n!}[2n(2n-1)\ldots(n+1)] = \frac{1}{2^n n!}[2n(2n-1)\ldots(n+1)]\frac{n(n-1)\ldots 1}{n(n-1)\ldots 1}$$

$$= \frac{1}{2^n n!}\frac{(2n!)}{n!}.$$

In order to prove the three term recursion formula we consider the expression $nP_n(x) - (2n - 1)xP_{n-1}(x)$. While each term is a polynomial of degree n, the leading order terms cancel. We first look at the coefficients of the leading order term in the second term. It is

$$(2n-1)\frac{1}{2^{n-1}(n-1)!}\frac{(2n-2)!}{(n-1)!} = \frac{1}{2^{n-1}(n-1)!}\frac{(2n-1)!}{(n-1)!}$$

The coefficients of the leading term for $nP_n(x)$ can be written as

$$n\frac{1}{2^n n!}\frac{(2n)!}{n!} = n\left[\frac{2n}{2n^2}\right]\frac{1}{2^{n-1}(n-1)!}\frac{(2n-1)!}{(n-1)!}$$

After some simple cancellations in the first factors, we see that the leading order terms cancel. The next terms will be of degree $n - 2$. This is because the P_n's are either even or odd functions, thus only containing even, or odd, powers of x. We conclude that

$$nP_n(x) - (2n - 1)xP_{n-1}(x) = \text{polynomial of degree } n - 2.$$

Therefore, since the Legendre polynomials form a basis, we can write this polynomial as a linear combination of of Legendre polynomials.

$$nP_n(x) - (2n-1)xP_{n-1}(x) = c_0P_0(x) + c_1P_1(x) + ... + c_{n-2}P_{n-2}(x).$$

Multiplying by $P_m(x)$ for $m = 0, 1, ..., n-3$, and integrating from -1 to 1, we obtain

$$0 = c_m\|P_m\|^2$$

using orthogonality. Thus, all of these c_m's are zero, leaving

$$nP_n(x) - (2n-1)xP_{n-1}(x) = c_{n-2}P_{n-2}(x).$$

The final coefficients can be found by using the normalization condition, $P_n(1) = 1$. Thus, $c_{n-2} = n - (2n-1) = -(n-1)$.

A second proof of the three term recursion formula can be obtained from the generating function of the Legendre polynomials. Many special functions have such generating functions. For Legendre polynomials the generating function is given by

$$g(x, t) = \frac{1}{\sqrt{1-2xt+t^2}} = \sum_{n=0}^{\infty} P_n(x)t^n, \quad |x| < 1, |t| < 1.$$

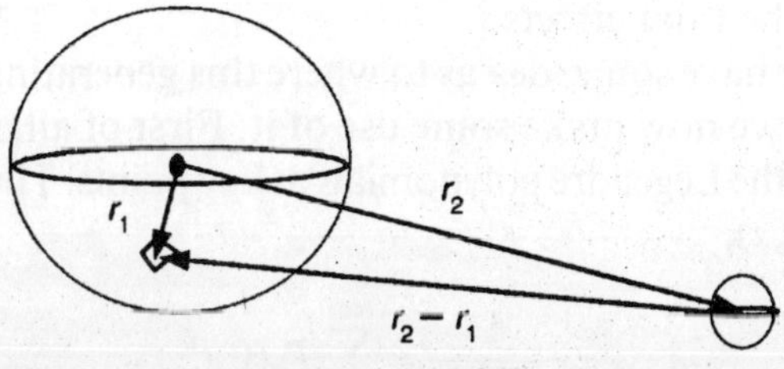

Fig. The Position Vectors used to Describe the Tidal force on the Earth Due to the Moon

This generating function occurs often in applications. In particular, it arises in potential theory, such as electromagnetic or gravitational potentials. These potential functions are $\frac{1}{r}$ type functions. For example, the gravitational potential between the Earth and the moon is proportional to the reciprocal of the magnitude of the difference between their positions relative to some coordinate system.

An even better example, would be to place the origin at the centre of the Earth and consider the forces on the non-pointlike Earth due to the moon. Consider a piece of the Earth at position r_1 and the moon at position r_2. The tidal potential Φ is given as

$$\Phi \propto \frac{1}{|r_2 - r_1|} = \frac{1}{\sqrt{(r_2-r_1)\cdot(r_2-r_1)}} = \frac{1}{\sqrt{r_1^2 - 2r_1r_2\cos\theta + r_2^2}}$$

where μ is the angle between r_1 and r_2. Typically, one of the position vectors is larger than the other. In this case, we have $r_1 \ll r_2$. So, one can write

$$\Phi \propto \frac{1}{\sqrt{r_1^2 - 2r_1r_2\cos\theta + r_2^2}} = \frac{1}{r_2}\frac{1}{\sqrt{1 - 2\frac{r_1}{r_2}\cos\theta + \left(\frac{r_1}{r_2}\right)^2}}$$

Now, define $x = \cos\theta$ and $t = \frac{r_1}{r_2}$. We then have the tidal potential is proportional to the generating function for the Legendre polynomials! So, we can write the tidal potential as

$$\Phi \propto \frac{1}{r_2}\sum_{n=0}^{\infty} P_n \cos(\theta)\left(\frac{r_1}{r_2}\right)^n.$$

The first term in the expansion will give the usual force between the Earth and the moon as point masses, or spheres. The next terms will give expressions for the tidal effects.

Now that we have some idea as to where this generating function might have originated, we now make some use of it. First of all, it can be used to provide values of the Legendre polynomials at key points. Thus, $P_n(0)$ is found by looking at $g(0, t)$,

$$g(0, t) = \frac{1}{\sqrt{1+t^2}} = \sum_{n=0}^{\infty} P_n(0)t^n.$$

However, we can use the binomial expansion to find our result. Namely, we have

$$\frac{1}{\sqrt{1+t^2}} = 1 - \frac{1}{2}t^2 + \frac{3}{8}t^4 + \ldots$$

Comparing these expansions, we have the $P_n(0) = 0$ for n odd and one can $P_{2n}(0) = (-1)^n \frac{(2n-1)!!}{(2n)!!}$ for n even. [note that the double factorial is defined by $n!! = (n-2)(n-2)!!$. So, $5!! = 5(3)(1)$ and $6!! = 6(4)(2)$.]

A simpler evaluation is to find $P_n(-1)$. In this case we have

$$g(-1, t) = \frac{1}{\sqrt{1+2t+t^2}} = \frac{1}{1+t} = 1 - t + t^2 - t^3 + \ldots$$

Therefore, $P_n(-1) = (-1)^n$.

We can also use the generating function to find recursion relations. To prove the three term recursion that we introduced above, we need only differentiate the generating function with respect to t in equation and rearrange the result. First note that

$$\frac{\partial g}{\partial t} = \frac{x-t}{(1-2xt+t^2)^{3/2}} = \frac{x-t}{1-2xt+t^2} g(x,t).$$

Combining this with

$$\frac{\partial g}{\partial t} = \sum_{n=0}^{\infty} nP_n(x)t^{n-1},$$

we have

$$(x-t)g(x,t) = (1-2xt+t^2)\sum_{n=0}^{\infty} nP_n(x)t^{n-1}.$$

Inserting the series expression for $g(x, t)$ and distributing the sum on the right side, we obtain

$$(x-t)\sum_{n=0}^{\infty} nP_n(x)t^n = \sum_{n=0}^{\infty} nP_n(x)t^{n-1} - \sum_{n=0}^{\infty} 2nxP_n(x)t^n + \sum_{n=0}^{\infty} nP_n(x)t^{n+1}$$

Rearranging leads to three separate sums:

$$\sum_{n=0}^{\infty} nP_n(x)t^{n-1} - \sum_{n=0}^{\infty} (2n+1)xP_n(x)t^n + \sum_{n=0}^{\infty} (n+1)P_n(x)t^{n+1} = 0.$$

Each term contains powers of t that we would like to combine into a single sum. This is done by reindexing. For the first sum, we could use the new index $k = n - 1$. Then,

$$\sum_{n=0}^{\infty} nP_n(x)t^{n-1} = \sum_{k=-1}^{\infty} (k+1)P_{k+1}(x)t^k.$$

These different indices are just another way of writing out the terms. Note that

$$\sum_{n=0}^{\infty} nP_n(x)t^{n-1} = 0 + P_1(x) + 2P_2(x)t + 3P_3(x)t^2 + \ldots$$

and

$$\sum_{k=-1}^{\infty} (k+1)P_{k+1}(x)t^k = 0 + P_1(x) + 2P_2(x)t + 3P_3(x)t^2 + \ldots$$

actually give the same sum. The indices are sometimes referred to dummy indices because they do not show up in the expanded expressions and can be replaced with another letter.

If we want to do so, we could now replace all of the k's with n's. The second sum in equation just needs the replacement $n = k$ and the last sum we reindex using $k = n + 1$. Therefore, equation becomes

$$\sum_{k=-1}^{\infty}(k+1)P_{k+1}(x)t^k - \sum_{k=1}^{\infty}(2k+1)xP_k(x)t^k + \sum_{k=1}^{\infty}kP_{k-1}(x)t^k = 0.$$

We can now combine all of the terms, noting the $k = -1$ term is zero and the $k = 0$ terms give

$$P_1(x) - xP_0(x) = 0.$$

Therefore, for $k > 0$,

$$\sum_{k=1}^{\infty}[(k+1)P_{k+1}(x)-(2k+1)xP_k(x)+kP_{k-1}(x)]t^k = 0.$$

Since this is true for all t, the coefficients s of the t^k's are zero, or

$$(k + 1)P_{k+1}(x) - (2k + 1)xP_k(x) + kP_{k-1}(x) = 0, k = 1, 2,....$$

There are other recursion relations. For example,

$$P\ 0n+1(x) - P'_{n-1}(x) = (2n + 1)Pn(x).$$

This can be proven using the generating function by differentiating $g(x, t)$ with respect to x and rearranging the resulting infinite series just as in this last manipulation.

Another use of the generating function is to obtain the normalization constant. Namely, $\|P_n\|^2$. Squaring the generating function, we have

$$\frac{1}{1-2xt+t^2} = \left[\sum_{n=0}^{\infty}P_n(x)t^n\right]^2 = \sum_{n=0}^{\infty}\sum_{m=0}^{\infty}P_n(x)P_m(x)t^{n+m}$$

Integrating from -1 to 1 and using orthogonality, we have

$$\int_{-1}^{1}\frac{dx}{1-2xt+t^2} = \sum_{n=0}^{\infty}\sum_{m=0}^{\infty}\int_{-1}^{1}t^{n+m}P_n(x)P_m(x)dx$$

$$= \sum_{n=0}^{\infty}\int_{-1}^{1}t^{2n}P_n^2(x)dx.$$

However

$$\int_{-1}^{1}\frac{dx}{1-2xt+t^2} = \frac{1}{t}\ln\left(\frac{1+t}{1-t}\right).$$

One can expand this expression about $t = 0$ to obtain

$$\frac{1}{t}\ln\left(\frac{1+t}{1-t}\right) = \sum_{n=0}^{\infty} \frac{2}{2n+1} t^{2n}.$$

Comparing this result with equation we find that

$$\|P_n\|^2 = \int_{-1}^{1} P_n^2(x)\,dx = \frac{2}{2n+1}.$$

Finally, we can expand functions in this orthogonal basis. This is just a generalized Fourier series. A Fourier-Legendre series expansion for $f(x)$ on (–1, 1) takes the form

$$f(x) \sim \sum_{n=0}^{\infty} c_n P_n(x).$$

As with Fourier trigonometric series, we can determine the coefficients s by multiplying both sides by $P_m(x)$ and integrating. Orthogonality give the usual form for the generalized Fourier coefficients s. In this case, we have

$$c_n = \frac{< f, P_n >}{\| P_n \|^2}.$$

We have just found $\|P_n\|^2 = \dfrac{2}{2n+1}$. Therefore, the Fourier-Legendre coefficients are

$$c_n = \frac{2n+1}{2} \int_{-1}^{1} f(x) P_n(x)\,dx.$$

Example: Expand $f(x) = x^3$ in a Fourier-Legendre Series.
We need to compute

$$c_n = \frac{2n+1}{2} \int_{-1}^{1} x^3 P_n(x)\,dx.$$

We first note that

$$\int_{-1}^{1} x^m P_n(x)\,dx = 0 \quad \text{for } m < n.$$

This is proven using Rodrigues formula in equation. We have

$$\int_{-1}^{1} x^m P_n(x)\,dx = \frac{1}{2^n n!} \int_{-1}^{1} x^m \frac{d^n}{dx^n}(x^2-1)^n\,dx.$$

For $m < n$, we integrate by parts m-times and use the facts that $P_n(1) = 1$

and $P_n(-1) = (-1)^n$. The right hand side vanishes. As a result, we will have that $c_n = 0$ for $n > 3$ in this example.

This leaves the computation of c_0, c_1, c_2 and c_3. Since x^3 is an odd function and P_0 and P_2 are even functions, $c_0 = 0$ and $c_2 = 0$. This leaves us with only two coefficients s to compute. These are

$$c_1 = \frac{3}{2}\int_{-1}^{1} x^4\, dx = \frac{3}{5}$$

$$c_3 = \frac{7}{2}\int_{-1}^{1} x^3 \left[\frac{1}{2}(5x^3 - 3x)\right] dx = \frac{3}{5}.$$

Thus,

$$x_3 = \frac{3}{5}P_1(x) + \frac{2}{5}P_3(x).$$

Of course, this is simple to check.

$$\frac{3}{5}P_1(x) + \frac{2}{5}P_3(x) = \frac{3}{5}x + \frac{2}{5}\left[\frac{1}{2}(5x^3 - 3x)\right] = x^3.$$

Well, maybe we could have guessed this without doing any integration. Let's see. Since $f(x) = x^3$ has degree three, we do not expect the expansion in Legendre polynomials for $f(x)$ to have polynomials of order greater than three. So, we assume from the beginning that

$$f(x) = c_0P_0(x) + c_1P_1(x) + c_2P_2(x) + c_3P_3(x).$$

Then,

$$x^3 = c_0 + c_1x + \frac{1}{2}c_2(3x^2 - 1) + \frac{1}{2}c_2(5x^3 - 3x).$$

However, there are no quadratic terms on the left side, so $c_2 = 0$. Then, there are no constant terms left except c_0. So, $c_0 = 0$. This leaves

$$x^3 = c_1x + \frac{1}{2}c_2(5x^3 - 3x)$$

$$= \left(c_1 - \frac{3}{2}c_2\right) + \frac{5}{2}c_2x^3$$

Equating coefficients s of the remaining like terms, we have that $c_2 = \frac{5}{2}$ and $c_1 = \frac{3}{2}\ c_2 = \frac{3}{5}$.

Example: Expand the Heaviside function in a Fourier-Legendre Series.

In this case, we cannot find the expansion coefficients s without some integration. We have to compute

$$c_n = \frac{2n+1}{2}\int_{-1}^{1} f(x)P_n(x)dx = \frac{2n+1}{2}\int_{0}^{1} P_n(x)dx$$

We can make use of the formula for $n > 1$. Then,

$$c_n = \frac{1}{2}\int_0^1 [P'_{n+1}(x) - P'_{n-1}(x)]dx = \frac{1}{2}[P_{n-1}(0) - P_{n+1}(0)].$$

For $n = 0$, we have

$$c_0 = \frac{1}{2}\int_0^1 dx = \frac{1}{2}.$$

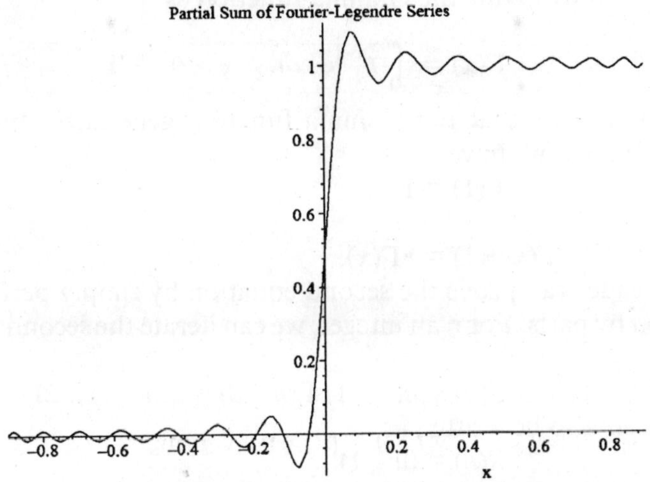

Fig. Sum of first 21 terms for Fourier-Legendre series Expansion of Heaviside Function

Then we have the expansion

$$f(x) \sim \frac{1}{2} + \frac{1}{2}\sum_{n=1}^{\infty}[P_{n-1}(0) - P_{n+1}(0)]P_n(x)$$

which can be written as

$$f(x) \sim \frac{1}{2} + \frac{1}{2}\sum_{n=1}^{\infty}[P_{2n-2}(0) - P_{2n}(0)]P_{2n-1}(x)$$

$$= \frac{1}{2} + \frac{1}{2}\sum_{n=1}^{\infty}\left[(-1)^{n-1}\frac{(2n-3)!!}{(2n-2)!!} - (-1)^n\frac{(2n-1)!!}{(2n)!!}\right] P_{2n-1}(x)$$

$$= \frac{1}{2} - \frac{1}{2}\sum_{n=1}^{\infty}(-1)^n\frac{(2n-3)!!}{(2n-2)!!}\left[1 + \frac{2n-1}{2n}\right] P_{2n-1}(x)$$

$$= \frac{1}{2} - \frac{1}{2}\sum_{n=1}^{\infty}(-1)^n\frac{(2n-3)!!}{(2n-2)!!}\frac{4n-1}{2n} P_{2n-1}(x).$$

The sum of the first 21 terms are shown.

SPHERICAL HARMONICS

Gamma Function

Another function that often occurs in the study of special functions is the Gamma function. The Gamma function is the natural generalization of the factorial function.

For $x > 0$ we define the Gamma function as

$$\Gamma(x) = \int_0^\infty t^{x-1}e^{-t}dt, \quad x > 0$$

We first show that the Gamma function generalizes the factorial function. In fact, we have

$$\Gamma(1) = 1$$

and

$$\Gamma(x + 1) = x\Gamma(x).$$

The reader can prove the second equation by simply performing an integration by parts. For n an integer, we can iterate the second expression to obtain

$$\Gamma(n + 1) = n\Gamma(n) = n(n - 1)\,\Gamma(n - 2) = n(n - 1)\ldots 2\Gamma(1) = n!.$$

This can also be written as

$$\Gamma(n) = (n - 1)!.$$

We can also define the Gamma function for negative, non-integer values of x. We first note that by iteration on $n \in Z^+$, we have

$$\Gamma(x + n) = (x + n - 1)\ldots(x + 1)x\Gamma(x), \; x < 0, \; x + n > 0.$$

Solving for $\Gamma(x)$, we then find

$$\Gamma(x) = \frac{\Gamma(x+n)}{(x+n-1)\cdots(x+1)x}, \quad x < 0, x + n > 0.$$

Note that the Gamma function is undefined at zero and the negative integers.

Another useful formula is

$$\Gamma\left(\frac{1}{2}\right) = \sqrt{\pi}.$$

It is simply found as

$$\Gamma\left(\frac{1}{2}\right) = \int_0^\infty t^{-\frac{1}{2}}e^{-t}dt$$

Letting $t = z^2$, we have

$$\Gamma\left(\frac{1}{2}\right) = 2\int_0^\infty e^{-z^2}dz.$$

Due to the symmetry of the integrand, we obtain the classic integral

$$\Gamma\left(\frac{1}{2}\right) = \int_{-\infty}^{\infty} e^{-z^2}\, dz,$$

which we had integrated when we computed the Fourier transform of a Gaussian. Recall that

$$\int_{-\infty}^{\infty} e^{-z^2}\, dz = \sqrt{\pi}.$$

Therefore, we have confirmed that $\Gamma\left(\frac{1}{2}\right) = \sqrt{\pi}$.

We have seen that the factorial function can be written in terms of Gamma functions. One can also relate the odd double factorials in terms of the Gamma function. First, we note that

$$(2n)!! = 2^n n!, \quad (2n+1)!! = \frac{(2n+1)!}{2^n n!}.$$

In particular, one can prove

$$\Gamma\left(n+\frac{1}{2}\right) = \frac{(2n-1)!!}{2^n}\sqrt{\pi}.$$

Formally, this givesμ

$$\left(\frac{1}{2}\right)! = \Gamma\left(\frac{3}{2}\right) = \frac{\sqrt{\pi}}{2}.$$

Another useful relation is

$$\Gamma(x)\Gamma(1-x) = \frac{\pi}{\sin \pi x}.$$

This result can be proven using complex variable methods.

BESSEL FUNCTIONS

Another important differential equation that arises in many physics applications is

$$x^2 y'' + xy' + (x^2 - p^2)y = 0.$$

This equation is readily put into self-adjoint form as

$$(xy')' + \left(x + \frac{p^2}{x}\right)y = 0.$$

This equation was solved in the first course on differential equations using power series methods, namely by using the Frobenius Method. One assumes a series solution of the form

$$y(x) = \sum_{n=0}^{\infty} a_n x^{n+s}$$

where one seeks allowed values of the constant s and a recursion relation for the coefficients s, a_n. One finds that $s = \pm p$ and

$$a_n = \frac{a_{n-2}}{(n+s)^2 - p^2}, \quad n \geq 2.$$

One solution is the Bessel function of the first kind of order p, given as

$$y(x) = J_p(x) = \sum_{n=0}^{\infty} \frac{(-1)^n}{\Gamma(n+1)\Gamma(n+p+1)} \left(\frac{x}{2}\right)^{2n+p}$$

The first few Bessel functions of the first kind of integer order. Note that these functions can be described as decaying oscillatory functions.

A second linearly independent solution is obtained for p not an integer as $J_{-p}(x)$. However, for p an integer, the $\Gamma(n + p + 1)$ factor leads to evaluations of the Gamma function at zero, or negative integers, when p is negative. Thus, the above series is not defined in these cases.

Another method for obtaining a second linearly independent solution is through a linear combination of $J_p(x)$ and $J_{-p}(x)$ as

$$N_p(x) = \frac{\cos \pi p J_p(x) - J_{-p}(x)}{\sin \pi p}.$$

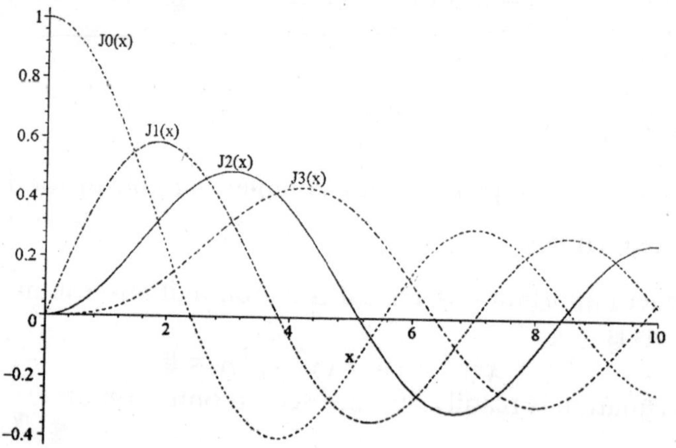

Fig. Plots of the Bessel functions $J_0(x)$, $J_1(x)$, $J_2(x)$, and $J_3(x)$

These functions are called the Neumann functions, or Bessel functions of the second kind of order p.

First few Bessel functions of the second kind of integer order. Note that these functions are also decaying oscillatory functions. However, they are singular at $x = 0$. In many applications these functions do not satisfy the boudary condition that one desires a bounded solution at $x = 0$. For example, one standard problem is to describe the oscillations of a circular drumhead. In this case the Bessel functions describe the radial part of the solution and one does not expect a singular solution at the centre of the drum.

Bessel functions satisfy a variety of properties, which we will only list at this time for Bessel functions of the first kind.

Derivative Identities

$$\frac{d}{dx}[x^p J_p(x)] = x^p J_{p-1}(x)$$

$$\frac{d}{dx}[x^{-p} J_{p-1}(x)] = x^{-p} J_{p+1}(x) .$$

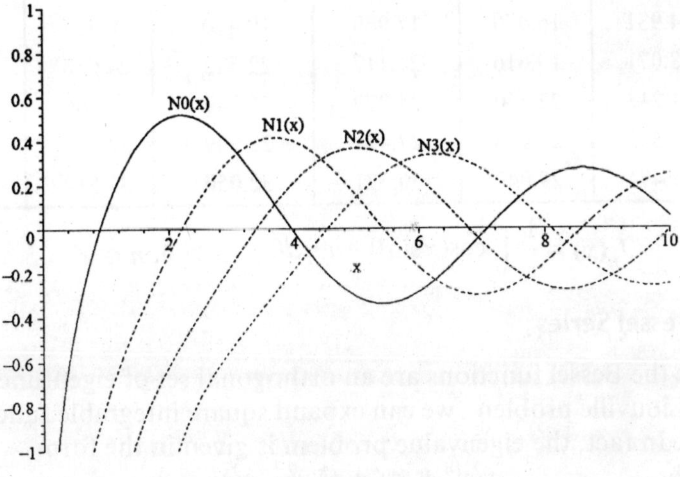

Fig. Plots of the Neumann Functions $N_0(x)$, $N_1(x)$, $N_2(x)$, and $N_3(x)$

Recursion Formulae

$$J_{p-1}(x) - J_{p+1}(x) = \frac{2p}{x} J_{p-1}(x) .$$

$$J_{p-1}(x) - J_{p+1}(x) = 2J_p'(x).$$

Orthogonality

$$\int_0^a xJ(j_{pn}x)J_p(j_{pm}x)\,dx = \frac{a^2}{2}\left[J_{p+1}(j_{pn}a)\right]^2 \delta_{n,m}$$

where j_{pn} is the nth root of $J_p(x)$, $J_p\,(j_{pn}) = 0$.

Generating Function

$$e^{x\left(t-\frac{1}{t}\right)/2} = \sum_{n=-\infty}^{\infty} J_n(x)t^n, \quad x>0, t \neq 0$$

Integral Representation

Table: The Zeros of Bessel Functions

p	$n=0$	$n=1$	$n=2$	$n=3$	$n=4$	$n=5$
1	2.405	3.832	5.135	6.379	7.586	8.780
2	5.520	7.016	8.147	9.760	11.064	12.339
3	8.654	10.173	11.620	13.017	14.373	15.700
4	11.792	13.323	14.796	16.224	17.616	18.982
5	14.931	16.470	17.960	19.410	20.827	22.220
6	18.071	19.616	21.117	22.583	24.018	25.431
7	21.212	22.760	24.270	25.749	27.200	28.628
8	24.353	25.903	27.421	28.909	30.371	31.813
9	27.494	29.047	30.571	32.050	33.512	34.983

$$J_n(x) = \frac{1}{\pi}\int_0^{\pi} \cos(x\sin\theta - n\theta)\,d\theta, \quad x > 0, n \in Z.$$

Fourier-Bessel Series

Since the Bessel functions are an orthogonal set of eigenfunctions of a Sturm-Liouville problem, we can expand square integrable functions in this basis. In fact, the eigenvalue problem is given in the form

$$x^2y'' + xy' + (\lambda x^2 - p^2)y = 0.$$

The solutions are then of the form J_p(Ex), as can be shown by making the substitution $t = \lambda x$ in the differential equation.

Furthermore, if $0 < x < a$, and one solves the differential equation with boundary conditions that $y(x)$ is bounded at $x = 0$ and $y(a) = 0$, then one can show that

$$f(x) = \sum_{n=1}^{\infty} c_n J_p(j_{pn}x)$$

where the Fourier-Bessel coefficients s are found using the ortho-gonality relation as

$$c_n = \frac{2}{a^2\left[J_{P+1}(j_{pn}a)\right]^2}\int_0^a xf(x)J_p(j_{pn}x)\,dx.$$

Example: Expand $f(x) = 1$ for $0 < x < 1$ in a Fourier-Bessel series of the form

$$f(x) = \sum_{n=1}^{\infty} c_n J_0(j_{pn} x)$$

We need only compute the Fourier-Bessel coefficients s in equation:

$$c_n = \frac{2}{[J_1(j_{0n})]^2} \int_0^1 x J_0(j_{0n} x)\, dx.$$

From Equation we have

$$\begin{aligned}\int_0^1 x J_0(j_{0n} x)\, dx &= \frac{1}{j_{0n}^2} \int_0^{j_{0n}} y J_0(y)\, dy \\ &= \frac{1}{j_{0n}^2} \int_0^{j_{0n}} \frac{d}{dy}[y J_1(y)]\, dy \\ &= \frac{1}{j_{0n}^2} [y J_1(y)]_0^{j_{0n}} \\ &= \frac{1}{j_{0n}} J_1(j_{0n}).\end{aligned}$$

As a result, we have found that the desired Fourier-Bessel expansion is

$$1 = 2\sum_{n=1}^{\infty} \frac{J_0(j_{0n} x)}{j_{0n} J_1(j_{0n})}, \quad 0 < x < 1.$$

The partial sum for the first fifty terms of this series. We note the slow convergence due to the Gibbs phenomenon.

Index

G

H

I

K

L

M

N

O

P

R

S